IUTAM Symposium on Mechanics of Martensitic Phase Transformation in Solids

SOLID MECHANICS AND ITS APPLICATIONS

Volume 101

Series Editor: G.M.L. GLADWELL

Department of Civil Engineering

University of Waterloo

Waterloo, Ontario, Canada N2L 3GI

Aims and Scope of the Series

The fundamental questions arising in mechanics are: *Why?*, *How?*, and *How much?* The aim of this series is to provide lucid accounts written by authoritative researchers giving vision and insight in answering these questions on the subject of mechanics as it relates to solids.

The scope of the series covers the entire spectrum of solid mechanics. Thus it includes the foundation of mechanics; variational formulations; computational mechanics; statics, kinematics and dynamics of rigid and elastic bodies: vibrations of solids and structures; dynamical systems and chaos; the theories of elasticity, plasticity and viscoelasticity; composite materials; rods, beams, shells and membranes; structural control and stability; soils, rocks and geomechanics; fracture; tribology; experimental mechanics; biomechanics and machine design.

The median level of presentation is the first year graduate student. Some texts are monographs defining the current state of the field; others are accessible to final year undergraduates; but essentially the emphasis is on readability and clarity.

For a list of related mechanics titles, see final pages.

IUTAM Symposium on

Mechanics of Martensitic Phase Transformation in Solids

Proceedings of the IUTAM Symposium held in Hong Kong, China, 11-15 June 2001

Edited by

Q.P. SUN

Department of Mechanical Engineering,
The Hong Kong University of Science and Technology,
Hong Kong, China, and
Department of Engineering Mechanics,
Tsinghua University, Beijng, China

KLUWER ACADEMIC PUBLISHERS
DORDRECHT / BOSTON / LONDON

A C.I.P. Catalogue record for this book is available from the Library of Congress.

ISBN 1-4020-0741-8

Published by Kluwer Academic Publishers,
P.O. Box 17, 3300 AA Dordrecht, The Netherlands.

Sold and distributed in North, Central and South America
by Kluwer Academic Publishers,
101 Philip Drive, Norwell, MA 02061, U.S.A.

In all other countries, sold and distributed
by Kluwer Academic Publishers,
P.O. Box 322, 3300 AH Dordrecht, The Netherlands.

Printed on acid-free paper

Printed in the Netherlands.

CONTENTS

(Participants and organizers of the IUTAM Symposium on *Mechanics of Martensitic Phase Transformation in Solids* at the Conference Hall of the University Center in the campus of Hong Kong University of Science and Technology on 14 June 2001)

Preface

Phase transition phenomena are of vital interest to physicists, materials scientists, and engineers who need to understand the behavior of solids during various phase transformations. Due to their academic and technological importance, martensitic phase transformations and the related mechanics research have received great attention and intensive investigation by the worldwide mechanics community over the past decade. The wide application of smart materials and Nano-/Microsystems (such as shape memory alloys and polymers, TRIP steels, Micro-electro-mechanical systems) has greatly promoted the development of this interdisciplinary research area. As one of the most active areas of research in theoretical and applied mechanics, the mechanics of phase transitions in solids is continually growing and new results in both fundamental and applied research are being constantly reported.

The objective of this first IUTAM Symposium held on this emerging research area was to provide a forum at which to discuss the latest research advances and future trends. It was the intention of the Scientific Committee to invite leading scientists and researchers from materials science, physics and mechanics together to present and to discuss different aspects of this active multidisciplinary field. During this five-day symposium held in Hong Kong, 33 invited lectures covering seven different topics were presented and intensively discussed. The seven topics were:

1. Microstructure and deformation of interfaces and moving boundaries;
2. Material instability and its propagation during martensite phase transformation;
3. Micromechanics description of constitutive behavior of representative volume element;
4. Interactions between plasticity (dislocation) and phase transformation;
5. Martensitic phase transformation in thin films;
6. Size effect and scaling in martensitic phase transformations;
7. Engineering applications of phase transformations in bio-materials and medical devices, intelligent materials and structures, TRIP steels, MEMS, shape memory alloys and shape memory polymers, etc.

In addition, one Round-Table discussion on future research directions was organized during the symposium. The open and friendly environment during the meeting provided an excellent opportunity for stimulating discussions and intensive exchange of ideas among all participants. An extended abstract booklet was distributed at the Symposium. Also local research graduate students participated in the meeting and were involved in the organization of the Symposium and, as a result, they benefited a lot from this symposium. Clearly, the major objective of this IUTAM symposium was achieved.

The Scientific Committee selected the distinguished participants to be invited and papers to be presented at the symposium. A total of 42 scientists from 14 countries and regions participated in the meeting as summarized below (not including local graduate students and research staffs):

Austria	1	Australia	1	China-mainland	4
Czech Republic	1	Estonia	2	France	5
Germany	2	Hong Kong	8	Italy	1
Japan	6	Singapore	2	Taiwan	1
UK	1	USA	8		

This volume collects the 29 written contributions to the Symposium from the invited speakers. The time and effort spent by these authors in participating in the meeting and preparing their manuscripts for this book is greatly appreciated. As in every successful scientific conference, financial support was an essential part of this symposium. Thanks are due to the IUTAM Bureau, the Research Grants Council of Hong Kong SAR, the National Natural Science Foundation of China, the US Army Research Office-Far East, Hong Kong University of Science and Technology, and the Hong Kong Society of Theoretical and Applied Mechanics for sponsoring the meeting and providing partial funding to bring international participants to Hong Kong.

Finally the editor would like to thank all the speakers and contributors to this symposium for their invaluable contributions. The editor also wishes to thank the Organizing Committee for their dedicated work. Special thanks goes to Prof. M. Cherkaoui as the Co-chairman and Ms. Ellie Ho as Conference Secretary, who both ensured the smooth course of the program. Special thanks also are due to Kluwer Academic Publishers, Dordrecht, for their effort and cooperation to produce this attractive Symposium Proceedings.

Qing-Ping SUN
Hong Kong, China
March 2002

International Scientific Committee of the Symposium

Q. P. Sun (Chairman, Hong Kong, China),
M. Berveiller (France),
O. T. Bruhns (Germany),
K. C. Hwang (China),
R. D. James (USA),
G. B. Olson (USA),
J. Ortin (Spain),
K. Tanaka (Japan),
J. Engelbrecht (Estonia).

Local Organizing Committee

Q. P. Sun (Chairman, Hong Kong, China),
M. Cherkaoui (Co-Chairman, Univ. of Metz, France)
Ellie Ho (Secretary, HKUST)

Sponsors

— The International Union of Theoretical and Applied Mechanics
— Research Grant Council of Hong Kong SAR, China
— The US Amy Research Office-Far East
— National Natural Science Foundation of China
— Kluwer Academic Publishers, Dordrecht
— School of Engineering, Hong Kong university of Science and Technology
— Hong Kong Society of Theoretical and Applied Mechanics
— Hysitron, Inc. USA

List of Participants

Name:	Address:
Antretter Th.	Institute of Mechanics, Montanuniversitä Leoben, Franz-Josef-Straße 18, A-8700 Leoben, Austria
Ball J.M.	Mathematical Institute, 24-29 St Giles', Oxford OX1 3LB, U.K.
Berezovski A.	Department of Mechanics and Applied Mathematics, Institute of Cybernetics at Tallinn Technical University, Akadeemia tee 21, 12618, Tallinn, Estonia
Bernardini Davide	Dipartimento di Ingegneria Strutturale e Geotecnica, Universita di Roma "La Sapienza", Via Eudossiana 18, 00184 Roma, Italy
Berveiller M.	Laboratoire de Physique et Mécanique des Matériaux, ISGMP-CNRS Ile du Saulcy –57045 Metz, France
Bhattacharya Kaushik	Applied Mechanics and Mechanical Engineering, 104-44 California Institute of Technology, Pasadena, CA 91125, USA
Bruhns O.T.	Institute of Mechanics, Ruhr-University Bochum, D-44780 Bochum, Germany
Cao Wenwu	Materials Research Laboratory, The Pennsylvania State University, University Park, Pennsylvania 16802, USA
Cherkaoui M.	Laboratoire de Physique et Mécanique des Matériaux, ISGMP-CNRS Ile du Saulcy –57045 Metz, France
Dai Hui-Hui	Department of Machematics, City University of Hong Kong, Tat Chee Avenue, Kowloon, Hong Kong, China
Engelbrecht J.	Department of Mechanics and Applied Mathematics, Institute of Cybernetics at Tallinn Technical University, Akadeemia tee 21, 12618, Tallinn, Estonia
Fang Dai-Ning	Department of Engineering Mechanics, Tsinghua University, Beijing 100084, China
Favier Denis	Laboratoire Sols, Solides Structures, CNRS – Univ. Joseph Fourier –Inst. Nat. Pol. de Grenoble, B.P.53, 38041 Grenoble Cedex 9, France
Huang W.M.	Centre for Advanced Numerical Engineering Simulations, Nanyang Technological University, Nanyang Avenue, 639798, Singapore
Hwang Keh-Chih	Department of Engineering Mechanics, Tsinghua University, Beijing 100084, China

Lexcellent Christian	Laboratoire de Mécanique Appliquée R. Chaléat (U.MR CNRS 6604), Université de Franche-Comté, 25030 Besançon CEDEX, France
Liu Yinong	Department of Mechanical and Materials Engineering, University of Western Australia, Crawley, WA 6009, Australia
Liu Yong	School of Mechanical & Production Engineering, Nanyang Technological University, Singapore
Messner C.	Lehrstuhl für Mechanik mit staatlichem Materialprüfamt für den Maschinenbau, und Christian Doppler Labor für moderne Mehrphasenstähle, Technische Universität München, Germany.
Miyazaki Shuichi	Institute of Materials Science, University of Tsukuba, Tsukuba, Ibaraki 305-8573, Japan
Patoor, E.	Laboratoire de Physique et Mécanique des Matériaux (UMR CNRS 7554), Institut Supérieur de Génie Mécanique et Productique, ENSAM, 4, rue Augustin Fresnel, 57078 Metz, France
Ren Xiaobing	National Research Institute for Metals, Tsukuba, Japan
Roytburd A.	Material and Nuclear Engineering, University of Maryland, College Park, MD 20742, U.S.A.
Sehitoglu Huseyin	University of Illinois, Department of Mechanical and Industrial Engineering, 1206 W. Green St., Urbana, IL 61801, USA
Shu Y.C.	Institute of Applied Mechanics, National Taiwan University, Taipei, Taiwan
Šittner P.	Institute of Physics, Na Slovance 2, Prague 8, 18221, Czech Republic
Slutsker Julia	Materials Science and Engineering Laboratory, National Institute of Standards and Technology, Gaithersburg, MD 20899, USA
Sun Q. P.	Department of Mechanical Engineering, The Hong Kong University of Science and Technology, Kowloon, Hong Kong, China
Tobushi H.	Aichi Institute of Technology, 1247 Yachigusa, Yagusa-cho Toyota, 470-0392 Japan
Tokuda Masataka	Department of Mechanical Engineering, Mie University, Kamihama 1515 Tsu, 514-8507 Japan
Tomita Yoshihiro	Faculty of Engineering, Kobe University, Nada Kobe 657-8501, Japan
Tong Pin	Department of Mechanical Engineering, Hong Kong University of Science and Technology, Hong Kong, China

Tsai Hungyu	Department of Materials Science and Mechanics, Michigan State University, East Lansing, MI 48824, U.S.A.
Wang Biao	School of MPE, Nanyang Technological University, Singapore
Yang W.	Department of Engineering Mechanics, Tsinghua University, Beijing 100084, China
Yu H. Y.	US Army Research Office – Far East, Minato-Ku Tokyo, 106-0032, Japan

List of Contributors

Name:	Address:
Ahaluwalia Rajeev	Materials Research Laboratory, The Pennsylvania State University, USA
Antretter Th.	Institute of Mechanics, Montanuniversitä Leoben, Franz-Josef-Straße 18, A-8700 Leoben, Austria
Ball J. M.	Mathematical Institute, 24-29 St Giles', Oxford OX1 3LB, U.K.
Berezovski A.	Department of Mechanics and Applied Mathematics, Institute of Cybernetics at Tallinn Technical University, Akadeemia tee 21, 12618, Tallinn, Estonia
Bernardini Davide	Dipartimento di Ingegneria Strutturale e Geotecnica, Universita di Roma "La Sapienza", Via Eudossiana 18, 00184 Roma, Italy
Berveiller M.	Laboratoire de Physique et Mécanique des Matériaux, ISGMP-CNRS Ile du Saulcy –57045 Metz, France
Bhattacharya Kaushik	Applied Mechanics and Mechanical Engineering, 104-44 California Institute of Technology, Pasadena, CA 91125, USA
Bouvet Christophe	Laboratoire de Mécanique Appliquée R. Chaléat (U.MR CNRS 6604), Université de Franche-Comté, 25030 Besançon CEDEX – France
Bruhns O.T.	Institute of Mechanics, Ruhr-University Bochum, D-44780 Bochum, Germany
Burcsu E.	Applied Mechanics and Mechanical Engineering, 104-44 California Institute of Technology, Pasadena, CA 91125, USA
Cailletaud G.	Centre des Matériaux, Ecole des Mines, B.P. 87, F-91003 Evry Cedex, France
Calloch Sylvain	Laboratoire de Mécanique et Technologie, E.N.S. de CACHAN / C.N.R.S. / Université PARIS 6, 61, Avenue du Président WILSON - 94235 CACHAN CEDEX – France
Cao Wenwu	Materials Research Laboratory, The Pennsylvania State University, University Park, Pennsylvania 16802, USA
Cherkaoui M.	Laboratoire de Physique et Mécanique des Matériaux, ISGMP-CNRS Ile du Saulcy –57045 Metz, France
Chirani S. Arbab	Laboratoire de Physique et Mécanique des Matériaux (UMR CNRS 7554), Institut Supérieur de Génie Mécanique et Productique, ENSAM, 4, rue Augustin Fresnel, 57078 Metz, France

Cihak U	Institute of Mechanics, Montanuniversitä Leoben, Franz-Josef-Straße 18, A-8700 Leoben, Austria
Cui Y.Q.	Department of Engineering Mechanics, Tsinghua University, Beijing 100084, China
Chumlyakov Yuriy	Siberian Physical-Technical Institute, Revolution Sq. 1, Tomsk, 634050, Russia
Endo M.	Aichi Institute of Technology, 1247 Yachigusa, Yagusa-cho Toyota, 470-0392 Japan
Engelbrecht J.	Department of Mechanics and Applied Mathematics, Institute of Cybernetics at Tallinn Technical University, Akadeemia tee 21, 12618, Tallinn, Estonia
Fan Xinjian	Department of Materials Science and Mechanics, Michigan State University, East Lansing, MI 48824, U.S.A.
Fang Dai-Ning	Department of Engineering Mechanics, Tsinghua University, Beijing 100084, China
Fang F.	Department of Engineering Mechanics, Tsinghua University, Beijing 100084, China
Favier Denis	Laboratoire Sols, Solides Structures, CNRS – Univ. Joseph Fourier – Inst. Nat. Pol. de Grenoble, B.P.53, 38041 Grenoble Cedex 9, France
Fischer F.D.	Institute of Mechanics, Montanuniversitä Leoben, Franz-Josef-Straße 18, A-8700 Leoben, Austria
Fratzl P.	Institute of Metal Physics, Montanuniversität Leoben, Jahnstraße 12, A-8700 Leoben, Austria
Gall Ken	Department of Mechanical Engineering, University of Colorado, Boulder, Colorado 80309, USA
Hayashi S.	Mitsubishi Heavy Industries, Ltd., Nagoya, Japan
Hirano M.	Institute of Materials Science, University of Tsukuba, Tsukuba, Ibaraki 305-8573, Japan
Huang W.M.	Centre for Advanced Numerical Engineering Simulations, Nanyang Technological University, Nanyang Avenue, 639798 Singapore
Hwang Keh-Chih	Department of Engineering Mechanics, Tsinghua University, Beijing 100084, China
Inaba Tadashi	Department of Mechanical Engineering, Mie University, Kamihama 1515 Tsu 514-8507 Japan

Iwamoto Takeshi	Faculty of Engineering, Hiroshima University, Higashi-Hiroshima, Hiroshima 739-8527, Japan
Karaman Ibrahim	University of Illinois, Department of Mechanical and Industrial Engineering, 1206 W. Green St., Urbana, IL 61801, USA
Lam D. C. C.	Department of Mechanical Engineering, Hong Kong University of Science and Technology, Hong Kong, China
Lexcellent Christian	Laboratoire de Mécanique Appliquée R. Chaléat (U.MR CNRS 6604), Université de Franche-Comté, 25030 Besançon CEDEX – France
Liew K.M.	Centre for Advanced Numerical Engineering Simulations, Nanyang Technological University, Nanyang Avenue, 639798 Singapore
Liu B.	Department of Engineering Mechanics, Tsinghua University, Beijing 100084, China
Liu Yinong	Department of Mechanical and Materials Engineering, University of Western Australia, Crawley, WA 6009, Australia
Liu Yong	School of Mechanical & Production Engineering, Nanyang Technological University, Singapore
Lukáš P.	Nuclear Physics Institute, 250 68 Řež, Czech Republic
Maier Hans J.	University of Paderborn, Lehrstuhl f. Werkstoffkunde, D-33095 Paderborn, Germany
Maugin G.A.	Université Pierre et Marie Curie, Laboratoire de Modélisation en, Mécanique, UMR 7607, Tour 66, 4 Place Jussieu, Case 162, 75252, Paris Cédex 05, France
Messner C.	Lehrstuhl für Mechanik mit staatlichem Materialprüfamt für den Maschinenbau, und Christian Doppler Labor für moderne Mehrphasenstähle, Technische Universität München, Germany.
Miyazaki Shuichi	Institute of Materials Science, University of Tsukuba, Tsukuba, Ibaraki 305-8573, Japan
Niclaeys C.	Laboratoire de Physique et Mécanique des Matériaux (UMR CNRS 7554), Institut Supérieur de Génie Mécanique et Productique, ENSAM, 4, rue Augustin Fresnel, 57078 Metz, France.
Novák V.	Institute of Physics, Na Slovance 2, Prague 8, 18221, Czech Republic
Oberste-Brandenburg C.	Institute of Mechanics, Ruhr-University Bochum, D-44780 Bochum, Germany

Okumura K.	Aichi Institute of Technology, 1247 Yachigusa, Yagusa-cho Toyota, 470-0392 Japan
Orgeas Laurent	Laboratoire Sols, Solides Structures, CNRS – Univ. Joseph Fourier – Inst. Nat. Pol. de Grenoble, B.P.53, 38041 Grenoble Cedex 9, France
Ortner B.	Institute of Metal Physics, Montanuniversität Leoben, Jahnstraße 12, A-8700 Leoben, Austria
Patoor E.	Laboratoire de Physique et Mécanique des Matériaux (UMR CNRS 7554), Institut Supérieur de Génie Mécanique et Productique, ENSAM, 4, rue Augustin Fresnel, 57078 Metz, France
Ravichandran G.	Applied Mechanics and Mechanical Engineering, 104-44 California Institute of Technology, Pasadena, CA 91125, USA
Reisner G.	Andritz AG, Graz, Austria.
Rio Gérard	Laboratoire de Génie Mécanique et Matériaux, Université de Bretagne Sud, 2 Rue le Coat Saint Haouens, 56325 Lorient, France
Roytburd A.	Material and Nuclear Engineering, University of Maryland, College Park, MD 20742, U.S.A.
Schryvers D.	EMAT, University of Antwerp, RUCA, Groenenborgerlaan 171, B-2020, Antwerpen, Belgium
Sehitoglu Huseyin	University of Illinois, Department of Mechanical and Industrial Engineering, 1206 W. Green St., Urbana, IL 61801, USA
Shu Y.C.	Institute of Applied Mechanics, National Taiwan University, Taipei, Taiwan
Šittner P.	Institute of Physics, Na Slovance 2, Prague 8, 18221, Czech Republic
Slutsker Julia	Materials Science and Engineering Laboratory, National Institute of Standards and Technology, Gaithersburg, MD 20899, USA
Sun Q. P.	Department of Mechanical Engineering, The Hong Kong University of Science and Technology, Kowloon, Hong Kong, China
Sogino Soichi	Department of Mechanical Engineering, Mie University, Kamihama 1515 Tsu 514-8507 Japan
Tanaka K.	Department of Aerospace Engineering, Tokyo Metropolitan Institute of Technology, Asahigaoka 6-6, J-191 Hino/Tokyo, Japan

Tobushi H.	Aichi Institute of Technology, 1247 Yachigusa, Yagusa-cho Toyota, 470-0392 Japan
Tokuda Masataka	Department of Mechanical Engineering, Mie University, Kamihama 1515 Tsu, 514-8507 Japan
Tomita Yoshihiro	Faculty of Engineering, Kobe University, Nada Kobe 657-8501, Japan
Tong P.	Department of Mechanical Engineering, Hong Kong University of Science and Technology, Hong Kong, China
Tsai Hungyu	Department of Materials Science and Mechanics, Michigan State University, East Lansing, MI 48824, U.S.A.
Vivet A.	Laboratoire Universitaire de Recherche Scientifique d'Alencon IUT Alencon, Universite de Caen Bsse-Normandie, 61250 DAMIGNY, France
Wang Biao	School of MPE, Nanyang Technological University, Singapore
Wang H.T.	Department of Engineering Mechanics, Tsinghua University, Beijing 100084, China
Werner E.	Lehrstuhl für Mechanik mit staatlichem Materialprüfamt für den Maschinenbau, und Christian Doppler Labor für moderne Mehrphasenstähle, Technische Universität München, Germany.
WU Xiaoxia	Department of Mechanical Engineering, The Hong Kong University of Science and Technology, Hong Kong
Xiao Zhongmin	School of MPE, Nanyang Technological University, Singapore
Yamamoto T.	Institute of Materials Science, University of Tsukuba, Tsukuba, Ibaraki 305-8573, Japan
Yang W.	Department of Engineering Mechanics, Tsinghua University, Beijing 100084, China
Zhang Xiangyang	University of Illinois, Department of Mechanical and Industrial Engineering, 1206 W. Green St., Urbana, IL 61801, USA
Zhu J.J.	Centre for Advanced Numerical Engineering Simulations, Nanyang Technological University, Nanyang Avenue, 639798, Singapore
Zineb T. Ben	Laboratoire de Physique et Mécanique des Matériaux (UMR CNRS 7554), Institut Supérieur de Génie Mécanique et Productique, ENSAM, 4, rue Augustin Fresnel, 57078 Metz, France

Session Program

Opening Session:

Opening and Welcome address by the Chairman of the Symposium *Q. P. SUN*

Welcome Address by: *Prof. Nelson Cue, Vice-President of the University (HKUST)*
Prof. Juri Engelbrecht, IUTAM representative

Session 1

Chairman: *Q-P SUN*

Keynote lecture: Materials with TRIP Effects: Mechanisms and Modelling
M. Berveiller, M. Cherkaoui, Laboratoire de Physique et Mécanique des Matériaux, CNRS, Metz, France

Session 2

Chairman: *M. Cherkaoui*

Constitutive Equation and Computational Prediction of Deformation Behavior of TRIP Steels Under Monotonic and Cyclic Loading
Yoshihiro Tomita, Kobe University, Japan
Takeshi Iwamoto, Hiroshima University, Japan

A Thermodynamic Approach to Modeling of Stress-Induced Phase-Transition Front Propagation in Solids
A.Berezovski, J.Engelbrecht, Institute of Cybernetics at Tallinn Technical University, Estonia; G.A.Maugin, Université Pierre et Marie Curie, France

The Formation of Macrotwins in NiAl Martensite
J.M. Ball, Mathematical Institute, UK
D. Schryvers, University of Antwerp, Belgium

Session 3

Chairman: *K. Bhattacharya*

On the Deformation Mechanism of Shape Memory Alloys
YONG Liu, Nanyang Technological University, Singapore

Nonlinear Electromechanical Deformation and Fracture Related to Domain Swithing in Ferroelectric Ceramics
Dai-Ning Fang, Keh-Chih Hwang, Tsinghua University, China

Session 4

Chairman: *Y. Tomita*

Two-way Shape Memory Effect Obtained by Training of Combined Cyclic Loading
Masataka Tokuda, Soichi Sogino, Tadashi Inaba, Mie University, Japan

Session 5

Chairman: *A. Roytburd*

Keynote lecture: Domain Patterns, Texture and Macroscopic Electro-mechanical Properties of Ferroelectrics
Kaushik Bhattacharya, California Institute of Technology, USA

Multiaxial Mechanical Behavior of a Cu-Al-Be Shape Memory Alloy: Experimental Investigations and Phenomenological Modeling
Christian Lexcellent, Christophe Bouvet, Université de Franche-Comté, France
Sylvain Calloch, Université PARIS, France

Session 6

Chairman: *M. Tokuda*

Thermomechanical Properties of Polyurethane-Shape Memory Polymer

H. Tobushi, K. Okumura, M. Endo, Aichi Institute of Technology, Japan
S. Hayashi, Mitsubishi Heavy Industries, Ltd., Japan

A New Model For The Volume Fraction of Martensitic Transformations
H. Y. Yu, US Army Research Office, Far East

Exotic Multi-scale Phenomena on Shape Memory Alloys Associated with the Interaction of Point Defects with Martensitic Transformation
Xiaobing Ren, National Research Institute for Metals, Japan

Session 7
Chairman: *H. Sehitoglu*

The Transformation Induced Plasticity (TRIP) in a Martensitically Transforming Steel Subjected to a General Loading Path
Th. Antretter, F.D. Fischer, P. Fratzl, B. Ortner, Austria
G. Cailletaud, France, K. Tanaka, Japan

Free Energy of Martensitic Transformation – From Discrete to Homogeneous Microstructure
C. Messner, E. Werner, Technische Universität München, Germany
Q.P. Sun, HKUST, Hong Kong, China, G. Reisner, Austria

Session 8
Chairman: *E. Patoor*

Single Crystal Deformation Mechanisms in NiTi, NiTiCu and Fe- Based Shape Memory Alloys
Huseyin Sehitoglu et al., University of Illinois, USA

Dependence of Energy Release Rate on the Propagation Speed of Martensitic Transformation in Materials
Biao Wang, Zhongmin Xiao, Nanyang Technological University, Singapore

Round Table Discussion on Future Research Directions
Chairmans: *A Roytburd and E. Patoor*

Session 9
Chairman: *M. Berveiller*

Keynote lecture: Martenitic Transformation in Constrained Films
A. Roytburd, University of Maryland, College Park, USA

Session 10
Chairman: *J. Ball*

Influence of Microstructural Parameters on Shape Memory Alloys Behavior
E. Patoor, C. Niclaeys, S. Arbab Chirani, T. Ben Zineb
Institut Supérieur de Génie Mécanique et Productique, France

Elastic Deformations in Shape Memory Alloy Fiber Reinforced Composites
Hungyu Tsai, Xinjian Fan, Michigan State University, USA

Modeling of Superelastic Adaptive Composites
Julia Slutsker, National Institute of Standards and Technology, USA
Alexander Roytburd, University of Maryland, USA

Session 11
Chairman: *B. Wang*

Influence of Plasticity on Transformation Behaviour of Martensite in NiTi
Yinong Liu, University of Western Australia, Australia

On the Description of Transformation in DO_3 To $18R$ Shape Memory Alloys Using Group Theory
W.M. Huang, J.J. Zhu, K.M. Liew, Nanyang Technological University, Singapore

Session 12
Chairman: *K. C. Hwang*
Pressurized Heterogeneous Thin Films of Martensitic Materials
Y.C. Shu, National Taiwan University, Taiwan
Load Partitioning Among Grains of Martensitically Transforming Shape Memory Alloy Polycrystals Studied by In-situ Neutron Diffraction
P. Šittner, V.Novák , Institute of Physics, Czech Republic
P. Lukáš, Nuclear Physics Institute, Czech Republic
M.Daymond, Rutherford Appleton Laboratory, UK

Session 13
Chairman: *W. Yang*
Keynote lecture: Ti-Ni Sputter-deposited SMA Thin Films and Their Applications
Shuichi Miyazaki, University of Tsukuba, Japan
On The Description of Martensitic Phase Transformations Using Tensorial Transformation Kinetics
O.T. Bruhns, C. Oberste-Brandenburg, Ruhr-University Bochum, Germany

Session 15
Chairman: *S. Miyazaki*
Mechanical Instability of NiTi in Tension, Compression and Shear
Denis Favier, Laurent Orgeas, CNRS – Univ. Joseph Fourier –Inst. Nat. Pol. de Grenoble, France
Yinong Liu, University of Western Australia, Australia
Gérard Rio, Université de Bretagne Sud, France
Phase Transition Phenomena of a Hyperelastic Rod Subjected to Static Load
Hui-Hui Dai, Qinsheng Bi, City University of Hong Kong, Hong Kong
Theoretical Study of Heterogeneous Nucleation in Martensitic Phase Transition
Wenwu Cao, Rajeev Ahaluwalia, The Pennsylvania State University, USA

Session 16
Chairman: *O. T. Bruhns*
Phase Transformation of Thin Wires in Tension
Pin Tong, D. C. C. Lam, Q. P. Sun, The Hong Kong University of Science and Technology, China
Unconventional Domain Bands Near a Crack Tip in Ferroelectrics
W. Yang, F. Fang, H.T. Wang, Y.Q. Cui, Tsinghua University, China

Session 17
Chairman: *P. Sittner*
A Macroscopic Model for Microscopically Heterogeneous Shape Memory Alloys
DAVIDE Bernardini, Universita di Roma "La Sapienza", Italy
Effect of Microstructure on the Deformation of Nano-grained Polycrystalline NiTi Wires, Thin Sheets and Microtubes
Q.P. Sun, The Hong Kong University of Science and Technology, China

MATERIALS WITH TRIP EFFECT: MECHANISMS AND MODELING

M. CHERKAOUI AND M. BERVEILLER
Laboratoire de Physique et Mécanique des Matériaux
ISGMP-CNRS, Ile du Saulcy –57045 Metz, France
e-mail : cherk@lpmm.univ-metz.fr

Abstract

The TRansformation Induced Plasticity (TRIP) phenomenon is clearly displayed when an inelastic strain occurs under an applied stress lower than the yield stress of the weaker phase. TRIP effect is attributed to the interaction between two basic mechanisms, namely the Magee and Greenwood Johnson. A quantitative description of such mechanisms is experimentally possible through non-proportional loading paths corresponding to the "creep" test. A micromechanical model is presented and tested under such conditions.

1. Introduction

The TRIP effect results from the competition between two basic mechanisms:
- Martensitic transformation occurs with 24 possible variants, each variant showing a distinct lattice orientation. At the mesoscale level, the mechanical description of each variant is performed by a given transformation strain ε^{trI} involving both a volumic part δ and a shear component g. In general, only preferred variants are developed during a thermomechanical loading; the applied as well as internal stresses play the role of variant selection. This mechanism is often called the "Magee" effect (1966) (referred here as the first mechanism).
- The accommodation process of the transformation strain by an additional plastic flow in parent and product phases. The according mechanism is known as the "Greenwood-Johnson" effect (1965) (referred here as the second mechanism).

The competition between these mechanisms depends on the progress of the transformation, which depends principally on thermomechanical loading conditions (stress, strain and temperature). An accurate experimental quantification of the interaction between the first and the second mechanisms is obtained by non-proportional loading paths namely the "creep" tests (cooling tests under various applied stresses). Here, we present those of the Nancy group (Gautier et al., 1988) performed on a Fe-20Ni-0.5C TRIP steel with a M^s temperature of $-34°C$. This group investigated

Q.P. Sun (ed.), IUTAM Symposium on Mechanics of Martensitic Phase Transformation in Solids, 1–8.

the evolution of the TRIP strain ε^{tp} versus the progress of the transformation by providing quantitative measurements of the martensitic volume fraction f during the thermomechanical loading.
From theoretical point of view, most of attempts follow a macroscopic description of the presented experimental results incorporating the concept of plastic accommodation of the transformation dilatation in the classical phenomenological plasticity. This concept was first adopted by Greenwood and Johnson (1965) and revised later by Abrassard (1972) and Leblond (Leblond et al., 1989) by successively including the martensitic volume fraction and stress deviator effects. The first mechanism and the transformation shear component are ignored in their analysis.
The aim of the present paper is to deal with the problem of the interaction between the first and second mechanisms, and their effect on the TRIP strain. In particular, an accurate description of the orientation effect should be accomplished by taking into account the typical microstructure developed during the phase transition, namely the 24 possible martensitic variants. An attempt is made in the present work by developing a micromechanic based thermodynamics model. The framework follows a double-scale transition method:
The first scale transition leads to the constitutive equation of an austenitic single crystal from a multi-variant description of martensitic phase transformation. The intragranular stresses emerging from the phase transition and the plastic flow of both phases are evaluated by introducing the concept of moving inelastic discontinuities. The interaction between the first and second mechanisms is discussed at this level.
The second scale transition allows us to obtain the overall behavior of the polycrystalline TRIP steels by using the self-consistent approximation. This method is a classical one; it will not be presented in this paper.
The thermodynamic aspects of the concept of moving inelastic discontinuities were published by the author in a previous paper (see Cherkaoui et al., 1998). Here, we focus our discussion on the ability of the model to take into account the interactions between the two mechanisms. The numerical calculations performed on a polycrystalline TRIP steel are mainly devoted to non-proportional loading paths.

2. Micromechanical Modeling

Inelastic strain called transformation strain, produced by martensitic transformation, results from two contributions: the Bain strain and the lattice invariant strain. Wechsler, Liebermann and Read (1953) were the first who formulated a geometrical theory giving the possible crystallographic habit planes. This theory is based on the concept of inelastic compatibility of the transformation strain $\varepsilon^{tr}(r)$ assumed to be uniform within an elementary transformed volume (plate or lath). Due to the high symmetry of the austenitic lattice, several transformation strains are possible, so that different martensitic variants may be formed with the corresponding habit planes. Transformation strains are given by

$$\varepsilon_{ij}^{trI} = \frac{1}{2} g\left(M_i^I N_j^I + M_j^I N_i^I\right) = g R_{ij}^I \quad (1)$$

where $\vec{N}$ and $\vec{M}$ are the habit plane normal and the direction of the transformation, respectively. The local transformation strain $\varepsilon^{\mathrm{tr}}(r)$ is then piecewise uniform and written as

$$\varepsilon^{\mathrm{tr}}(r)=\sum_{I=1}^{N}\varepsilon^{\mathrm{tr}I}\theta^{I}(r) \tag{2}$$

where $\theta^{I}(r)$ are the Heaviside step functions for the different transformed domains I and N their number.

For a single crystal with volume V subjected at its external boundary ∂V to a displacement $u_i=E_{ij}x_j$, and undergoing an inelastic strain field $\varepsilon^{\mathrm{tr}}(r)+\varepsilon^{\mathrm{p}}(r)$ within V, the total macroscopic E strain is

$$E=\frac{1}{V}\int_{V}\left(\varepsilon^{\mathrm{e}}(r)+\varepsilon^{\mathrm{tr}}(r)+\varepsilon^{\mathrm{p}}(r)\right)dV=\left\langle\varepsilon^{\mathrm{e}}(r)+\varepsilon^{\mathrm{tr}}(r)+\varepsilon^{\mathrm{p}}(r)\right\rangle \tag{3}$$

where the elastic strain field $\varepsilon^{\mathrm{e}}(r)$ arises from both E and the incompatibility of fields $\varepsilon^{\mathrm{tr}}(r)$ and $\varepsilon^{\mathrm{p}}(r)$.

For homogeneous elasticity with its elastic constant C, one obtains directly the elastic part E^{e} of E as follows:

$$E=C^{-1}:\Sigma+E^{\mathrm{tp}} \tag{4}$$

where $\Sigma=\langle\sigma(r)\rangle$, and $E^{\mathrm{tp}}=\left\langle\varepsilon^{\mathrm{tr}}(r)+\varepsilon^{\mathrm{p}}(r)\right\rangle$ corresponds to the macroscopic inelastic strain arising from the phase transformation and the plastic flow.

In the case of TRIP steels, the plastic strain $\varepsilon^{\mathrm{p}}(r)$ describes simultaneously the plastic strain of an elementary volume element at austenitic and martensitic states. For the TRIP steels, the reorientation and inverse transformation being insignificant mechanisms, the progression of the transformation can be described by the instantaneous growth of new plates or laths in the austenitic phase. Therefore, one obtains the expression

$$\dot{E}^{\mathrm{tp}}=(1-f)\dot{\bar{\varepsilon}}^{\mathrm{pA}}+\sum_{I}f^{I}\dot{\bar{\varepsilon}}^{\mathrm{M}I}+\sum_{I}\varepsilon^{\mathrm{tr}I}\dot{f}^{I} \tag{5}$$

- The first term describes the average plastic flow in the residual austenitic phase.
- The second term corresponds to the plastic flow in the pre-existing martensitic phase.
- The last term expresses the formation of new plates or laths.

Recently, the author and coworkers (Cherkaoui et al., 1998) extended the Eshelby's (1970) pioneering work on the energy momentum tensor to derive the thermodynamic driving force acting on a moving boundary point between the product and the parent phases. That is

$$F = -\frac{1}{2}\left(\sigma^{+} + \sigma^{-}\right)\left[\varepsilon^{\mathrm{tp}}\right] + [\varphi] \tag{6}$$

where $\left[\varepsilon^{\mathrm{tp}}\right]$ and $[\varphi]$ denote the jumps of the inelastic strain and the chemical energy, respectively. σ^{+} and σ^{-} correspond to the limiting values of the local stress field for each side of the moving interface, which can be linked with the help of interface operators $Q(C,n)$

$$\sigma^{+} = \sigma^{-} - Q(C,n)[\varepsilon^{\mathrm{tp}}] \tag{7}$$

depending on the unit normal to interface and elastic constant of the material. With (7), (6) can be rewritten as

$$F = \left(\frac{1}{2}Q(C,n) - \sigma^{-}\right)\left[\varepsilon^{\mathrm{tp}}\right] + [\varphi] \tag{8}$$

The introduction of (8) as a thermodynamic driving force assumes that the interface normal velocity is taken as the internal variable characterizing the phase transition. It is also noticed that equation (8) requires the determination of the local stress and strain fields in each point of the moving interface. Therefore, the determination of (8) for any topology of the interface could be obtained only by strong numerical calculations. In order to go a step further in the scale transition, an alternative relation of (8) is obtained by introducing the concept of an ellipsoidal growing. The former hypothesis assumes that the moving interface represents an ellipsoidal growing domain. This is in correlation with the microstructure developed during martensitic transformation corresponding to the formation of microdomains with of a typical morphology of plate or lath.

Furthermore, as we discussed in the previous kinematic study, the martensitic transformation progresses by the nucleation and instantaneous growth of new domains. This is related to a high interface velocity between the parent and product phases, leading consequently to the following expression for the local jump $[\varepsilon^{\mathrm{tp}}]$ of the inelastic strain

$$[\varepsilon^{\mathrm{tp}}]^{I} = \varepsilon^{\mathrm{p}+} - \left(\varepsilon^{\mathrm{p}-} + \varepsilon^{\mathrm{tr}I}\right) = -\varepsilon^{\mathrm{tr}I} \tag{9}$$

In other words, the instantaneous growth of a martensitic domain allows the continuity of the plastic strain over its boundary. In connection with (5), it implies that the volume fractions f^{I} of each possible martensitic variant are the internal variables characterizing the phase transition at the mesoscale, the integration of (8) along the ellipsoidal domain interface and thanks to (9), one obtain the driving force F^I for the "flow" $\dot{f}^I$ of a martensitic variant due to the formation of a new domain. That is

$$F^I = \sigma^{-} : \varepsilon^{\mathrm{tr}I} - B\left(T - T^{\circ}\right) + \frac{1}{2}\varepsilon^{\mathrm{tr}I} : C : \left(I - S^I\right)\varepsilon^{\mathrm{tr}I} + \frac{1}{2}\varepsilon^{\mathrm{tr}I} : C : \frac{\dot{S}^I}{\dot{f}^I} : \varepsilon^{\mathrm{tr}I} f^I, \tag{10}$$

I is the identity tensor and S^I is the Eshelby tensor depending on the aspect ratio as well as on the orientation of the microdomain; $\dot{S}^I$ correspond to the variation of S^I due to the ellipsoidal growth. For the sake of simplicity a homothetic growth is assumed giving $\dot{S}^I = 0$. The term $B\left(T - T^{\circ}\right)$ is a linear approximation of the chemical energy jump $[\varphi]$. B is a material constant and T° the equilibrium temperature. It should be noticed that the derivation of (10) requires the hypothesis of a uniform stress field σ^{-} within the growing ellipsoidal domain.

Eqn (10) gives the thermodynamic driving force for nucleation and growth of martensitic microdomains belonging to different martensitic variants; it requires the knowledge of stress inside the growing domains. In addition to the overall applied stress, this stress field contains several contributions due to different couplings between plasticity and phase transformation at the microscale. Several ways for the determination of internal stresses are discussed by the authors from a micromechanic point of view in a previous work (see Cherkaoui et al., 2000).

3. Numerical results for polycrystalline TRIP steels

In the previous sections, the constitutive equations of an austenitic single crystal have been established within micromechanic and thermodynamic frameworks. This constitutes the principal contribution of the theoretical approach since the second step corresponding to a meso-macro scale transition is a classical one. In fact, eqns (5) and (10) are implemented in an elastoplastic type self-consistent scheme taking into account the statistical disorder of the polycrystalline TRIP steel. This scale transition method has been extensively presented and discussed in previous studies.

Numerical results for the behavior of a polycrystalline TRIP steel obtained by this approach are presented in this section. The obtained results are compared with the experimental data of Gautier et al. (1988) performed on a Fe-Ni-C base alloy. For the numerical simulation, an aggregate of 100 grains is employed to describe the polycrystalline structure. The grains are assumed to be spherical in shape with identical size. The elasticity of the material is assumed to be isotropic and homogeneous defined by the shear modulus $(\mu = 80\,\mathrm{GPa})$ and the Poisson's ratio $(\nu = 0.3)$. Before presenting the numerical results, it will be necessary to distinguish between the material parameters and model parameters:

The material parameters are B, M^s, T^o, g and α the aspect ratio of martensitic plates as well as the classical plasticity constants describing the hardening of austenite. For the sake of simplicity, the plasticity of martensite is neglected.
Furthermore, the Schmidt tensors P^{A^n} in austenitic phase are deduced from classical plasticity of FCC metals while the transformation tensors R^I $(I=1,24)$ are determined from the crystallographic theory of Wechsler et al. (1953).
The different parameters are identified by a tensile test at a given temperature. The model is then used to perform non-proportional loading paths corresponding to cooling tests at various applied stresses. In figure 1, we plotted the obtained results of the evolution of the TRIP strain as function of the total volume fraction of martensite, whose evolution during the thermomechanical load is displayed in figure 2.
It is clearly seen that the TRIP strain depends on the progress of the transformation and the applied stress level. We can distinguish in Fig. 1 two ranges; the sloop $d\varepsilon^{tp}/df$ is maximal at the beginning of the transformation and decreases as the transformation progresses. This is essentially related to the contribution of the first mechanism governed by the accommodation condition of the internal stresses emerged from the phase transformation. In the case of steels, the relaxation of these stresses is performed by plastic deformation of the austenite and the formation of "self-accommodating" groups of martensite.
The first mechanism is dominant at the beginning of the transformation where the preferred martensitic variants are mainly oriented by the applied stress leading to a maximum deformation in its direction. Note that in this transformation range, the orientation of martensitic variants increases when the applied stress reaches a value near to the yield stress of the austenite (here about 180 MPa). One can conclude that the relaxation process occurs by plastic deformation of the austenite rather than the formation of "self-accommodating" groups. In other words, when the applied stress approaches the yield stress of the austenite, the two mechanisms are active, in such a manner that the second mechanism favors the first one.
As the transformation progresses, the martensitic variants are selected by the internal stress state (self-accommodating process), the resulted deformation in the direction of the applied stress is low. In this range, the main contribution of the TRIP strain in the direction of the applied stress comes from the second mechanism.
A good agreement with the experimental results of Gautier et al. is observed. Therefore, the presented model provides a good description of the interaction between the orientation and accommodation effects.

4. Conclusions

The behavior of the TRIP steels is governed by the competition between the orientation and accommodation effects. Most of the previous studies neglected the orientation effect and therefore provided a partial description of the TRIP effect. The typical microstructure developed at the grain level during martensitic transformation requires a multiple scale transition to account for the orientation effect. This is accomplished in this work using the concept of moving inelastic discontinuities.

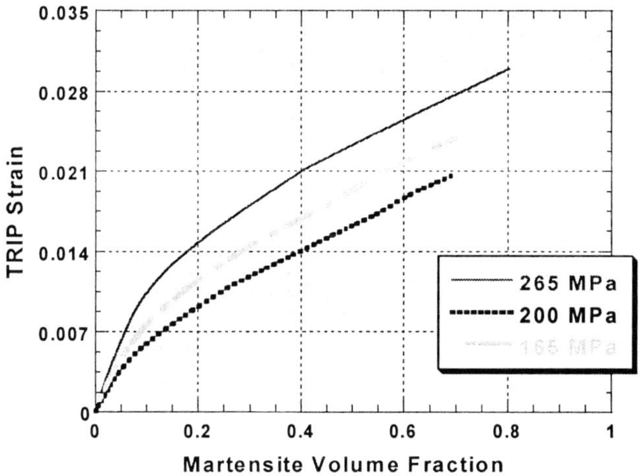

Figure 1 Evolution of the TRIP strain as a function of the martensite content during a « creep » test. Theoretical results.

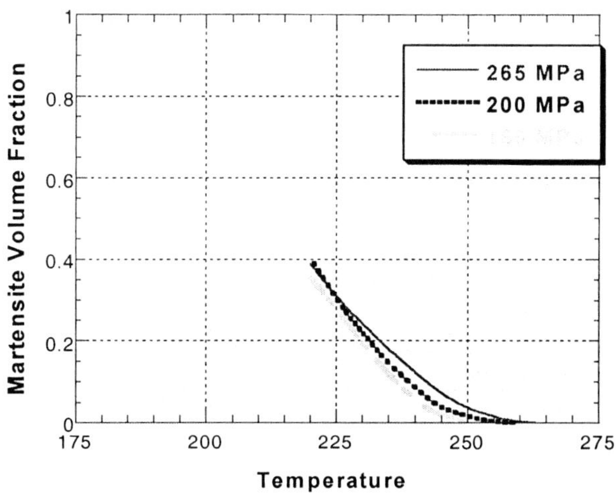

Figure 2 Evolution of the martensite content during a « creep » test. Theoretical results.

5. References

Abrassart, F., (1972), Thèse d'état Nancy.
Cherkaoui, M., Berveiller, M. and Sabar, H. (1998), Micromechanical Modeling of Martensitic TRansformation Induced Plasticity (TRIP) in Austenitic Single Crystals, *Int. J. Plasticity.* **14**, pp. 597-626.
Eshelby, J.D. (1970), Energy Relations and the energy-Momentum Tensor in Continuum Mechanics. Inelastic Behavior of Solids, (Eds. Kanninen, M.F., Alder, W.F., Rosenfield, A.R., and Joffee, R.I.), Mc Graw-Hill, New York, pp. 77-115.
Gautier, E. and Simon, A. (1988), In Phase Transformation 87. (Edited by LORIMER G. W.), Institute of Metals, London, pp. 285-287
Greenwood, G.W. and Johnson, R.H. (1965), The Deformation of Metals Under Small Stresses During Phase Transformation, *Proc. Roy. Soc.* **A 283**, 403.
Leblond, J.B., Devaux, J. and Devaux, J.C. (1989), Mathematical Modelling of Transformation Plasticity in Steels, I: Case of Ideal-Plastic Phases, *Int. J. Plasticity.* **5**, pp. 551-572.
Magee, C.L. (1966) ,Transformation Kinetics, Microplasticity and Aging of Martensite in Fe-31 Ni. Ph. D. Thesis, Carnegie Institute of Technology, Pittsburgh, PA.
Wechsler, M.S., Lieberman, D.S. and Read, T.A. (1953), On the Theory of the Formation of Martensite, *AIME Trans. J. Metals* **197**, pp. 1503-1515.

CONSTITUTIVE EQUATION AND COMPUTATIONAL PREDICTION OF DEFORMATION BEHAVIOR OF TRIP STEELS UNDER MONOTONIC AND CYCLIC LOADING

YOSHIHIRO TOMITA* TAKESHI IWAMOTO**
*Graduate School of Science and Technology, Kobe University, Rokkodai, Nada, Kobe 657-8501, JAPAN
**Faculty of Engineering, Hiroshima University, Kagamiyama, Higashi-hiroshima, Hiroshima 739-8527, JAPAN

ABSTRACT

A constitutive equation accounting for strain rate, temperature and applied stress system dependences of strain-induced martensitic transformation is given. A series of computational predictions of monotonic and cyclic deformation behavior, including uniaxial and proportional loading of typical 304 austenitic stainless steel, have been performed under the environmental temperatures of 77K and 298K. The effects of stress range, prestrain, temperature and applied strain system on such responses of transformation induced plasticity (TRIP) steels as the evolution of the martensitic phase, the accumulated plastic strain, and the asymptotic nature of the stress-strain relationship with an increasing number of cycles are clarified. The predictability of the present constitutive model is verified by experimental results. Furthermore, the monotonic and cyclic deformation behaviors of TRIP steel bars with a ringed notch are investigated.

1. Introduction

A TRIP (transformation induced plasticity) steel possesses advantageous mechanical properties including high strength, ductility and toughness due to the effect of the strain-induced martensitic transformation, which strongly depends on temperature, stress, strain and strain rate. Therefore, computational simulations using appropriate constitutive equations [1]-[4] are indispensable for predicting the deformation processes, as well as the final strength, of the product.

Here, in order to clarify the applicability of the constitutive equations, a series of computational simulations for uniform, as well as nonuniform, deformation behavior has been performed, and the applicability of the computational simulation for the

Q.P. Sun (ed.), IUTAM Symposium on Mechanics of Martensitic Phase Transformation in Solids, 9–18.

prediction of experimentally observed results [5] has been examined. These include the macro- to microscale deformation and the distribution of the volume fraction of the martensitic phase under macroscopically uniform and nonuniform deformations.

Subsequently, the approved computational simulation method is applied for estimating of deformation processes of cyclic axial, shear and proportional loading. Computationally predicted results are compared with carefully obtained experimental results [6], [7]. Furthermore, the deformation behavior of ringed-notched specimens subjected to cyclic loading has been clarified.

2. Constitutive Equations

Olson and Cohen [1] established a model for strain-induced martensitic transformation kinetics, which can express the temperature dependence of the transformation phenomenon. Stringfellow et al. [2] generalized the Olson and Cohen model so as to include the stress state and the contribution of the martensite phase to the strength. Tomita and Iwamoto [3] modified the two models to account for the experimental finding that the mode of the deformation behavior is controlled by the shear band mode as the strain rate increases [8]. The rate of increase of the volume fraction of martensite, $\dot{f}^{\alpha'}$, is given by

$$\dot{f}^{\alpha'} = (1 - f^{\alpha'})(A\dot{\bar{\varepsilon}}_{\gamma}^{pslip} + B\dot{g}), \tag{1}$$

$$A = \alpha p n \eta \left(f^{sb}\right)^{n-1}\left(1 - f^{sb}\right), \quad B = \eta \frac{dp}{dg}(f^{sb})^{n} H(\dot{g}),$$

where $\dot{\bar{\varepsilon}}_{\gamma}^{pslip}$ is the equivalent strain rate of slip deformation in austenite, f^{sb} is the volume fraction of the shear band, p is the probability that an intersection forms a martensitic embryo [2], [3], g is the driving force for martensitic transformation, $H(\dot{g})$ is the Heaviside step function with respect to $\dot{g}$ which describes the irreversible process of the martensitic transformation, and n and η are geometric constants. α is a parameter related to the stacking fault energy and is a function of temperature T [1], strain rate $\dot{\bar{\varepsilon}}_{\gamma}^{pslip}$ [3] and the stress triaxiality parameter $\Sigma = \sigma_{ii} / \left(3\bar{\sigma}\right)$ [4] as

$$\alpha = \left(\alpha_1 T^2 + \alpha_2 T + \alpha_3 - \alpha_4 \Sigma\right)\left(\frac{\dot{\bar{\varepsilon}}_{\gamma}^{pslip}}{\dot{\varepsilon}_y}\right)^{M}, \tag{2}$$

where M is the strain rate sensitivity exponent, $\alpha_1 - \alpha_4$ are material parameters and $\dot{\varepsilon}_y$ is the reference strain rate.

For the loading function, to account for the experimentally observed result [4] that uniaxial tension and compression curves for austenite and martensitic phases in TRIP steels are different, the second and third invariants, J_2 and J_3 of deviatoric stress σ'_{ij} are introduced as

$$f(J_2, J_3) = J_2 - \kappa \frac{J_3}{J_2^{1/2}} - \frac{1}{3}\bar{\sigma}^2 = 0, \tag{3}$$

where $\bar{\sigma}$ is the equivalent stress and κ is the parameter which reflects the development of the microstructure and texture inside the materials. The simplified evolution equation for κ proposed by Miller and McDowell [9] is

$$\dot{\kappa} = C_\kappa (1 - \kappa) \dot{\bar{\varepsilon}}^{pslip}, \tag{4}$$

where C_κ is a constant.

The corresponding constitutive equation for the plastic strain rate can be derived along the same line as discussed elsewhere [3], [4] and the constitutive equation for the stress rate can be established by introducing the plastic strain rate into the thermoelastic constitutive equation [10]. The constitutive equation for a two-phase composite material of austenite with the volume fraction of the martensitic phase $f^{\alpha'}$ is established in a manner similar to that described by Stringfellow et al. [2]. The forward gradient method is introduced to improve computational efficiency. The constitutive parameters employed are those indicated in previous works [3], [4].

A computational strategy for the simulation of the deformation behavior of TRIP steels has been established with an updated Lagrangian formulation of the virtual work principle and the finite element method [10], along with the constitutive equation.

3. Predictability of Deformation Behavior of TRIP Steels

3.1 COMPUTATIONAL CONDITIONS

Initially, the homogeneous deformation behaviors of a typical stainless steel, SUS304, are considered. The plane strain rectangular blocks under the actions of the axial strain rate $\dot{\varepsilon}$ and the shear strain rate $\dot{\gamma}$ with an environmental temperature of $T_{env} = 77, 298K$ are investigated. For proportional straining, as indicated in Fig. 1(a), the relationship $\dot{\gamma}/\sqrt{3} = \alpha\dot{\varepsilon}$ holds with $\dot{\varepsilon} = 5 \times 10^{-4}/s$. The heat generation due to irreversible work is small [3] therefore the temperature change is neglected. The angle defining the direction is expressed as α = Tan θ through the relationship θ = Tan^{-1} $[\gamma/(\sqrt{3}\varepsilon)]$. The corresponding equivalent plastic strain rate under the plane strain condition becomes $\dot{\bar{\varepsilon}}^p = \sqrt{4\dot{\varepsilon}^{p^2}/3 + \dot{\gamma}^{p^2}/3}$. In order to illustrate the cyclic behavior, as shown in Fig. 1(a), the equivalent strain rate may have negative values depending on the signs of $\dot{\varepsilon}^p$ and $\dot{\gamma}^p$. The corresponding equivalent stress becomes $\bar{\sigma} = \sqrt{3\sigma^2/4 + 3\tau^2}$.

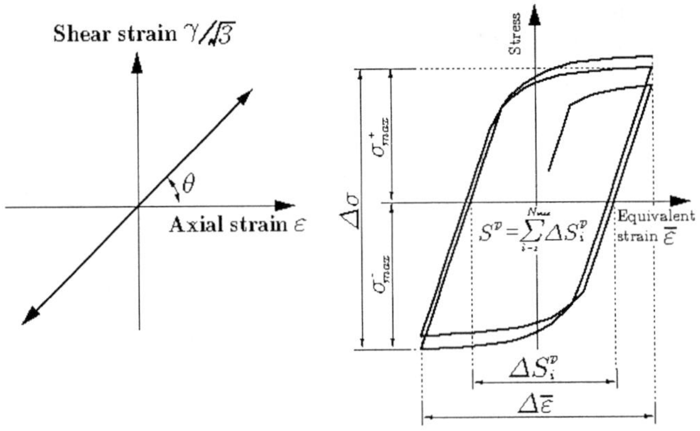

(a) (b)

Figure 1 Strain trajectory and definition of notations

In order to explain the characteristic feature of the deformation processes under the cyclic loading condition, the typical cyclic equivalent stress-strain relationship [7] is shown in Fig. 1(b). In this figure, $\Delta\sigma$ is the stress range corresponding to the strain range $\Delta\overline{\varepsilon}$, σ^+_{max} and σ^-_{max} are the absolute maximum values of stress attained during a deformation cycle, ΔS^P_i is the magnitude of strain difference at zero stress, as indicated in the figure, and S^P is the accumulation of the strain difference up to the Nth cycle. These notations will be used for the explanation of the cyclic deformation behavior under different loading conditions. Thus the equivalent stress range is evaluated by $\Delta\overline{\sigma}=\sqrt{3\Delta\sigma^2/4+3\Delta\tau^2}$. Similar calculations have been performed for the case with prestrain $\overline{\varepsilon}_{pre}$ = 0.03, 0.04 and 0.08 for monotonic and cyclic deformations, respectively.

3.2 MONOTONIC UNIFORM DEFORMATION

Figure 2 shows the predicted (a) equivalent stress- and (b) volume fraction of the martensitic phase-equivalent plastic strain relationship for different proportional straining with $\theta=0$ to 180 degrees under relatively small strain, the magnitude of which is comparable to that in the case of cyclic loading that will be discussed in 3.3. The results clearly reveal the stress state dependence of the evolution of the martensitic phase, which results in the difference in the magnitude of deformation resistance of the materials. Namely, the evolution of the martensitic phase is strongly dependent on the triaxiality parameter Σ in Eq. (2) and decreases from $1/\sqrt{3}$ to $-1/\sqrt{3}$ depending on the direction from 0 to 180 degrees. Therefore, the amount of evolution of the martensitic phase is inverse proportion. The tendency observed in Fig. 2(b) corresponds well to the experimentally obtained results [11]. However, a quantitative comparison is required for further identification of the material constants introduced in the present constitutive equation.

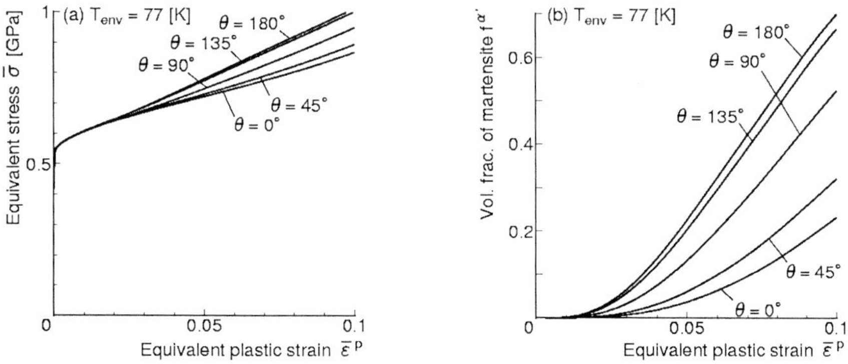

(a) *equivalent stress-equivalent plastic strain relations* (b) *volume fraction of martensitic phase-equivalent plastic strain relations*

Figure2 Deformation behavior of TRIP steels without pre-strain

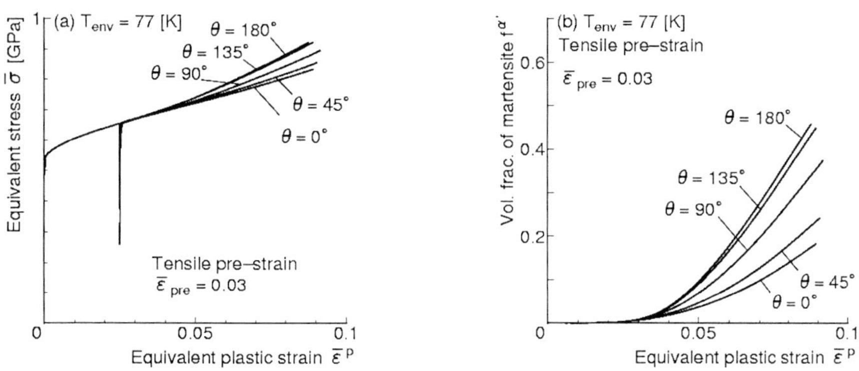

(a) *equivalent stress-equivalent plastic strain relations* (b) *volume fraction of martensitic phase-equivalent plastic strain relations*

Figure 3 Deformation behavior of TRIP steels with tensile prestrain

Figure 3 illustrates (a) equivalent stress- and (b) volume fraction of martensitic phase-plastic strain relationship with tensile prestrain at environmental temperature T_{env} =77K. The same tendency, observed in the case, without prestrain is observed but is slightly suppressed, which is attributable to the reduction of the fraction of hardening due to martensitic transformation after prestrain. Therefore, the hardening after prestraining depends on the magnitude and type of prestrain.

3.3 MONOTONIC NONUNIFORM DEFORMATION

The proposed computational strategy has been applied to the prediction of the nonuniform deformation behavior of TRIP steels bars with a ringed notch under tension and the capability of the method has been discussed.

Figure 4 illustrates the distribution of the volume fraction of martensite around the notch tip deformed under uniaxial tension at 77K. Spreading of the martensite phase around the notch tip is observed in Fig. 4. It is found that the high-volume-fraction area spreads in accordance with strain localization, extending to the maximum shearing direction, which corresponds well with the experimentally estimated results based on the newly developed method [5]. Blunting of the notch due to the transformation is observed in Fig. 4. Therefore the development of the martensite phase may relieve the strain localization around the notch tip and subsequently, improvement of the toughness may occur. Figure 5 shows the volume fraction of the martensite along the radial direction. Due to the microscopically nonuniform distribution of martensite, the experimental results fluctuate depending on the indentation loads. However, the average value corresponds well with the computationally predicted results.

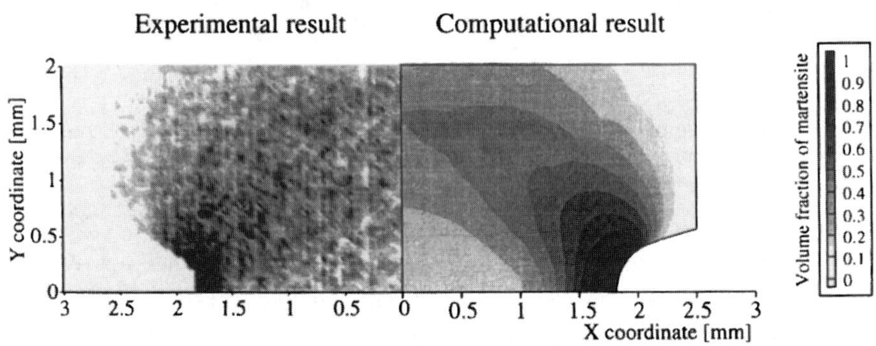

Figure4 Martensitic volume fraction distribution around the notch tip estimated experimentally and predicted by computational simulation for end displacement $\Delta l = 0.3\,mm$.

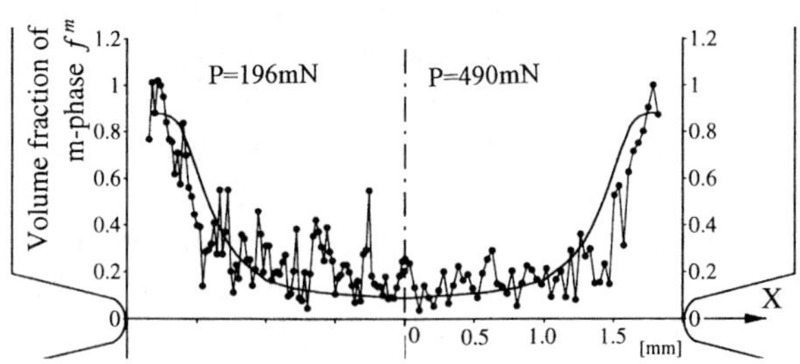

Figure 5 Martensitic volume fraction distributions along the radial direction on the notched section. Dotted line indicates experimental results due to microhardness testing device with indentation loads 196mN and 490mN, and solid line stands for that due to computational simulation.

3.4 CYCLIC UNIFORM DEFORMATION

The computational simulation of the plane strain block deformed under cyclic loading with and without prestrain at environmental temperatures of 77K and 298K, was performed. A strain range of 0.03 was selected for the simulation.

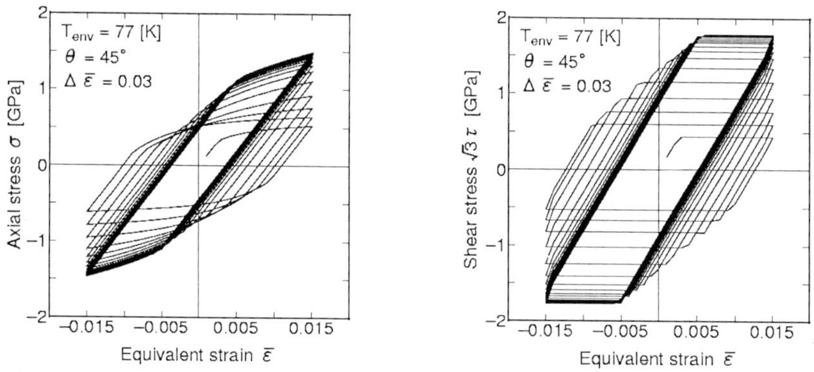

(a) *axial stress-equivalent stress relations* (b) *shear stress-equivalent strain relations*
Figure 6 Cyclic deformation behaviors under combined proportional straining without pre-strain

Figure 6 shows (a) axial stress- and (b) shear stress-strain relationship under proportional straining with $\theta = 45^\circ$ at 77K. Due to the martensitic transformation, hardening, which causes an increase in the stress range $\Delta\sigma$, and a monotonic decrease in the width of the hysteresis loop ΔS_i^p with respect to strain, are observed. Figure 7 illustrates (a) the equivalent stress range $\Delta\bar{\sigma}$ and (b) volume fraction of the martensitic phase $f^{\alpha'}$ with respect to the number of cycles N applied under axial, proportional and shear straining with strain range $\Delta\varepsilon$ = 0.03 at 77 and 298K. As expected, a rapid saturation of the volume fraction of the martensitic phase is attained at a lower temperature, which is manifested as the saturation of the stress range $\Delta\bar{\sigma}$ with few cycles. On the other hand, the saturation of the stress range $\Delta\bar{\sigma}$ was not observed at 40 cycles of loading in the high temperature range. The results obtained are similar to those experimentally obtained for 304 austenitic stainless steel [6]. It should be noted that the saturation values of the volume fraction of the martensitic phase for cyclic axial and proportional loading are higher than that for monotonic deformation, which is attributable to a different probability of martensitic transformation at the intersection of shear bands, as indicated by Eq. (1). That is, in cyclic loading, the transformation favorably occurs during the compression process. The stress-dependent features of cyclic deformation processes can be observed in Fig. 6(a), $\sigma^+_{max} \geq \sigma^-_{max}$, which are also attributable to the dependence of the martensitic transformation on the stress, as shown by Eq. (2). Thus, the typical nature of the cyclic deformation processes observed in

TRIP steels is attributable to the martensitic transformation due to plastic deformation depending on the type of straining.

In order to clarify the effect of prestrain on the cyclic deformation behavior of TRIP steels, the computational predictions have been performed with various magnitudes of prestrain. Similar to the asymptotic nature observed in Fig. 6 for the case without prestrain, a gradual reduction in the width of the stress-strain relationship ΔS_i^p, an increase in the stress range $\Delta\sigma$, and their saturation with the number of cycles were observed. It has been clarified that the rate of accumulation of plastic strain decreases as the number of cycles increases and asymptotically approaches a specified value, which is due to the martensitic transformation of the austenitic phase and is attributable to the constant saturation value of the martensitic phase for the case with different prestrains. On the other hand, the stress range $\Delta\sigma$ increases with higher accumulation of plastic strain S^p, whereas the rate of increase is suppressed with prestrain ε_{pre}. This can be explained by the fact that the fraction of hardening due to martensitic transformation after prestraining decreases as the magnitude of prestrain increases. This characteristic feature of prestrain was observed in the uniaxial cyclic loading case and is compatible with that obtained in experiments [6].

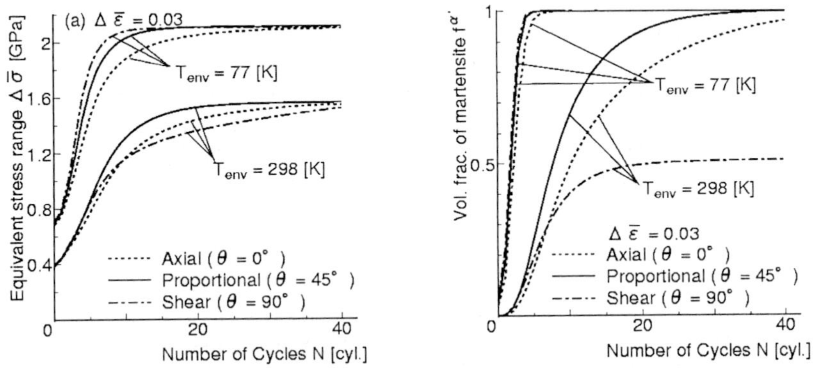

(a) *equivalent stress range-number of cycles relations* (b) *volume fraction of martensitic phase-number of cycles relations*

Figure 7 Cyclic deformation behaviors under axial, shear and proportional straining without prestrain

3.5 CYCLIC NONUNIFORM DEFORMATION BEHAVIOR

Here we consider the cyclic loading process of ringed-notched specimens that were used in the experiment described in 3.3 at an environmental temperature of 77K. Since the deformation was localized to the notched region, the saturation of the volume fraction of the martensitic phase was restricted to a small area. Therefore, the asymptotic nature of the load amplitude versus the number of cycles was observed in the early stage of the cycles, and the hysteresis loop in load-displacement relationship was very small and converges rapidly to the quasi-elastic deformation process. Figure

8(a) shows the evolution of the volume fraction of the martensitic phase for various numbers of cycles. The area with a high volume fraction of the martensitic phase starts from the notch root and penetrates into the interior of the specimen at approximately 45 degrees with respect to the tension axis. This martensitic transformation causes the distribution effect on the deformed region. Due to the high transforming probability of the intersection of microshear bands to a martensitic embryo, the transformation in the case of compression is promoted further than that in the case of tension. Therefore, the evolution of the volume fraction of the martensitic phase under the cyclic loading process is mainly promoted during the compressive loading process.

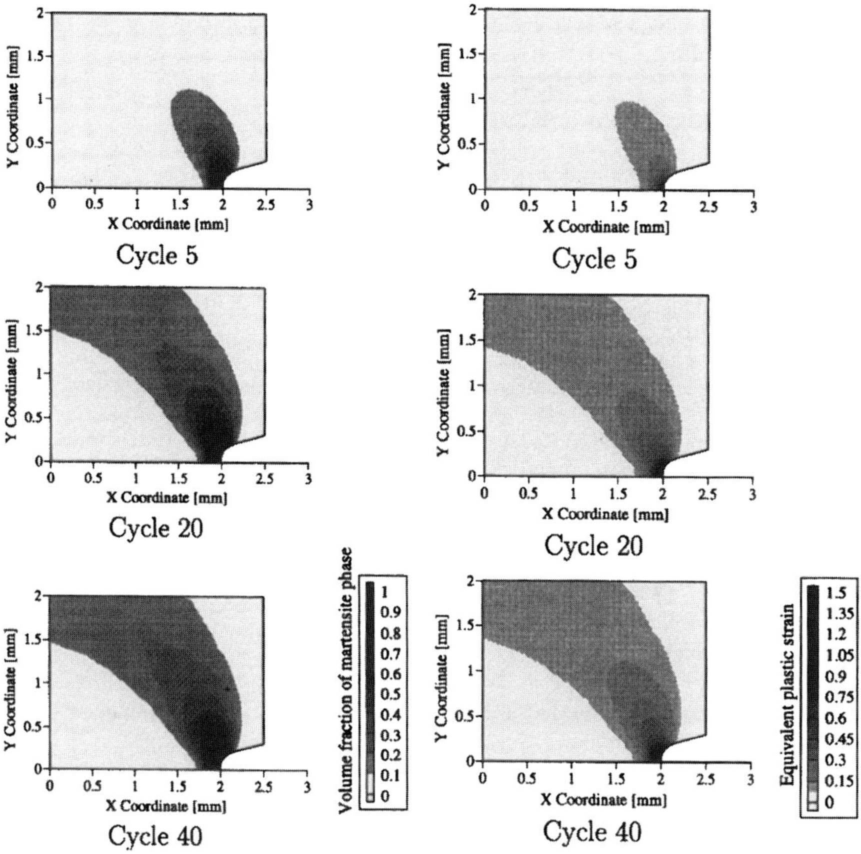

(a) *volume fraction of martensitic phase* (b) *equivalent plastic strain distribution*

Figure 8 Local deformation behavior of ringed notch specimen.

4. Conclusions

The constitutive equation that models the martensitic transformation dependence on the strain rate, temperature and applied stress system was applied to clarify the deformation behavior of cyclic axial, shear and proportional loading of TRIP steels with and without prestrain ε_{pre}. The effect of stress $\Delta\sigma$ and strain $\Delta\varepsilon$ ranges, temperature T and prestrain ε_{pre} on the stress-strain relationship and the saturation nature of stress range $\Delta\sigma$ and strain range $\Delta\varepsilon$ in the stress-strain relationship were clarified. The computational simulation of the deformation behavior of TRIP steel bars with the ringed notch under uniaxial tension at 77K was performed. Since the validity of the predictability of monotonic/cyclic deformation behavior of TRIP steels for uniform stress cases and monotonic deformation of ringed-notched bars was verified, the deformation behavior thus predicted for the ringed-notched specimen under cyclic loading should provide useful information on the deformation behavior of TRIP steels under complex loading.

Acknowledgements

Financial support from the Ministry of Education of Japan is gratefully acknowledged.

References

1. Olson, G. B., Cohen, M. (1975) Kinematics of strain-induced martensitic nucleation. *Metall Trans*. A **6**, 791-795.
2. Stringfellow, R. G, Parks, D. M, Olson, G. B. (1992) A constitutive model for transformation plasticity accompanying strain-induced martensitic transformation in metastable austenitic steels. *Acta Metall*. **40**, 1703-1716.
3. Tomita, Y, Iwamoto, T. (1995) Constitutive modeling of TRIP steel and its application to the improvement of mechanical properties. *Int. J. Mech. Sci.* **37**, 1295-1305.
4. Iwamoto, T, Tsuta, T, Tomita, Y. (1998) Investigation on deformation mode dependence of strain-induced martensitic transformation in TRIP steels and modeling of transformation kinetics. *Int. J. Mech. Sci.* **40**, 173-182.
5. Shibutani, Y., Taniyama, A., Tomita, Y. and Adachi, T. (1997) Measurement of local strain-induced martensitic phase transformation by microhardness, *JSMS Japan*, **46**, 893-899.
6. Tanaka, N, Shirasawa, Y, Niitsu, Y, Ikegami, K. (1991) Experimental investigation of plastic deformation of SUS304 stainless steel at 77K, *Trans. JSME* A, **57**, 2775-2781. (in Japanese)
7. Yokotsuka, T, Ikegami, K. (1999) Cyclic plastic deformation at low temperature in liquid nitrogen and room temperature subsequent to pre – strain at low temperature in liquid nitrogen and room temperature. *J. Soc. Mat. Sci., Japan* **48**, 38-43. (in Japanese)
8. Hecker, S .S, Stout, M, G, Staudhammer, K. P, Smith, J. L. (1982) Effect of strain state and strain rate on deformation induced transformation in 304 stainless steel: part I. magnetic measurements and mechanical behavior. *Metall. Trans*. A**13**, 619-626.
9. Miller, M. P, McDowell, D. L. (1996) Modeling large strain multiaxial effects in FCC polycrystals. *Int. J. Plast*. **12**, 875-902.
10. Tomita, Y, Shindo, A, Sasayama, S. (1990) Plane strain tension of thermo-elasto- viscoplastic blocks. *Int. J. Mech. Sci.* **32**, 613- 622.
11. Shirasawa, Y, Ikegami, K, Niitsu, Y. (1993) Experimental investigation of cyclic plastic deformation of stainless steel 304 at low temperature. *Trans. JSME*.A, **59**, 2427-2434. (in Japanese)

A THERMODYNAMIC APPROACH TO MODELING OF STRESS-INDUCED PHASE-TRANSITION FRONT PROPAGATION IN SOLIDS

A. BEREZOVSKI, J. ENGELBRECHT
Department of Mechanics and Applied Mathematics
Institute of Cybernetics at Tallinn Technical University
Akadeemia tee 21, 12618 Tallinn, Estonia

AND

G.A. MAUGIN
Laboratoire de Modélisation en Mécanique
Université Pierre et Marie Curie
UMR 7607, Tour 66, 4 Place Jussieu, Case 162
75252, Paris Cédex 05, France

1. Introduction

A number of continual models for describing the dynamics of phase boundaries has been developed in order to understand evolving phase transitions (cf., for example, [1]). All these models take the basic balance laws of continuum mechanics as a starting point. However, the propagation speed of phase-transition fronts is not completely determined by the balance of momentum in the dynamic problems involving phase boundaries. As a result, an additional constitutive criterion is needed for the determination of the dynamics of phase boundaries. In the sharp-interface theory, phase boundaries are treated as discontinuity surfaces of zero thickness. The canonical formalism of continuum mechanics with a full exploitation of the balance of so called pseudo- or canonical momentum leads to the balance of "material" forces at the phase-transition front. The surface "balance" equation plays an essential role in the description of phase-transition front propagation. However, in all the theories, the constitutive relation for free energy is assumed to be able to describe the states in both sides of the phase-transition front simultaneously.

Q.P. Sun (ed.), IUTAM Symposium on Mechanics of Martensitic Phase Transformation in Solids, 19–26.

Here we consider a simpler situation, when each phase has its own prescribed classical constitutive equation for the thermoelastic free energy. In this case, the driving force and the entropy flux should be computed by means of the surface "balance" equation. However, for the computation of the phase-transition front propagation, the equations are not closed, because a criterion of the progress of the front is additionally needed. In this paper, a thermodynamic criterion is proposed, based on the stability conditions for complex thermodynamic systems. The conditions of the stability are derived and expressed in terms of contact quantities, which are introduced in the framework of the thermodynamics of discrete systems in order to describe the nonequilibrium states of discrete elements. The contact quantities at the phase boundary are determined by means of the above-mentioned balance of material forces at the interface and the jump relation for the entropy. As a result, an efficient and robust numerical method capable of solving dynamical phase transition problems is developed. As a first attempt, the numerical simulation of stress-induced austenitic-martensitic phase-transition front propagation in the one-dimensional case is performed. The analysis of the two-dimensional case is in progress.

2. Balance laws for linear thermoelasticity

We consider both two phases of the material as thermoelastic heat conductors. If geometrical nonlinearities are neglected, then the main *bulk* equations of inhomogeneous linear isotropic thermoelasticity (in the absence of body forces) can be written as the following three equations:

$$\rho_0(\mathbf{x})\frac{\partial v_i}{\partial t} - \frac{\partial \sigma_{ij}}{\partial x_j} = 0, \tag{1}$$

$$\frac{\partial \sigma_{ij}}{\partial t} = \lambda(\mathbf{x})\frac{\partial v_k}{\partial x_k}\delta_{ij} + \mu(\mathbf{x})\left(\frac{\partial v_i}{\partial x_j} + \frac{\partial v_j}{\partial x_i}\right) + m(\mathbf{x})\frac{\partial \theta}{\partial t}\delta_{ij}, \tag{2}$$

$$C(\mathbf{x})\frac{\partial \theta}{\partial t} = \frac{\partial}{\partial x_i}\left(k(\mathbf{x})\frac{\partial \theta}{\partial x_i}\right) + m(\mathbf{x})\frac{\partial v_k}{\partial x_k}, \tag{3}$$

of which the second one is none other than the time derivative of the Duhamel-Neumann thermoelastic constitutive equation [2]. Here t is time, x_j are spatial coordinates, v_i are components of the velocity vector, σ_{ij} is the Cauchy stress tensor, ρ_0 is the density, θ is temperature, $C(\mathbf{x}) = \rho_0 c$, c is the specific heat at constant stress, k is the thermal conductivity. The *dilatation coefficient* α is related to the thermoelastic coefficient m, and the Lamé coefficients λ and μ by $m = -\alpha(3\lambda + 2\mu)$. The indicated explicit dependence on the point $\mathbf{x}$ means that the body is materially inhomogeneous in general.

The system of equations (1)-(3) is a system of hyperbolic conservation laws with source terms in a form that is suitable for a numerical solution. Well developed numerical methods exist for solving of this system of equations including the case of inhomogeneous media like, for example, the wave-propagation algorithm [3],[4]. However, in the case of moving phase transition fronts some additional considerations are needed.

3. Jump relations

In order to consider the possible irreversible transformation of a phase into another one, the separation between the two phases is idealized as a sharp, discontinuity surface $\mathcal{S}$ across which some of the fields suffer finite discontinuity jumps. It is assumed that the phase transition fronts are *homothermal* (no jump in temperature) and *coherent* (they present no defects such as dislocations). Thus, we have the following continuity conditions [5]:

$$[\mathbf{V}] = \mathbf{0}, \qquad [\theta] = \mathbf{0} \qquad at \;\; \mathcal{S}. \tag{4}$$

Here $[A]$ denotes the jump of a discontinuous field A across $\mathcal{S}$.

The material velocity $\mathbf{V}$ is defined by means of the inverse mapping $\mathbf{X} = \chi^{-1}(\mathbf{x}, t)$, where $\mathbf{X}$ denotes the material points. Jump relations associated with the conservation laws in the bulk are formulated according to the theory of *weak solutions* of hyperbolic systems [6]-[7]

$$\tilde{V}_N[\rho_0 v_i] + N_j[\sigma_{ij}] = 0, \tag{5}$$

$$\tilde{V}_N[S] - N_i \left[\frac{k}{\theta}\frac{\partial \theta}{\partial x_i}\right] = \sigma_{\mathcal{S}} \geq 0. \tag{6}$$

where $\tilde{\mathbf{V}}$ is the material velocity of the geometrical points of $\mathcal{S}$, $\tilde{V}_N$ is the normal speed of the points of $\mathcal{S}$, N_i are components of the unit normal to $\mathcal{S}$, S is the entropy per unit volume, and $\sigma_{\mathcal{S}}$ is the entropy production at the interface. As it was shown in [5]-[7], the entropy production can be expressed in terms of the so-called "material" driving force $f_{\mathcal{S}}$

$$f_{\mathcal{S}}\tilde{V}_N = \theta_{\mathcal{S}}\sigma_{\mathcal{S}} \geq 0, \tag{7}$$

where $\theta_{\mathcal{S}}$ is the temperature at $\mathcal{S}$. In addition, the *balance of "material" forces* at the interface between phases should be satisfied [5]-[9]. In the considered case, this can be specified to the form

$$f_{\mathcal{S}} = -[W] + < \sigma_{ij} > [\varepsilon_{ij}], \tag{8}$$

where W is the free energy per unit volume and $< A >$ denotes the mean value of a discontinuous field A. The surface "balance" equation (8) follows from the balance law for pseudomomentum [5]-[7] and generalizes the equilibrium conditions at the phase transition front [10] to the dynamical case. Usually, a relation of constitutive type between the values of driving force and material velocity at the front should be written for the computation of the phase transition. Instead of that we use a thermodynamic considerations based on the thermodynamics of discrete systems [11].

4. Thermodynamic consistency conditions for thermoelastic media

The thermodynamic consistency conditions for simple thermodynamic systems are expressed in terms of internal energy [12]

$$\left[\left(\frac{\partial \hat{E} + \hat{E}^{int}}{\partial V}\right)_\theta\right] = 0, \qquad \left[\left(\frac{\partial \hat{E} + \hat{E}^{int}}{\partial V}\right)_p\right] = 0, \tag{9}$$

in the homogeneous and heterogeneous case, respectively. Here $\hat{E}$ is the internal energy, V is volume, and E^{int} is the so-called *interaction energy*, which appears in the description of the interaction of non-equilibrium discrete elements [12]. It should be noted that the value of $\hat{E}^{int}$ is undetermined yet. In the *thermoelastic case*, the thermodynamic derivatives which should be exploited instead of $\left(\frac{\partial \hat{E}}{\partial V}\right)_\theta$ and $\left(\frac{\partial \hat{E}}{\partial V}\right)_p$ are the following [13] (i and j are fixed):

$$\left(\frac{\partial E}{\partial \varepsilon_{ij}}\right)_\theta = -\theta\left(\frac{\partial \sigma_{ij}}{\partial \theta}\right)_{\varepsilon_{ij}} + \sigma_{ij}, \qquad \left(\frac{\partial E}{\partial \varepsilon_{ij}}\right)_{\sigma_{ij}} = \theta\left(\frac{\partial S}{\partial \varepsilon_{ij}}\right)_{\sigma_{ij}} + \sigma_{ij}. \tag{10}$$

Here E is the internal energy per unit volume.

For extending the concepts of the thermodynamics of discrete systems to the *thermoelastic case*, we divide the body into a *finite number of identical elements*. The state of each element is then identified with the thermodynamic state of a discrete system associated with that element; each element being assumed to be in local equilibrium. In thermoelasticity, in addition to the *contact temperature* Θ [11], which governs heat exchange, we must define a *contact dynamic stress tensor* Σ_{ij} since the state space includes the deformation. We have thus

$$\frac{\partial \varepsilon_{ij}}{\partial t}(\Sigma_{ij} - \sigma^*_{ij}) \geq 0, \qquad (\dot{Q} = 0). \tag{11}$$

Here σ^*_{ij} is the Cauchy stress tensor in the environment.

It is supposed that the introduced contact quantities are connected with the energy of interaction in the similar way as in Eqs (10), i.e.

$$\left(\frac{\partial E^{int}}{\partial \varepsilon_{ij}}\right)_\theta = -\Theta\left(\frac{\partial \Sigma_{ij}}{\partial \theta}\right)_{\varepsilon_{ij}} + \Sigma_{ij}, \left(\frac{\partial E^{int}}{\partial \varepsilon_{ij}}\right)_{\sigma_{ij}} = \Theta\left(\frac{\partial S^{int}}{\partial \varepsilon_{ij}}\right)_{\sigma_{ij}} + \Sigma_{ij}. \tag{12}$$

In fact, the latter two equations are the relations of a constitutive type for the introduced contact quantities. It should be noted that the interaction entropy S^{int} is still undetermined yet. Just these conditions will be applied to determine the values of the contact quantities at the phase boundary.

5. Contact quantities at the phase boundary

Suppose that the interface between two thermoelastic phases is placed between elements numbered $(p-1\,q)$ and (pq). We propose to apply the heterogeneous consistency conditions for the calculation of the contact stresses at the phase boundary. In the two-dimensional case, we can rewrite the heterogeneous consistency conditions in the form

$$[\sigma_{ij}] + \theta_S \left[\left(\frac{\partial S^{int}}{\partial \varepsilon_{ij}} \right)_\sigma \right] + (\Sigma^l_{ij})_{pq} - (\Sigma^r_{ij})_{p-1\,q} = 0. \tag{13}$$

Further, we suppose that the jump of the entropy of interaction is equal to the jump of entropy at the phase boundary

$$[S^{int}] = [S]. \tag{14}$$

It is obvious that for zero entropy of interaction the heterogeneous consistency conditions can be reduced to those in homogeneous case and, therefore, no change is needed. However, the jump of entropy is nonzero at the phase boundary if phase transition occurs.

For the computation of the jump of entropy at the phase boundary we will exploit the jump relation corresponding to the balance of the entropy (6). In the two-dimensional homothermal case, we have

$$[S] = f_S/\theta_S, \tag{15}$$

where the driving force is determined by the balance of "material" forces (8) with the free energy in the two-dimensional thermoelastic case

$$W = \frac{1}{2}\left(\sigma_{ij}\varepsilon_{ij} - \frac{C}{\theta_0}(\theta - \theta_0)^2 - \alpha(3\lambda + 2\mu)(\theta - \theta_0)(\varepsilon_{11} + \varepsilon_{22}) \right). \tag{16}$$

Thus, the jump of the entropy of interaction is determined completely. Remaining relations follow from the coherency conditions for the material velocity (4). Obtained conditions form a linear system of equations which can be solved exactly, and we can update the state of the elements adjacent to the phase boundary by means of the wave-propagation algorithm [12]. It should be noted that the proposed new procedure should be applied at the phase boundary instead of standard one only after the initiation of the phase transition process. The material velocity at the phase boundary can be determined by means of jump relation for linear momentum (5). At last, the direction of the front propagation is determined by the positivity of the entropy production

$$\sigma_S = \frac{f_S \tilde{V}_N}{\theta_S} \geq 0. \tag{17}$$

Now all bulk quantities can be computed and the driving force and the material velocity at the phase boundary determined. However, these quantities manifest themselves only if the phase transition takes place.

6. A thermodynamic initiation criterion for the stress-induced phase transition

We propose to expect the initiation of the stress-induced phase transition if both heterogeneous and homogeneous consistency conditions are fulfilled at the phase boundary simultaneously. Let the heterogeneous consistency condition (13) be fulfilled. Then the homogeneous consistency condition can be represented at the phase boundary as follows

$$-\theta_S \left[\left(\frac{\partial \sigma_{ij}}{\partial \theta} \right)_{\varepsilon} \right] - \theta_S \left[\left(\frac{\partial S^{int}}{\partial \varepsilon_{ij}} \right)_{\sigma} \right] + \theta_S \left[\left(\frac{\partial \Sigma_{ij}}{\partial \theta} \right)_{\varepsilon} \right] = 0. \tag{18}$$

It should be noted that the combined consistency condition (18) must be fulfilled at least for one component of the stress tensor, because only one additional condition to be satisfied in order to determine the initiation of the phase transition. In particular, for the shear component of the stress tensor σ_{12} we have a simple relation of continuity of the shear stress at the phase boundary

$$[\sigma_{12}] = 0. \tag{19}$$

The expression (19) is inconvenient for the determination of the initiation of the phase transition, because it is fulfilled even in the absence of phase transitions and any other perturbations. Just therefore the combined consistency condition (18) for the normal components of stress tensor should be also checked. The combined consistency condition (18) for the normal component of the stress tensor σ_{11} leads to the following expression for the driving force f_S:

$$f_{critical} = -\frac{\theta_S^2}{4}[\alpha(3\lambda + 2\mu)] \left\langle \frac{\alpha(3\lambda + 2\mu)}{\lambda + \mu} \right\rangle -$$

$$-\frac{\theta_S}{8}[\sigma_{11}] \left\langle \frac{\alpha(3\lambda + 2\mu)}{\lambda + \mu} \right\rangle + \frac{\theta_S}{2}[(\lambda + \mu)(\varepsilon_{11} + \varepsilon_{22})] \left\langle \frac{\alpha(3\lambda + 2\mu)}{\lambda + \mu} \right\rangle. \tag{20}$$

Therefore, the proposed criterion for the initiation of the stress-induced phase transition is the following:

$$|f_S| \geq |f_{critical}|. \tag{21}$$

A similar condition can be obtained for the second normal component of the stress tensor.

7. Numerical results for stress-induced phase transition

As an example, an one-dimensional martensite-austenite phase-transition front propagation is simulated in a thermoelastic medium. The physical properties of the material correspond to the NiTi alloy [14].

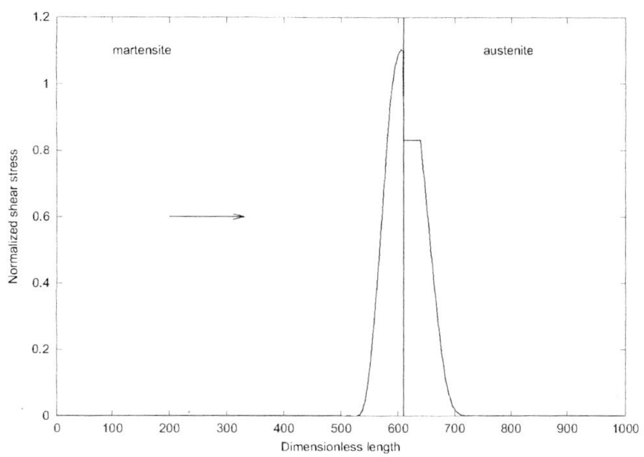

Figure 1. Interaction of shear wave with phase transition front.

It is supposed that the medium contains both martensitic and austenitic parts, which coexist at the temperature $\theta_S = \theta = 300\,K$. An isothermal loading of shear stress is applied at the left boundary of the martensitic region. The form of the hump is shown in the left part of Fig. 1. The location of the phase boundary (initially $x = 600$) is marked by a vertical line.

If the amplitude of loading is sufficiently large to overcome the critical value of the driving force, we have the interaction of the shear wave with the phase-transition front (Fig. 1). As one can see, the value of the stress remains constant after the interaction, while the location of the phase boundary is displaced into the austenitic region.

It should be noted that the displacement of the front location depends nonlinearly on the amplitude of the loading, because of the nonlinearity in the computation of the driving force.

Acknowledgment

Support of the Estonian Science Foundation under contract No.3203 (A.B.), of the NATO grant PST.CLG.976009 "Thermomechanics of Progress and Stability of Phase Interfaces (Crystals, Alloys)" (A.B., J.E and G.A.M.), and of the European Network TMR. 98-0229 on "Phase transitions in crystalline substances" (G.A.M.) is gratefully acknowledged.

References

1. Hou, T.Y, Rosakis, P. and LeFloch, P.: A level-set approach to the computation of twinning and phase-transition dynamics, *J. Comp. Physics* **150** (1999), 302-331.
2. Maugin, G.A. and Berezovski, A.: Material formulation of finite strain thermoelasticity and applications, *J. Thermal Stresses* **22** (1999), 421-449.
3. LeVeque, R.J.: Wave propagation algorithms for multidimensional hyperbolic systems, *J. Comp. Physics* **131** (1997), 327-353.
4. LeVeque, R.J.: Balancing source terms and flux gradients in high-resolution Godunov methods: the quasi-steady wave-propagation algorithm, *J. Comp. Physics* **148** (1998), 346-365.
5. Maugin, G.A.: On shock waves and phase-transition fronts in continua, *ARI* **50** (1998), 141-150.
6. Maugin, G.A.: Thermomechanics of inhomogeneous - heterogeneous systems: application to the irreversible progress of two- and three-dimensional defects, *ARI* **50** (1997), 41-56.
7. Maugin, G.A. and Trimarco, C.: Driving force on phase transition fronts in thermoelectroelastic crystals, *Math. Mech. Solids* **2** (1997), 199-214.
8. Abeyaratne, R. and Knowles, J.K.: On the driving traction acting on a surface of strain discontinuity in a continuum, *J. Mech. Phys. Solids* **38** (1990), 345-360.
9. Truskinovsky, L.: Dynamics of nonequilibrium phase boundaries in a heat conducting nonlinear elastic medium, *J. Appl. Math. Mech.* (PMM) **51** (1987), 777-784.
10. Grinfeld, M.A.: *Thermodynamic Methods in the Theory of Heterogeneous Systems*, Longman, London, 1991.
11. Muschik, W.: Fundamentals of non-equilibrium thermodynamics, in W. Muschik ed., *Non-Equilibrium Thermodynamics with Application to Solids*, Springer, Wien, 1993, pp. 1-63.
12. Berezovski,A., Engelbrecht, J. and Maugin, G.A.: Thermoelastic wave propagation in inhomogeneous media, *Arch. Appl. Mech.* **70** (2000), 694-706.
13. Callen, H.B. *Thermodynamics*, Wiley & Sons, New York, 1960.
14. Hodgson, D.E., Wu, M.H. and Biermann, R.J.: Shape memory alloys, in *Metals Handbook*, 10th Edition, **2**, ASM International, 1990, pp. 897-903.

THE FORMATION OF MACROTWINS IN NiAl MARTENSITE

J.M. BALL* (ball@maths.ox.ac.uk)
Mathematical Institute, University of Oxford, 24-29 St. Giles', Oxford OX1 3LB, U.K.

D. SCHRYVERS† (schryver@ruca.ua.ac.be)
EMAT, University of Antwerp, RUCA, Groenenborgerlaan 171, B-2020 Antwerpen, Belgium

1. Introduction.

Using high resolution electron microscopy, Boullay & Schryvers [6] observed and documented interesting macrotwin interfaces in bulk and splat-cooled $Ni_{65}Al_{35}$ polycrystals. This alloy undergoes a cubic (bcc) to tetragonal (bct) martensitic transformation. The macrotwin interfaces separate two different plates of martensite, each plate involving the same two martensitic variants but with microtwin interfaces that are nearly orthogonal. They measured the orientations both of the microtwin interfaces and of the rotations of the variants comprising the microtwins, in the vicinity of the macrotwin plane and at distances of up to several hundred nanometres from this plane. In addition, they described atomic scale details of the macrotwin region, in particular identifying two different ways in which the microtwins meet at the macrotwin plane, one in which they appear to cross, and the other in which they meet in a stepped configuration (see Fig. 1). In the bulk samples, which form the main focus of this paper, the grain-size is large (of the order of 1mm) compared to that of the macrotwins. In the splat-cooled samples it is smaller (of the order of 1μm).

It is natural to believe that these macrotwins arise from coalescence of two different martensitic plates nucleated in the austenite at different points, each having a different habit plane separating the martensite from the austenite, that propagate towards each other, eliminating the austenite

* Research supported by EC TMR Contract FMRX-CT98-0229 on 'Phase Transitions in Crystalline Solids'.

† Research supported by the above EC contract and the IUAP project of the Belgium Federal government on "Reduced Dimensionality Systems" (P4/10).

Q.P. Sun (ed.), IUTAM Symposium on Mechanics of Martensitic Phase Transformation in Solids, 27–36.

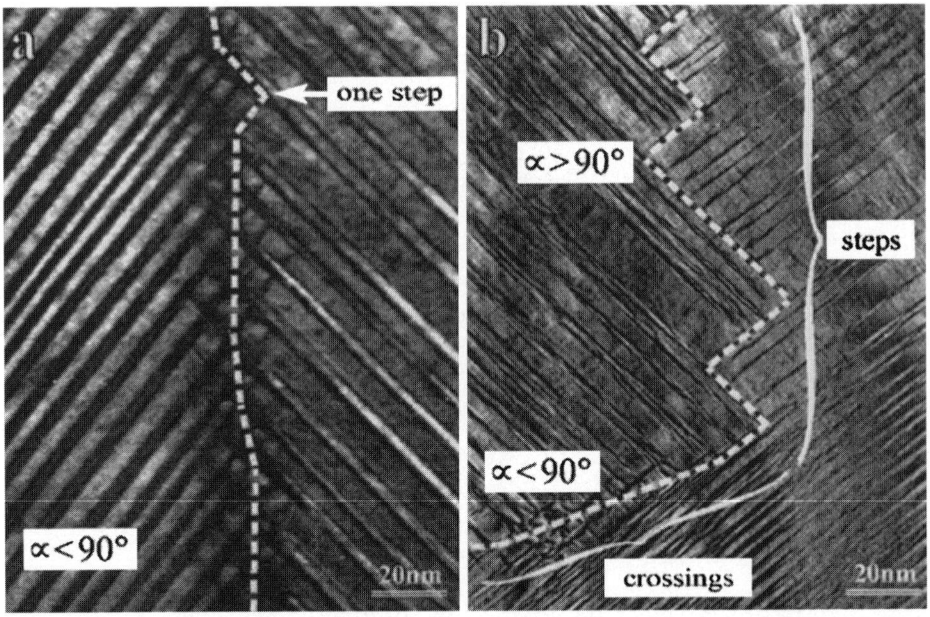

Figure 1. Low magnification images of macrotwin boundaries of (a) crossing, and (b) step type. The angles α indicated are those between the microtwin interfaces in each plate.

between them. However for the bulk and splat-cooled samples used in the experiments the transformation is too fast to observe the sequence of events leading to the final martensitic microstructure, and so indirect evidence is needed to confirm the above scenario.

Some such evidence is provided by correlating details of the observed microstructures with predictions of the nonlinear elasticity model of martensitic transformations (see Ball & James [2], [3]), a model that incorporates the crystallographic theory of martensite (Wechsler, Lieberman & Read [13]) into a much more general framework. (For corresponding linearized models see [8], [9], [10], [11], [12], [5].) We treat the sample as if it were a single crystal, ignoring constraints imposed by neighbouring grains. In terms of the kind of calculations made, this is justified provided the microstructures have close to zero energy (the zero of energy being taken to be that of a pure variant of martensite or of undistorted austenite, these energies being equal at the transformation temperature θ_c).

A calculation of Bhattacharya [4] related to the so-called *wedge microstructure* shows that it is impossible for two different martensitic plates, involving the same two martensitic variants and having different habit plane normals, to be compatible at zero energy at θ_c across any macrotwin plane. This means that if the observed macrotwins arise by coalescence of two such plates, then stresses must be accommodated in the vicinity of the macrotwin boundaries. Nevertheless one can calculate (see Section 3) the relative orientations of the microtwin planes and of the different variants in the two plates prior to meeting of the habit planes, as well as the relative volume fractions of the two variants in each plate. These volume fractions are predicted by the theory to be either the same in each plate, or reversed. In all the examples of macrotwins observed by Boullay & Schryvers the volume fractions in the two plates were found to be almost the same and to be close to that predicted by the theory for a single plate in contact with the austenite, suggesting that these volume fractions remain more or less unchanged as the plates coalesce.

On the other hand one can do a second calculation (see Section 4) to determine whether two plates involving the same two martensitic variants and having the same volume fractions can be compatible at zero energy across a macrotwin interface, without requiring, as in the Bhattacharya calculation, that these plates are also compatible with the austenite. This calculation shows that the macrotwin normal must be a prior [100] cubic plane, and delivers relative orientations of the microtwin planes and variants that are different from those of the first calculation.

It turns out that both these calculations seem to be relevant. The first predicts with satisfactory accuracy many details of the experimental data at a sufficiently large distance (more than 500nm) from the macrotwin interface, while for the case of the crossing-type macrotwins, the second correctly predicts the trends in the relative orientations that are observed in an intermediate region between 50 and 500nm from the macrotwin interface. In this intermediate region the microtwin interfaces curve, this curvature being accompanied by a corresponding rotation of the crystal lattice. In the region very close to the macrotwin interface, from 0 to 50nm from it, there is still an apparent macroscopic curvature of the microtwin interfaces, but at the atomic level this is seen to be achieved without lattice rotation, the microtwin interfaces consisting of planar sections having the same orientation and separated by nearly equidistant atomic scale ledges. Presumably lattice curvature is the energetically optimal way to relieve small stresses, while ledges, for which the energy is concentrated in dislocations, are best for relieving the larger stresses close to the macrotwin interface. The situation for the stepped-type macrotwins closer to the macrotwin interface is somewhat different, and is described in more detail in [7].

In this paper we give precise statements and a few details of these calculations, the results of which are used for the experimental comparisons in [6], and which led to some of the observations made there. A more comprehensive treatment will appear elsewhere.

2. The nonlinear elasticity model of martensitic transformations

This model is based on an elastic free-energy density $\varphi(F,\theta)$, in terms of which the total free-energy of a single crystal at temperature θ is given by

$$I_\theta(y) = \int_\Omega \varphi(\nabla y(x),\theta)\,dx. \tag{2.1}$$

In (2.1) $y(x) = (y_1(x), y_2(x), y_3(x))$ denotes the deformed position of the material point of the crystal occupying the point $x = (x_1, x_2, x_3)$ in the region Ω occupied by the crystal in a reference configuration, which we take to be undistorted austenite at the transformation temperature θ_c, and $\nabla y(x)$ denotes the deformation gradient, namely the 3×3 matrix $(\partial y_i/\partial x_j)$. The free-energy density is assumed to be frame-indifferent, that is

$$\varphi(RF,\theta) = \varphi(F,\theta) \tag{2.2}$$

for all F in the set $M_+^{3\times 3}$ of 3×3 matrices with positive determinant, and all R in the set $SO(3)$ of rotations. Further, φ is assumed to satisfy the material symmetry condition

$$\varphi(FQ,\theta) = \varphi(F,\theta) \tag{2.3}$$

for all rotations Q in the point group $\mathcal{S}$ of the crystal, in our case given by $\mathcal{S} = \mathcal{P}^{24}$, the set of rotations of a cube into itself. By adding a suitable function of θ to φ we may assume that $\min \varphi(\cdot,\theta) = 0$, so that the set $K(\theta)$ of energy-minimizing deformation gradients is given by

$$K(\theta) = \{F \in M_+^{3\times 3} : \varphi(F,\theta) = 0\}$$

By (2.2), (2.3) we have that

$$K(\theta) = RK(\theta)Q \qquad \text{for all } R \in SO(3), Q \in \mathcal{S}.$$

More specifically, for a cubic-to-tetragonal transformation, $K(\theta_c)$ is assumed to have the form

$$K(\theta_c) = SO(3) \cup \bigcup_{i=1}^{3} SO(3)U_i,$$

where $U_1 = \mathrm{diag}\,(\eta_3,\eta_1,\eta_1)$, $U_2 = \mathrm{diag}\,(\eta_1,\eta_3,\eta_1)$, and $U_3 = \mathrm{diag}\,(\eta_1,\eta_1,\eta_3)$ are the transformation strains of the three martensitic variants, and where $\eta_1 > 0, \eta_3 > 0$ are the deformation parameters.

We can identify zero-energy microstructures with sequences of deformations $y^{(j)}$ such that $I_{\theta_c}(y^{(j)}) \to 0$ as $j \to \infty$, or with the *Young measures* $(\nu_x)_{x\in\Omega}$ corresponding to the deformation gradients $Dy^{(j)}$ of such sequences. For each $x \in \Omega$, ν_x is a probability measure on $M_+^{3\times 3}$ describing the limiting probability of finding particular values of $Dy^{(j)}(z)$ for very large j when z is sampled at random from a small ball $B(x,r)$ of radius r about the point x (see [1]). More precisely, for a subset E of $M_+^{3\times 3}$

$$\nu_x(E) = \lim_{r\to 0}\lim_{j\to\infty} \frac{\text{volume } \{z \in B(x,r) : Dy^{(j)}(z) \in E\}}{\text{volume } B(x,r)}.$$

Since $I_{\theta_c}(y^{(j)}) \to 0$, we have that ν_x is supported in $K(\theta_c)$ for each x; that is, the limiting probability of finding a value for the deformation gradient outside $K(\theta_c)$ is zero.

As is well known, twins correspond to rank-one connections

$$A - B = a \otimes n \tag{2.4}$$

between two of the martensitic energy wells $SO(3)U_i$, A and B being the deformation gradients on opposite sides of the twin plane, which has normal n. Taking without loss of generality twins involving the first two variants and $B = U_1$, the two possible twins with $A \in SO(3)U_2$ are given by

$$a = \sqrt{2}\frac{\eta_3^2 - \eta_1^2}{\eta_1^2 + \eta_3^2}(-\eta_3, \kappa\eta_1, 0), \;\; n = \frac{1}{\sqrt{2}}(1, \kappa, 0), \tag{2.5}$$

where $\kappa = \pm 1$. The angle of rotation γ between variants across a microtwin plane is given by

$$\gamma = \cos^{-1}\frac{2\eta_1\eta_3}{\eta_1^2 + \eta_3^2}. \tag{2.6}$$

A martensitic plate with twins A, B in the volume fraction λ to $1-\lambda$ corresponds to the Young measure

$$\nu_x = \lambda\delta_A + (1-\lambda)\delta_B,$$

where δ_G denotes the Dirac mass at G defined by $\delta_G(E) = 1$ if $G \in E$, $\delta_G(E) = 0$ otherwise. The corresponding macroscopic deformation gradient is $F = \lambda A + (1-\lambda)B$.

In order for such a plate to be compatible with undistorted austenite across the habit plane $\{x \cdot m = k\}$ the equation

$$\lambda A + (1-\lambda)B = \mathbf{1} + b \otimes m \tag{2.7}$$

must hold for some vector b, this being necessary and sufficient for there to exist a sequence $y^{(j)}$ with gradient $Dy^{(j)}$ having Young measure ν_x given by

$$\nu_x = \begin{cases} \delta_{\mathbf{1}} & \text{if } x \cdot m < k, \\ \lambda\delta_A + (1-\lambda)\delta_B & \text{if } x \cdot m > k. \end{cases}$$

The solutions $A \in SO(3)U_i$, $B \in SO(3)U_j$, $\lambda \in [0,1]$, b, m of (2.4), (2.7) are given by the formulae of the crystallographic theory of martensite [13]. Taking $i = 1, j = 2$ we have that $A = QU_1$, $B = Q(U_1 + a \otimes n)$ with $Q \in SO(3)$ and a, n given by (2.5), that $\lambda = \lambda^*$ or $1 - \lambda^*$, where

$$\lambda^* = \frac{1}{2}\left(1 - \sqrt{\frac{2(\eta_3^2 - 1)(\eta_1^2 - 1)(\eta_1^2 + \eta_3^2)}{(\eta_3^2 - \eta_1^2)^2} + 1}\right),$$

and that

$$m = (\frac{1}{2}\chi(\delta + \tau), \frac{1}{2}\chi\kappa(\tau - \delta), 1), \tag{2.8}$$

$$b = (\frac{1}{2}\chi\zeta(\delta + \tau), \frac{1}{2}\chi\zeta\kappa(\tau - \delta), \beta), \tag{2.9}$$

for $\lambda = \lambda^*$, with the sign before τ changed if $\lambda = 1 - \lambda^*$. Here

$$\delta = [(\eta_3^2 + \eta_1^2 - 2)(1 - \eta_1^2)^{-1}]^{\frac{1}{2}}, \tag{2.10}$$

$$\tau = [(2\eta_1^2\eta_3^2 - \eta_1^2 - \eta_3^2)(1 - \eta_1^2)^{-1}]^{\frac{1}{2}}, \tag{2.11}$$

$$\zeta = \frac{1 - \eta_1^2}{1 + \eta_3}, \ \beta = \frac{\eta_3(\eta_1^2 - 1)}{1 + \eta_3}, \ \chi = \pm 1. \tag{2.12}$$

These solutions exist provided the inequalities

$$\eta_1^2 + \eta_3^2 < 2 \text{ if } \eta_1 > 1; \qquad \eta_1^{-2} + \eta_3^{-2} < 2 \text{ if } \eta_1 < 1 \tag{2.13}$$

hold. For details see [2].

3. Calculations for macrotwins (i)

Two different martensitic plates with macroscopic deformation gradients $\mathbf{1} + b \otimes m, \mathbf{1} + \bar{b} \otimes \bar{m}$ can only be compatible if $b \otimes m - \bar{b} \otimes \bar{m}$ has rank one, that is if b is parallel to $\bar{b}$ or m is parallel to $\bar{m}$. Using the formulae (2.8)-(2.12), Bhattacharya [4] showed that this cannot happen if the two plates comprise the same two variants, and that it can happen for plates which together use all three variants if and only if

$$\eta_1^2 = \frac{(1 - \eta_3^2)^2 + 4\eta_3^2(1 + \eta_3^2)}{(1 - \eta_3^2)^2 + 8\eta_3^4}, \tag{3.1}$$

in which case the macrotwin plane is $[110]_{B2}$ and the microtwin interfaces meet at an angle of about 120°. This case is observed in $Ni_{65}Al_{35}$, for which

(3.1) is nearly satisfied, but we do not consider it further in this paper. Instead we concentrate on the case of macrotwins involving only two variants, whose microtwin interfaces meet at about 90°, for which the calculation in [4] shows that compatibility cannot be achieved at zero energy, and thus that the deformation close to the macrotwin interface cannot be stress-free.

We take the viewing direction in the TEM images to be $e_3 = [001]_{B2}$. Since there is no austenite present it is not possible to check the accuracy of this assumption directly. The situation is also complicated by various imaging issues that are discussed more fully in [6]. Fortunately the results of the calculations are insensitive to changes in the viewing direction of a few degrees. To fix orientation we take the e_3 axis to be pointing into the sample plane, with the $e_2 = [010]_{B2}$ axis horizontal and the $e_1 = [100]_{B2}$ axis vertical. From (2.8) we see that e_3 is not parallel to the habit plane. As a consequence, the microtwin plane with undeformed normal $n = \frac{1}{\sqrt{2}}(1, \kappa, 0)$ is not parallel to the viewing direction, but has normal n' parallel to $(\mathbf{1} + b \otimes m)^{-T}n$, from which it follows that n' is obtained by rotating n about $n \wedge n'$ clockwise through the angle

$$\theta = \cos^{-1}\frac{1+\eta_3 z^2}{(1+\eta_3)z}, \qquad z = \sqrt{\frac{\eta_1^{-2}+\eta_3^{-2}}{2}}.$$

The angle of rotation of the trace of the microtwin plane in $\{x_3 = 0\}$ is different and is given by

$$\psi = \cos^{-1}\frac{\eta_1^2\eta_3 - \frac{1}{2}\zeta\tau^2}{\sqrt{\eta_1^4\eta_3^2 - \zeta\tau^2\eta_1^2\eta_3 + \frac{1}{4}\zeta^2\tau^2(\delta^2+\tau^2)}}, \tag{3.2}$$

the sense of rotation being clockwise in the case $\lambda = \lambda^*$ (corresponding to vol $U_1 >$ vol U_2) if $\kappa = 1$ and anticlockwise if $\kappa = -1$ (with the orientation described above).

The deformation parameters are measured from the HRTEM images assuming that the transformation strain is volume-preserving, giving values $\eta_1 = .93, \eta_3 = 1.15$ used below. These values satisfy (2.13). In particular, from (3.2) we get $\psi = 1.88°$.

Suppose that a macrotwin is formed by coalescence of two martensitic plates, Plate I and Plate II, comprising the variants U_1 and U_2 in the same relative volume fractions λ^* but using different microtwins, namely with normals $n = \frac{1}{\sqrt{2}}(1,1,0)$ for Plate I, and $n = \frac{1}{\sqrt{2}}(1,-1,0)$ for Plate II. We suppose that the macrotwin normal N is either e_1 or e_2. That these are preferred normals is motivated in Section 4.

The above calculations imply that for $N = e_1$ the microtwin traces in the two plates make an angle $\alpha = 90° + 2\psi = 93.76° > 90°$ with one another, while for $N = e_2$ we have $\alpha = 90° - 2\psi = 86.24° < 90°$.

The rotations of the variants in each plate can also be calculated. Consider, for example, the U_1 variant in a macrotwin with $\kappa = 1$, $\lambda = \lambda^*$. In the austenite a $[010]_{B2}$ plane intersects $\{x_3 = 0\}$ in a line $x_2 =$ constant, while in the U_1 variant of the macrotwin the corresponding line of intersection for a prior $[010]_{B2}$ plane is rotated anticlockwise through an angle of 4.23°. In the U_2 variant the lines of intersection of prior $[100]_{B2}$ planes with $\{x_3 = 0\}$ are rotated clockwise through 7.89° with respect to the corresponding lines for the austenite. If $\kappa = -1$ the sense of rotation is reversed.

4. Calculations for macrotwins (ii)

Although two martensitic plates cannot simultaneously be compatible with each other and the austenite at zero energy, they can be compatible with each other across a macrotwin plane if suitably rotated. As in the preceding section we consider the case of two plates with the variants U_1 and U_2 having the same relative volume fraction λ^*. Since the average deformation gradient corresponding to layering U_1 with $U_1 + a \otimes n = RU_2$ in the volume fraction λ^* to $1 - \lambda^*$ is $U_1 + \lambda^* a \otimes n$, we need to solve the equation

$$Q(U_1 + \lambda^* a_2 \otimes n_2) - (U_1 + \lambda^* a_1 \otimes n_1) = d \otimes N,$$

where a_1, n_1 and a_2, n_2 are given by (2.5) with $\kappa = 1$ and $\kappa = -1$ respectively, and where $Q \in SO(3)$ and d, N are to be determined. The solutions are given by

$$Q = \begin{pmatrix} \cos\varphi & -\sin\varphi & 0 \\ \sin\varphi & \cos\varphi & 0 \\ 0 & 0 & 1 \end{pmatrix}$$

with either

$$\varphi = \varphi_1 = 2\tan^{-1}\frac{\rho\eta_3}{(1+\rho)\eta_1}, \qquad N = e_1,$$

or

$$\varphi = \varphi_2 = 2\tan^{-1}\frac{\rho\eta_1}{(1-\rho)\eta_3}, \qquad N = e_2,$$

where

$$\rho = \lambda^* \frac{\eta_3^2 - \eta_1^2}{\eta_1^2 + \eta_3^2}.$$

Thus the possible macrotwin normals are $[100]_{B2}$ and $[010]_{B2}$, as observed. The angle α' between the microtwin planes in each plate is given by

$$\alpha' = 90° + \gamma - \varphi_1 = 92.43° \ \text{ if } N = e_1,$$
$$\alpha' = 90° - \gamma + \varphi_2 = 85.24° \ \text{ if } N = e_2.$$

5. Discussion

The observed α angles at distances of more than 500nm from the macrotwin interface are in the ranges 92°-97° for $[100]_{B2}$ macrotwins and 83°-86° for $[010]_{B2}$ macrotwins, corresponding to the predicted values of $\alpha = 93.76°$ and $\alpha = 86.24°$ respectively. As the macrotwin interface is approached the α angle is observed to decrease, with values α' in the 50-500nm region in the ranges 90°-95° for $[100]_{B2}$ macrotwins and 83°-85° for $[010]_{B2}$ macrotwins. The sense of rotation of the microtwin variants is as predicted.

Suppose Plate I has habit plane normal $m_1 = (\frac{1}{2}\chi(\delta+\tau), \frac{1}{2}\chi(\tau-\delta), 1)$, while Plate II has habit plane normal $m_2 = (\frac{1}{2}\bar{\chi}(\delta+\tau), -\frac{1}{2}\bar{\chi}(\tau-\delta), 1)$. Can we say which choices of the signs χ and $\bar{\chi}$ lead to a macrotwin normal $N = e_1$, and which to a macrotwin normal $N = e_2$? This is difficult to answer directly from the TEM observations because the angles calculated in Section 3 do not depend on these signs.

A possible mechanism for the selection of N is that when two austenite-martensite interfaces propagate towards each other, they contact first along the common line of their two habit-planes. Up to this time, the deformation has close to zero energy. The process by which the wedge-shaped regions between the plates close up is unclear, but the formation of the final macrotwin is presumably aided if the common line lies in the macrotwin plane. This is the case for $N = e_1$ if and only if $\chi = -\bar{\chi}$, and for $N = e_2$ if and only if $\chi = \bar{\chi}$. On the other hand, the calculations in Section 4 imply that given Plate I, there are two rotations R_1 and R_2 which make it compatible with Plate II across the macrotwin planes with normals $N = e_1$ and $N = e_2$ res[pectively. These rotations are quite close to one another, the angle of rotation of $R_1R_2^{-1}$ being $\varphi_1 - \varphi_2 = 2.33°$. The angle of rotation of R_1 is 1.75° if $\chi = \bar{\chi}$, and 7.99° if $\chi = -\bar{\chi}$, while that of R_2 is 1.63° if $\chi = \bar{\chi}$, and 7.99° if $\chi = -\bar{\chi}$. Whether the larger angles of rotation when $\chi = -\bar{\chi}$ make this case less likely is unclear. In one case (see [6]) two plates were observed to make contact across both e_1 and e_2 at either ends of a curved macrotwin interface.

In the less symmetrical situation in which the two plates are composed of the same two variants but with reversed volume fractions, using either the same or different microtwins, it is surprisingly still true that the common line of the two habit planes can lie in a zero-energy macrotwin plane. Research is in progress to understand whether there are features of the microstructure in these cases deducible from the nonlinear elasticity model which might explain either why they are unlikely to arise, or why they are difficult to see in TEM images.

To summarise, the observations are consistent with the hypothesis that the macrotwins indeed arise from coalescence of two plates with different habit planes, and that these plates remain relatively undistorted in the transformation process except near the macrotwin interface, where achieving compatibility necessarily introduces stresses.

Our work can be thought of as a small step towards a better understanding of pattern formation due to displacive phase transformations. Of course the generation of microstructure is a dynamic process, which one might hope to treat using appropriate dynamic equations at the continuum level. However it is not clear which dynamic equations in the bulk and for the motion of interfaces should be used, and for no such equations would we currently have much hope of making quantitative or qualitative predictions. Like many calculations using the static theory, our work has more the flavour of a consistency check for a scenario of pattern formation, rather than being a genuine prediction starting from the governing equations.

References

1. J.M. Ball. A version of the fundamental theorem for Young measures. In M. Rascle, D. Serre, and M. Slemrod, editors, *Proceedings of conference on 'Partial differential equations and continuum models of phase transitions'*, pages 3–16. Springer Lecture Notes in Physics. No. 359, 1989.
2. J.M. Ball and R.D. James. Fine phase mixtures as minimizers of energy. *Arch. Rat. Mech. Anal.*, 100:13–52, 1987.
3. J.M. Ball and R.D. James. Proposed experimental tests of a theory of fine microstructure, and the two-well problem. *Phil. Trans. Roy. Soc. London A*, 338:389–450, 1992.
4. K. Bhattacharya. Wedge-like microstructure in martensites. *Acta Metallurgica et Materialia*, 39:2431–2444, 1991.
5. K. Bhattacharya. Comparison of geometrically nonlinear and linear theories of martensitic transformation. *Cont. Mech. Thermodyn.*, 5:205–242, 1993.
6. Ph. Boullay, D. Schryvers, and J.M. Ball. Nano-structures at martensite macrotwin interfaces in bulk and splat-cooled $Ni_{65}Al_{35}$. To appear.
7. Ph. Boullay, D. Schryvers, and R.V. Kohn. Bending martensite needles in Ni-Al using 2D elasticity and HRTEM. Phys. Rev. B, in press.
8. A.G. Khachaturyan. Some questions concerning the theory of phase transformations in solids. *Soviet Physics - Solid State*, 8:2163–2168, 1967.
9. A.G. Khachaturyan. *Theory of Structural Transformations in Solids.* John Wiley, 1983.
10. A.G. Khachaturyan and G.A. Shatalov. Theory of macroscopic periodicity for a phase transition in the solid state. *Soviet Physics JETP*, 29:557–561, 1969.
11. A.L. Roitburd. *Kristallografiya*, page 567 ff., 1967. In Russian.
12. A.L. Roitburd. Martensitic transformation as a typical phase transformation in solids. *Solid State Physics*, 33:317–390, 1978.
13. M.S. Wechsler, D.S. Lieberman, and T.A. Read. On the theory of the formation of martensite. *Trans. AIME J. Metals*, 197:1503–1515, 1953.

ON THE DETWINNING MECHANISM IN SHAPE MEMORY ALLOYS

YONG LIU
School of Mechanical & Production Engineering
Nanyang Technological University,
Nanyang Avenue
Singapore 639798
Email: mliuy@ntu.edu.sg

1. Introduction

Why do some materials exhibit shape memory effect while others do not? Taking steel and NiTi, for example, in both materials a martensitic transformation takes place when the temperature is lowered, and a reverse transformation occurs when they are subsequently heated [1-6]. Upon heating, steels normally do not show a well defined shape memory effect while NiTi does. One of the major differences between these two materials is the deformation mechanism. When steel is stressed beyond its yield limit in martensite, dislocations are generated and are responsible for the observed plastic deformation. However, when a martensitic NiTi is stressed beyond its yield limit, a "detwinning" process is responsible for the observed inelastic deformations [7-12]. This inelastic deformation can reach about 6% strain without a significant increase in dislocation density. The critical stress for activating the detwinning process is lower than that for dislocation generation and, owing to an insignificant dislocation process, the associated shape change can be partially or even fully restored through a reverse phase transformation. This phenomenon is widely observed in shape memory alloys (SMAs). What makes SMAs so attractive is their unique combination of various novel properties including the shape memory effect, superelasticity, high damping capacity, good fatigue and wear resistance, high kinetic output per volume and, of significant importance, an excellent biocompatibility of NiTi. As listed in Figure 1, most of these properties are somehow related to the detwinning process. It is well known that, in SMAs, the thermally formed martensite consists of 100% lattice twins that existed in a self-accommodated state [8-10,13,14]. When under stress, the self-accommodated twins will reorient and "detwin" without changing the lattice symmetry. Accompanying the detwinning, a stress-plateau is generated at macroscopic scales, meaning that once the detwinning is started, it can progress under a constant load. Several factors can affect the detwinning process which will in turn affect the subsequent shape recovery. These factors include test temperature [4,15,16], annealing condition [17,18], grain size [19,20], stress mode [21-24], and microstructure especially textures [14,25-28]. In the following, the detwinning of SMAs, its importance in the process of shape recovery, and some influential factors will be discussed.

Q.P. Sun (ed.), IUTAM Symposium on Mechanics of Martensitic Phase Transformation in Solids, 37–44.

	Properties	Mechanisms
NiTi SMAs	Shape memory effect	Detwinning and thermally induced PT*
	High damping capacity	Detwinning and stress induced PT
	Wear resistance	Detwinning and stress induced PT
	Fatigue resistance	Detwinning and stress induced PT
	High kinetic output	Detwinning and thermally induced PT
	Superelasticity	Stress induced PT
	Biocompatibility	Inert titanium oxide layer

* PT - phase transformation

Figure 1. Shape memory alloys, especially NiTi have unique combination of various novel properties highly attractive for applications in biomedical, MEMS, sensors & actuators, energy dissipation and vibration suppression, etc. Current understanding shows that most of the properties are related to the deformation mechanisms of both detwinning and stress-induced phase transformation.

2. Deformation under monotonic tension

When a martensitic SMA is deformed under tension, it initially yields at a relatively low stress (~120 MPa for NiTi), followed by a stress-plateau extending to about 6% strain as schematically illustrated in Figure 2. Starting from the end of the stress-plateau, the deformation tendency of SMAs becomes similar to that of traditional structural materials. It was found that much of the "secret" of SMAs lies in the stress-plateau.

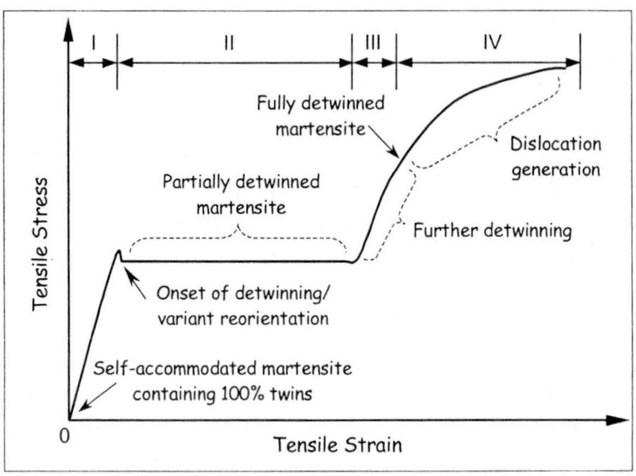

Figure 2. Current understanding of the martensitic deformation process in SMAs. Thermally formed martensite in SMAs consists of 100% lattice twins. Under tension, the lattice twins are detwinned leading to macroscopic deformations up to about 6% strain under constant load. Further deformation is realized through further detwinning and dislocation mechanisms.

The onset of the stress-drop has been generally recognized to be the onset of the martensite detwinning process, and the stress-plateau is due to a continuous detwinning process. Within the plateau region, detwinning proceeds unevenly throughout the testing sample [12,29]. In polycrystalline SMAs, the end of stress-plateau is not the end of the detwinning process. By further deformation into the region III, further detwinning of martensite twins which are less favorable to the loading direction can be expected. In this region, martensite detwinning is accompanied by a further increase in the stress. By further loading, plastic deformation will be introduced in the martensite plates having orientations unfavorable to the applied force [12].

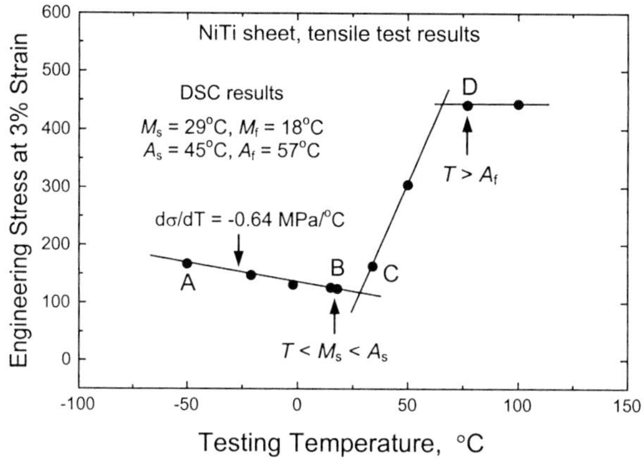

Figure 3. Stress of a NiTi sheet at 3% strain as a function of testing temperature. From A to B, a stress-plateau due to martensite detwinning is observed. The detwinning stress increases with decreasing temperature. Above C, no stress plateau exists while the yield stress increases with increasing testing temperature.

The detwinning stress is a function of testing temperature. As shown in Fig. 3, when tested below M_s, the detwinning stress increases linearly with lowering the temperature. Experimental results show that, at testing temperatures between points A and B in Figure 3, a stress-plateau occurred during tensile deformation. From temperature C above, no stress-plateau exists in the stress-strain curve. The yield stress increases with increasing temperature from C to D. The plateau stress has the lowest value at temperature B which might be the highest testing temperature before the reverse transformation starts. It needs to be mentioned that, in this NiTi sheet, no stress-induced phase transformation is observed within the testing temperature range.

3. Differences between tension and compression

For polycrystalline SMAs, the stress-strain curves are different for tension and compression, exhibiting an "asymmetry" as shown typically in Figure 4. In contrast to the case of tension, under compression the SMA is quickly strain hardened and no flat stress-plateau is observed. This result clearly suggests that, when application of SMAs is concerned, attention should be paid to their loading mode in operation from both the design and performance prediction points of view. The mechanism of the mechanical asymmetry is further revealed by TEM study of the microstructures developed during deformation [24]. In contrast to tension, a specimen deformed to 4% strain under compression has a high density of dislocations and less significant martensite detwinning. The different deformation mechanisms between tension and compression further lead to differences in the thermo-mechanical properties.

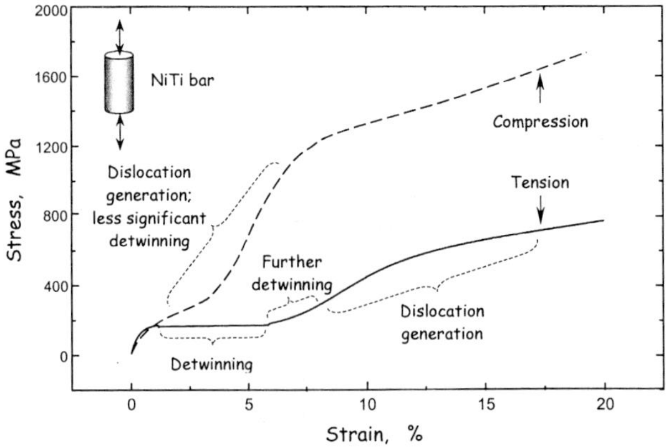

Figure 4. Differences in the martensitic deformation process between tension and compression. Microstructure study shows that tensile deformation leads to martensite detwinning, while compressive deformation results in formation of a high density of dislocations and less significant detwinning process.

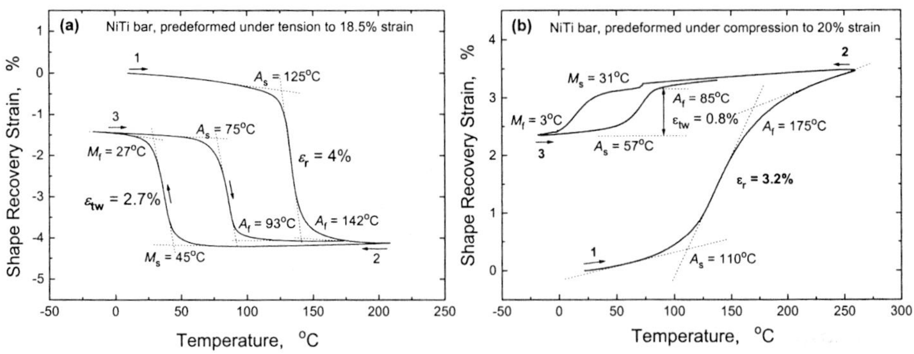

Figure. 5. Shape recovery and TWME developed by martensitic deformation under (a) tension to 18.5% strain and (b) compression to 20% strain. Differences in both shape recovery and TWME strain are found between tension and compression. The transformation temperatures are roughly estimated by a slope line method and can be different from the values determined by other methods, e.g. DSC. In the undeformed condition the phase transformation temperatures are $M_s = 59$°C, $M_f = 39$°C, $A_s = 74$°C and $A_f = 93$°C, respectively.

The shape recovery results obtained from two pre-deformed NiTi bar samples, one under tension to 18.5 % strain (Sample A) and the other under compression to 20% strain (Sample B) are highlighted in Figure 5. Both samples were heated to above 200°C and their shape recovery was recorded. Upon heating, Sample A contracts while Sample B elongates. For both samples, pre-deformation leads to a stabilization of the martensite phase. A two-way memory effect (TWME) was also developed with, however, different magnitudes. The magnitude of martensite stabilization and the reverse transformation features were also different between these two samples. During the 1st heating, Sample A recovered 4% while Sample B recovered about 3.5%. For Sample A, the reverse transformation began at about 125°C which is about 50°C higher than that for the undeformed condition, and the ΔM ($=M_s$-M_f) was about 18°C. For Sample B, the reverse transformation began at about 110°C; however, the ΔM was about 65°C. In addition, the

TWME strain in Sample A was 2.7% being much higher than in Sample B of 0.8%. The transformation hysteresis was much narrower in Sample A. The deformation mode not only affects the detwinning process and the resulting mechanical behavior, but also affects the subsequent shape recovery and transformation characteristics. Detwinning corresponds to a better shape recovery and higher TWME strain, while dislocation generation corresponds to a lower shape recovery strain and lower TWME strain.

4. Anisotropy due to textures

In most application, SMAs are used in polycrystalline state prepared by either rolling or drawing or other heavy deformation processes. Thus, textures exist in most of the SMAs. In general, having texture is desirable if the material is used properly. However, it can also lead to unwanted poor performance if the materials are used improperly. Various results [14,25-28] on textured SMA rolled sheet show a considerable difference in plateau-strain and shape recovery along different directions. The shape recovery strain is associated with the competition between detwinning and dislocation generation during deformation. The detwinning process is strongly texture dependent. Without texture, it is isotropic in polycrystalline materials. With texture, it is enhanced for certain directions and reduced for other directions, as is the shape recovery strain. The anisotropy in detwinning and thus the shape recovery due to texture provides a possibility of maximizing the potential of the SMAs through processing.

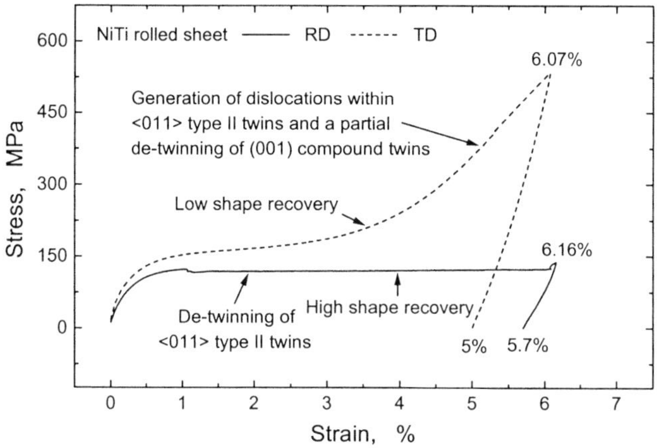

Figure 6. Anisotropy of martensitic deformation process in a textured rolled NiTi SMA sheet. Tension along RD leads to detwinning of <011> type II twins, while tension along TD leads to formation of dislocations and partial detwinning of (001) compound twins.

Figure 6 summarizes the texture effect on mechanical and thermomechanical behavior of a NiTi SMA sheet. Tension along the rolling direction (RD) leads to a stress-plateau, a well developed detwinning and higher shape recovery strain, while tension along the transverse direction (TD) results in strain hardening, dislocation generation and a lower shape recovery strain [14,27]. For a rolled sheet, the direction for which the highest shape recovery strain is obtained corresponds to the direction for which the longest plateau-strain is obtained during the martensitic deformation. On the other hand, a direction for which a poor shape recovery is obtained corresponds to a martensite stress-strain curve that shows significant strain hardening.

The relation between texture orientation and the shear direction (SD) of lattice twins plays a critical role in the overall performance of the SMAs [14]. It was found that the stress-plateau occurring during tension along RD is mainly related to detwinning of <011> type II twins, while the detwinning of (001) compound twins and dislocation generation are responsible for the observed non-flat stress-plateau and strain hardening along TD. This observation can be further understood based on Figure 7. For two major texture components, $(010)[001]_M$ and $(001)[010]_M$, the SD of type II twins make angles of 41.7° and 48.3° to the RD, while they make respectively 90.2° and 95° angles to the TD. The SD of compound twins makes angles of 6.56° and 0° to the TD, while it makes respectively 96.8° and 90° angles to the RD for the $(010)[001]_M$ and $(001)[010]_M$ textures. Thus, the type II twins are easy to detwin when loading along RD due to the higher resolved shear stress and they are difficult to detwin when loading along TD. However, for the compound twins, the situation is reversed; they will easily detwin along TD. This relation is responsible for the difference in detwinning along different directions and the respective stress-strain curves. This detwinning anisotropy is also believed to be responsible for the observed anisotropy in shape recovery strain.

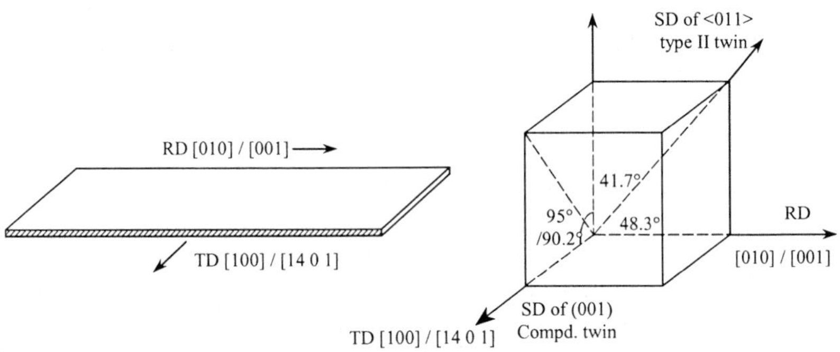

Figure 7. Schematic relationship among SD of twins, RD and TD for two martensite texture components, $(001)[010]_M$ and $(010)[001]_M$. SD of <011> type II twins is parallel to $[011]_M$ which has 48.3°/41.7° angle to RD and 95°/90.2° angle to TD for $(001)[010]_M/(010)[001]_M$ textures. SD of (001) compound twins is nearly parallel to the TD and perpendicular to the RD for both $(001)[010]_M$ and $(010)[001]_M$ textures.

5. Prediction of the detwinning anisotropy

The present author and collaborator have tried to predict the martensitic detwinning process by taking into consideration of textures, detwinning shear and loading direction [30]. Based on a detwinning model, an analysis was performed on the resolved shear stress of both type II and compound twins. By assuming several spread functions of ideal orientation distributions of the above mentioned two texture components the detwinning of both types of twin along different directions was examined. Theoretical analysis agrees that the detwinning anisotropy is responsible for the mechanical anisotropy. In general, the stronger the texture the stronger is the detwinning anisotropy.

The results further suggest that two types of detwinning processes exist, *domino detwinning* and *assisted detwinning*. *Domino detwinning* is a self-propagating process that requires only a critical driving force to proceed, while *assisted detwinning* requires a continuous increase in the driving force. *Domino detwinning* is responsible for the stress-plateau, while the *assisted detwinning* takes place at higher stresses beyond the stress-plateau region. Along with the *assisted detwinning* process, dislocation generation will take place as soon as a critical resolved shear stress is reached. In highly textured SMAs, the *domino detwinning* is strong along certain directions and longer plateau-strains can be expected. This prediction agrees well with the experimental observations [27]. Since a higher plateau-strain corresponds to a higher shape recovery strain, *domino detwinning* is thus expected to promote shape recovery. The results in Fig. 4 and Fig. 6 can thus be explained. Tension promotes *domino detwinning*, while compression results in *assisted detwinning*; tension along RD promotes *domino detwinning* while tension along TD leads to *assisted detwinning* in the materials presently studied.

Based on the above understanding, we may further consider a new approach to predict the texture dependency of shape recovery. The underlying principle is: *the shape recovery strain is intrinsically related to the magnitude of the martensite detwinning*. The martensitic deformation consists of mainly two mechanisms, *detwinning* and *dislocation generation*. Both mechanisms are operative during deformation but have an opposite effect on the subsequent shape recovery. Detwinning does not introduce "permanent" deformation, while dislocation generation does. Thus, upon reverse phase transformation, the residual strain associated with detwinning will be recovered while the residual strain associated with dislocation process will remain "permanently".

6. Rate dependence

The author and collaborators have studied the rate dependence of the martensitic deformation under both dynamic compression and dynamic tension [31,32]. Dynamic compression of a NiTi bar at strain rate up to 3000 s^{-1} shows no significant difference in deformation mechanism if compared to specimens deformed under quasi-static mode. Results [32] from dynamic tension at strain rates up to 300 s^{-1} show that the stress-plateau due to martensite detwinning still exists. Further experiments on shape recovery of the dynamically tensioned specimens suggest that a detwinning process has well developed in the specimens. This shows that the detwinning taking place at the early stage of the martensite deformation is not very sensitive to the strain rate. This result provides further experimental evidence for the *domino detwinning*.

Acknowledgements — Collaborations with Z.L. Xie, J. Van Humbeeck, L. Delaey, Q.S. Zheng and Y.L. Li among others helped us to further understand the deformation mechanisms in SMAs.

References

1. Kurdjumov, G.V.: The nature of martensitic transformations, *Journal of Metals* **July** (1959), 449-453.
2. Cohen, M.: The strengthening of steel, *Transactions of the Metallurgical Society of AIME* **224** (1962), 638-656.
3. Winchell, P.G., and Cohen, M.: The strength of martensite, *Transactions of the ASM* **55** (1962), 347-361.
4. Wayman, C.M.: The growth of martensite since E.C. Bain (1924) - some milestones, *Materials Science Forum* **56-58** (1990), 1-32.

5. Olson G.B., and Owen, W.S.: *Martensite*, ASM International, 1992.
6. Xie, Z.L., Liu, Y., and Hänninen, H.: Stabilization of retained austenite due to partial martensitic transformations, *Acta Metall. et Mater.* **42** (1994), 4117-4133.
7. Otsuka K., and Shimizu, K.: Pseudoelasticity and shape memory effects in alloys, *International Metals Reviews* **31, no. 3** (1986), 93-114.
8. Otsuka K., and Wayman, C.M.: *Shape Memory Materials*, Cambridge University Press, 1998.
9. Miyazaki, S., Otsuka, K., and Wayman, C.M.: The shape memory mechanism associated with the martensitic transformation in Ti-Ni alloys - I. Self-accommodation., *Acta Metall.* **37** (1989), 1873-1884.
10. Miyazaki, S., Otsuka, K., and Wayman, C.M.: The shape memory mechanism associated with the martensitic transformation in Ti-Ni alloys - II. Variant coalescence and shape recovery, *Acta Metall.* **37** (1989), 1885-1890.
11. Wayman, C.M.: Shape memory and related phenomena, *Progress in Materials Science* **36** (1992), 203-224.
12. Liu, Y., Xie, Z.L., Van Humbeeck, J., Delaey, L., and Liu, Y.N.: On the deformation of twinned domain in NiTi shape memory alloys, *Phil. Mag. A* **80** (2000), 1935-1953.
13. Knowles, K.M., and Smith, D.A.: The crystallography of the martensitic transformation in equiatomic nickel-titanium, *Acta Metall.* **29** (1981), 101-110.
14. Liu, Y., Xie, Z.L., Van Humbeeck, J., and Delaey, L.: Effect of texture orientation on the martensite deformation of NiTi shape memory alloy sheet, *Acta Mater.* **47** (1999), 645-660.
15. Cross, W.B., Kariotis, A.H., and Stimler, F.J.: *NASA CR-1433*, **09**/1969.
16. Saburi, T., Tatumi, T., and Nenno, S.: Effects of heat treatment on mechanical behavior of Ti-Ni alloys, *J. de Physique* **34, Suppl. 12** (1982), C4-261-266.
17. Liu Y.N., and McCormick, P.G.: Influence of heat treatment on the mechanical behaviour of a NiTi alloy, *ISIJ International* **29/5** (1989), 417-422.
18. Liu, Y., Van Humbeeck, J., Stalmans, R., and Delaey, L.: Some aspects of the properties of NiTi shape memory alloy, *J. of Alloys and Compounds* **247** (1997), 115-121.
19. Wayman, C.M.: Deformation mechanisms and other characteristics of shape memory alloys, in *Shape Memory Effects in Alloys*, edited by J. Perkins, The Metallurgical Society of AIME, Plenum Press, New York, 1975, 1-27.
20. Motohashi, Y., Sakuma, T., Suzuki, M., Hoshiya, T., and Ohsawa, K.: Effect of grain refinement on phase transformation characteristics and mechanical properties of Ti-Ni shape memory alloys, *Proc. of ICOMAT-92*, pp. 993-998, Monterey Institute for Advanced Studies, 1992.
21. Wasilewski, R.J.: The effects of applied stress on the martensitic transformation in TiNi, *Metall. Trans.* **2** (1971), 2973-2981.
22. Wasilewski, R.J.: The shape memory effect in TiNi - one aspect of stress-assisted martensitic transformation, in *Shape Memory Effects in Alloys*, edited by J. Perkins, The Metallurgical Society of AIME, Plenum Press, New York, 1975, 245-271.
23. Adler, P.H. Yu, W., Pelton, A.R., Zadno, R., Duerig, T.W., and Barresi, R.: *Scripta Metall.* **24** (1990), 943.
24. Liu, Y., Xie, Z.L., Van Humbeeck, J., and Delaey, L.: Asymmetry of stress-strain curves under tension and compression for NiTi SMAs*Acta Mater.* **46** (1998), 4325-4338.
25. Mulder, J.H., Thoma, P.E., and Beyer, J.: Anisotropy of the shape memory effect in tension of cold-rolled 50.8Ti49.2Ni (at.%) sheet, *Z. Metallkd.* **84/7** (1993), 501-508.
26. Inoue, H., Miwa, N., and Inakazu, N.: Texture and shape memory strain in TiNi alloy sheets, *Acta Mater.* **44** (1996), 4825-4834.
27. Zhao, L.: Texture development and anisotropic behaviour in a Ti-45Ni-5Cu (at.%) shape memory alloy, *Ph.D. thesis*, ISBN 90-9011069-0, University of Twente, Enschede, The Netherlands, 1997.
28. Melton, K.N.: Ni-Ti based shape memory alloys, in *Engineering Aspects of Shape Memory Alloys*, edited by T.W. Duerig, K. N. Melton, D. Stöckel and C. M. Wayman, Butterworth-Heinemann, London, 1990, 21-35.
29. Liu, Y.N., Liu, Y., and Van Humbeeck, J.: Lüders-like deformation associated with martensite reorientation in NiTi, *Scripta Mater.* **38** (1998), 1047-1055.
30. Zheng, Q.S. and Liu, Y.: Prediction of the detwinning anisotropy in textured NiTi shape memory alloy, *Philosophical Magazine A* (in press).
31. Liu, Y., Li, Y.L., Ramesh, K.T., and Van Humbeeck, J.: High rate deformation of martensitic NiTi shape memory alloy, *Scripta Mater.* **41** (1999), 89-95.
32. Liu, Y., Li, Y.L., Ramesh, K.T.: Rate dependence of deformation mechanisms in shape memory alloy, *Phil. Mag. A*, submitted.

NONLINEAR ELECTROMECHANICAL DEFORMATION AND FRACTURE RELATED TO DOMAIN SWITCHING IN FERROELECTRIC CERAMICS

D. N. Fang, B. Liu and K. C. Hwang
Department of Engineering Mechanics
Tsinghua University, Beijing 100084, China

Abstract

Domains exist in ferroelectric ceramics. External loads, such as electric field and stress, can cause domain switching. Domain switching always results in nonlinear ferroelectricity and ferroelasticity of ferroelectric ceramics. In this investigation, nonlinear electric-mechanical behavior related to the ferroelctric and ferroelastic domain switching is experimentally and theoretically studied. The domain orientation and its evolution are determined by the joint action of the domain switching driving force and the dissipation during domain switching. 90° and 180° domain switchings play different roles in ferroelectricity and ferroelasticity and have different switching dissipations associated with them. A nonlinear electromechanical coupling finite element model based on the domain switching theory is proposed and used for finite element analysis.

Keywords: ferroelectric material, domain switching, finite element method.

1. Introduction

Due to the property of mechanical-electrical coupling, ferroelectric ceramics have been widely used in key components of sensors, transducers and actuators [1-2]. However, both property degeneration and crack growth in the devices due to the nonlinear response hamper the reliable performance of ferroelectric ceramics in smart structures [3-4]. Domains exist in ferroelectric ceramics [1-3]. External loads, such as electric field and stress, can cause domain switching [3-5]. Domain switching, similar to phase transformation and reorientation, always results in nonlinear ferroelectricity and ferroelasticity of ferroelectric ceramics [3-5]. Ferroelectric ceramics are susceptible to brittle fracture that can lead to catastrophic failure [4]. Therefore, it is important to understand the deformation and fracture behavior of ferroelectric materials under coupled electromechanical fields [4]. One way to accomplish this is to resort to well-established linear elastic fracture mechanics [6-9]. On the other hand, a better understanding of the fracture behavior in PZT ceramics was obtained on the base of the domain switching mechanisms and related theory of nonlinear fracture mechanics [4, 10]. Since the crack tip stress fields are singular, nonlinear effect such as domain switching

Q.P. Sun (ed.), IUTAM Symposium on Mechanics of Martensitic Phase Transformation in Solids, 45–54.

exists near the crack tip even though the remote field is less than the coercive value. This implies that the structure reliability concerns of electromechanical devices call for a better understanding of nonlinear constitutive laws of ferroelectric ceramics as well as related domain-switching criteria. In this investigation, we theoretically, numerically and experimentally study nonlinear electric-mechanical behavior related to the ferroelectric and ferroelastic domain switching. A nonlinear electromechanical coupling finite element model based on the domain switching theory is proposed and used for finite element analysis.

2. Experimental Procedure

The material tested is a typical soft PZT-5H with large piezoelectric coefficients and dielectric permittivity [11]. Each grain of the ceramic is a single crystal with a tetragonal perovskite structure at room temperature. The specimens, with nominal dimensions 10mm×10mm×16mm, were cut from bulk ferroelectrics, and all faces were polished. The upper and bottom faces of the specimens with an area of 10mm×10mm were electroded with sputtered Ag. A servo-hydraulic loading fixture is set up for applying stress field. A silicon oil bath is mounted on the loading fixture to prevent high voltage arcing. As a simple mechanical load, the stress field is applied to a polarized sample in the x_3-direction parallel to the poling direction. Under combined electric-mechanical loads, samples are subjected to a cyclic electric field parallel to the x_3-direction under different constant compressive stresses. The load sensor output is recorded through an A/D circuit connected to the computer. Longitudinal and transverse strain gauges are bonded to the center of the area near the ground electrode to measure strains parallel and perpendicular to the polarization direction, i.e., x_3-direction. The charge per unit area on the electrode is equal to the normal component of the electric displacement. The charge on the electrode is measured by monitoring the voltage of a capacitor (10 μF) connected from the bottom electrode of the specimen to the ground. The voltage of the capacitor is monitored by means of a high input impedance electrometer and the A/D circuit. A high voltage source triangle wave (±30KV at from 1 HZ to 0.01HZ) is connected to the upper electrode of the specimen. The measured curves are presented in section 4 to make a comparison of the calculations and the experiments.

3. Mathematic Description

Based on the domain switching criteria proposed by some of the authors and their co-workers [12], the stress, strain, electric displacement and electric field, $\boldsymbol{\sigma}$, $\boldsymbol{\varepsilon}$, $\boldsymbol{D}$ and $\boldsymbol{E}$, respectively, obey the following constitutive relations:

$$\begin{aligned} \sigma_{ij} &= C^E_{ijmn}\varepsilon_{mn} - e_{nij}E_n - \hat{\sigma}_{ij} \\ D_i &= e_{imn}\varepsilon_{mn} + k^{\varepsilon}_{in}E_n - \hat{D}_i \end{aligned} \tag{1}$$

where $C^{E}_{ijmn}, e_{nij}, k^{\varepsilon}_{in}$ are the components of the elastic tensor measured at constant electric field, the piezoelectric tensor, and the permittivity tensor measured at constant strain, respectively. $\hat{\sigma}_{ij}$ and $\hat{D}_i$ are related to the spontaneous strain and electric displacement, ε^*_{ij} and D^*_i, respectively, as follows:

$$\hat{\sigma}_{ij} = C^{E}_{ijmn}\varepsilon^*{}_{mn}, \qquad \hat{D}_i = e_{imn}\varepsilon^*{}_{mn} - D^*_i \tag{2}$$

Assume that a ferroelectric ceramic is subjected to a field of displacement $\boldsymbol{u}$ and electric potential ϕ. Based on the requirement of geometric compatibility, the strain and the electric fields can be derived from the gradients as follows:

$$\varepsilon_{ij} = \frac{1}{2}\left(u_{i,j} + u_{j,i}\right), \qquad E_i = -\frac{\partial \phi}{\partial x_i} \tag{3}$$

The Gibbs' free energy (per unit volume) in a domain can be expressed as

$$g = -\left(\boldsymbol{\varepsilon}^* : \boldsymbol{\sigma} + \boldsymbol{D}^* \cdot \boldsymbol{E} + \frac{1}{2}\boldsymbol{\sigma} : \boldsymbol{M} : \boldsymbol{\sigma} + \frac{1}{2}\boldsymbol{E} \cdot \boldsymbol{k} \cdot \boldsymbol{E} + \boldsymbol{E} \cdot \boldsymbol{d} : \boldsymbol{\sigma}\right) \tag{4}$$

The domain switching driving force of 90° and 180° switching, F_{90} and F_{180}, respectively, are defined as follows:

$$F_{90}(\boldsymbol{R};\boldsymbol{\sigma}_{t+dt},\boldsymbol{E}_{t+dt}) = \max\{g(\boldsymbol{R};\boldsymbol{\sigma}_t,\boldsymbol{E}_t,S_t) - g(\boldsymbol{R};\boldsymbol{\sigma}_{t+dt},\boldsymbol{E}_{t+dt},S)\}_{S\in\{A^t_{90}\}} \tag{5}$$

$$F_{180}(\boldsymbol{R};\boldsymbol{\sigma}_{t+\mathrm{d}t},\boldsymbol{E}_{t+\mathrm{d}t}) = \max\{g(\boldsymbol{R};\boldsymbol{\sigma}_t,\boldsymbol{E}_t,S_t) - g(\boldsymbol{R};\boldsymbol{\sigma}_{t+\mathrm{d}t},\boldsymbol{E}_{t+\mathrm{d}t},S)\}_{S=A^t_{180}} \tag{6}$$

where $\boldsymbol{R} = R_{iJ}\bar{\boldsymbol{e}}_I\bar{\boldsymbol{e}}_J$ ($i = I$, $j = J$ sum for I, J) is called the domain orientation tensor, and the type of domains, S, is a function of $\boldsymbol{R}$, denoted by $S(\boldsymbol{R})$. The thresholds for 90° and 180° domain switching are different and are denoted as W^f_{90} and W^f_{180}, respectively. The domain switching criterion can be stated as follows:

a) If $F_{90}(\boldsymbol{R};\boldsymbol{\sigma}_{t+dt},\boldsymbol{E}_{t+dt}) \geq W^f_{90}$, then 90° domain switching occurs at orientation $\boldsymbol{R}$ and $S_{t+\mathrm{d}t} = S^p_{90}$, where S^p_{90} denotes the domain type S that satisfies the above definition of F_{90}. Otherwise, no domain switching occurs at orientation $\boldsymbol{R}$ and $S_{t+\mathrm{d}t} = S_t$.

b) If $F_{180}(\boldsymbol{R};\boldsymbol{\sigma}_{t+dt},\boldsymbol{E}_{t+dt}) \geq W^f_{180}$, then 180° domain switching occurs at orientation $\boldsymbol{R}$ and $S_{t+\mathrm{d}t} = S^p_{180}$, where S^p_{180} denotes A^t_{180}. Otherwise, no domain switching occurs at orientation $\boldsymbol{R}$ and $S_{t+\mathrm{d}t} = S_t$.

Based on the principle of virtual work, for the real solution of the system in domain V with its boundary ∂V, the following variational equation exists

$$\int_V \left[\left(\sigma_{ij,i} + f_j\right)\delta u_j + \left(D_{i,i} + p^e\right)\delta\phi\right] dV - \int_{S^\sigma}\left(\sigma_{ij}n_i - \bar{T}_j\right)\delta u_j dS - \int_{S^D}\left(D_i n_i - \bar{p}^e\right)\delta\phi dS = 0 \tag{7}$$

where f_j and p^e are the body force and electric charge, respectively. $\overline{p}^e$, $\overline{T}_j$ and n_i are the surface electric charge, surface force and normal of the surface, respectively; and S^σ, S^D refer to the stress and electric displacement boundaries, respectively. Thus, the potential of the system in domain V with its boundary ∂V is as follows:

$$\Pi = \int_V \left(\frac{1}{2} C^E{}_{ijkl}\varepsilon_{ij}\varepsilon_{kl} - e_{ijk}\varepsilon_{ij}E_k - \frac{1}{2} k^\varepsilon{}_{ij}E_iE_j \right) dV \\ - \int_V \left(\hat{\sigma}_{ij}\varepsilon_{ij} - \hat{D}_iE_i\right)dV - \int_V \left(f_iu_i - p^e\phi\right)dV - \int_{S^\sigma} \overline{T}_iu_i dS - \int_{S^D} \overline{p}^e\phi dS \tag{8}$$

Of all the fields $(\boldsymbol{u}, \phi)$ satisfying the mechanical and electric boundary conditions, the solution renders [12] $\delta\Pi = 0$. Suo, *et al.* [12] pointed out that since the electric enthalpy, $\Psi(\boldsymbol{\varepsilon}, \boldsymbol{E})$, is not positive definite, it is not necessary to minimize the solution field, Π. In addition, although the relevant energy function, Ψ, is not positive definite, uniqueness can be stated for a fairly general class of nonlinear materials in the case of small deformation. Thus, an isoparametric finite element, which is of displacement-electric potential type, can be developed. The details of the standard FE procedure will not be reiterated here. Note that the second term on the right hand side of Equation (5) contains the spontaneous strain $\boldsymbol{\varepsilon}^*$ and spontaneous electric displacement $\boldsymbol{D}^*$.

4. Results and Analysis

In the following, we shall compare calculations with results of experiments on the soft PZT-5H ceramic, which were conducted in our Failulre Mechanics Laboratory [11]. The material properties are listed in Table 1. The electric displacement vs. electric field curves and longitudinal strain vs. electric field curves under various compressive stresses are demonstrated in Fig. 1(a)~(f), respectively. The solid lines respresent theoretical results and the marked lines respresent experimental results. Fig. 1 shows that the model can characterize the electric-mechanical response and the effect on depolarization due to the compressive stress under the combined electric-mechanical loading. In the case that the specimen is subjected to the combined electric-mechanical loads, a constant compressive stress is imposed in the polarization direction parallel to the x_3 axis while the electric field is cyclically applied. Fig. 1(a), (c) and (e) illustrate that D_3, vs. E_3 hysteresis loops are developed at different constant compressive stress, indicating that there is a significant effect of the uniaxial compressive stress on the hysteresis behavior. The resulting hysteresis loops show a steady decrease in both the remanent polarization and the saturation polarization as the compressive stress is changed form 0 to -80 MPa. The slope of the loops at zero electrical field, which can be regarded as the permittivity, decreases when the magnitude of compressive stress increases. The coercive field almost linearly decreases as the uniaxial compressive stress increases. This is the so-called depolarization phenomenon[2, 5], that is, when the magnitude of compressive stress incresases, the remanent polarization decereases and the area encircled by the hysteresis decreases with the increase of compressive stress as well. Similarly, the butter-fly shape longitudinal strain vs. electric field curves are influenced by the external stress in such a way that they become flatter when the compresive stress increases. The longitudinal strain gradually becomes negative with reference to the unpolarized state when the compressive stress becomes larger. The range of the strain variation becomes smaller and smaller as the magnitude of compressive stress increases. The range turns out

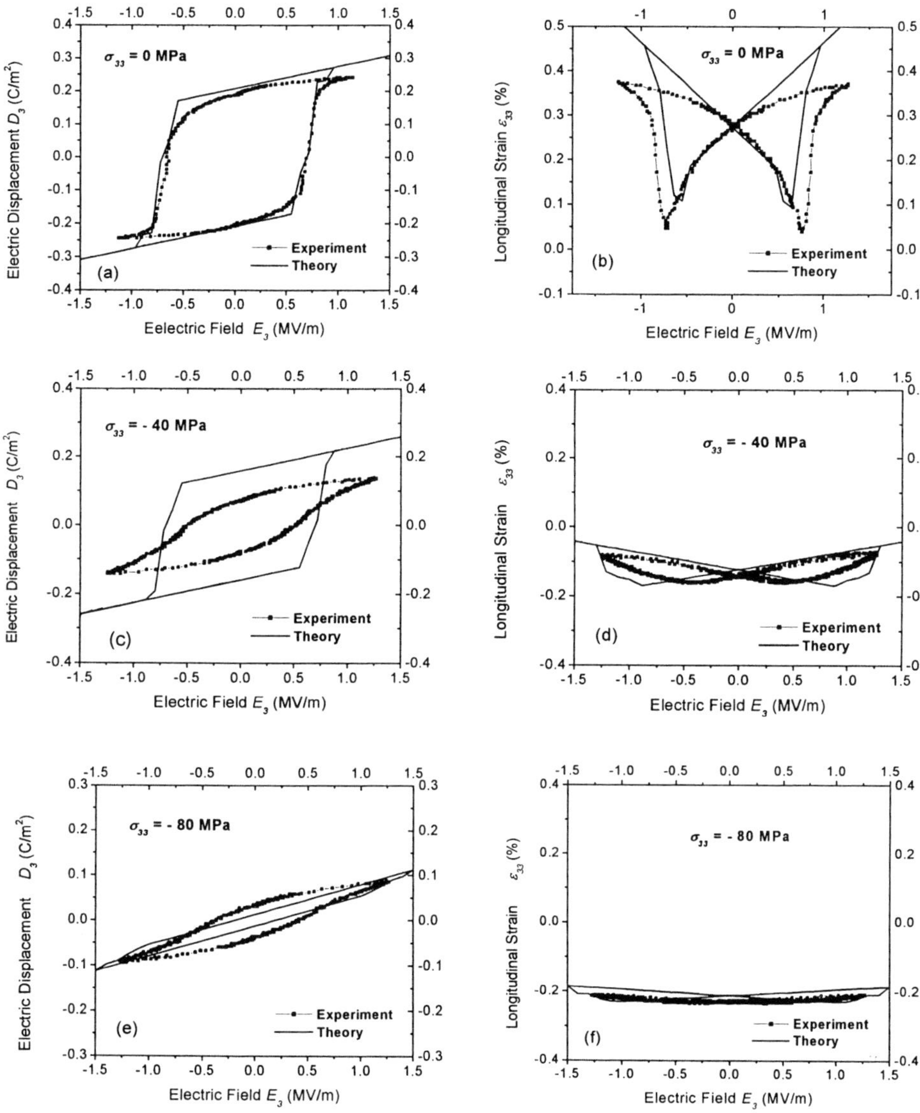

Fig.1 Comparison of theory and experiment[11] for the case of the combined electric-mechanical loading: the axial compressive stress equals (a) and (b) 0MPa; (c) and (d) -40MPa; (e) and (f) -80MPa, respectively.

to be almost zero when the stress goes over –80 MPa that is nearly the coercive stress field. This means that the domain switching becomes more difficult under a larger magnitude of compressive stress.

Note that the plane strain model was employed in all the FE calculations. In other words, all the numerical examples selected for the present study are two-dimensional problems. The material chosen for the analyses was a PZT-5H ceramic, for which the properties can be obtained from [11].

4.1 EXAMPLE 1: A BLOCK OF HOMOGENEOUS MATERIAL SUBJECTED TO UNIFORM ELECTRIC FIELD

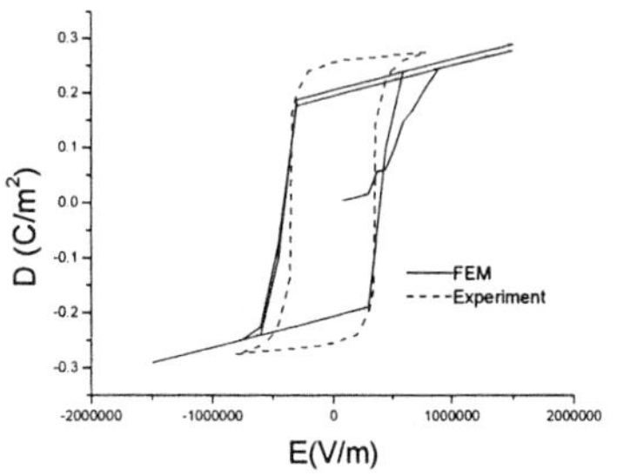

Fig. 2 D vs. E

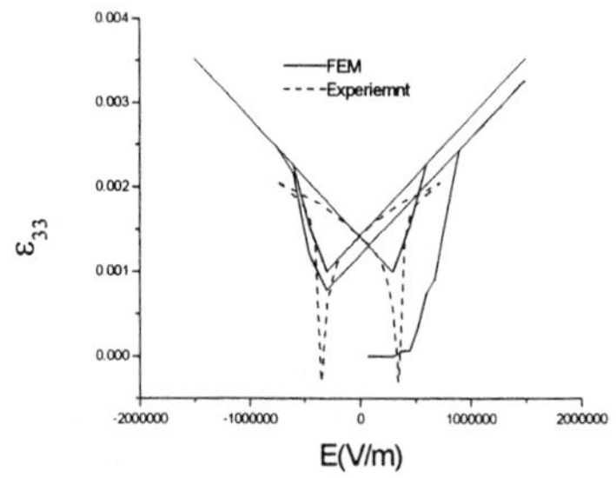

Fig. 3 ε_{33} vs. E

Consider a block of homogeneous material subjected to uniform electric field in the x_3-direction. The hysteresis loops of electric field versus electric displacement and strain versus electric field, obtained by the finite element and experimental methods, are shown in Figure 2 and 3. Note that the calculated results were obtained using the constitutive model proposed by the authors [12], and the experimental results were obtained from the work done by Fang and Li [11]. It can be found that the FE results, as shown in Figures 2 and 3, are in good agreement to the experiments.

4.2 EXAMPLE 2: A POLED PZT PLATE WITH A CAVITY SUBJECTED TO PRESCRIBED ELECTRIC LOADS

A central elliptic hole with $\alpha = 1/3$ is analyzed. A positive electric field E_0 is applied in the x_3-direction. Note that a positive electric field is defined, as the direction of an electric field is the same as that of the polarization of the material, which is the x_3-direction in the present case. On the free boundary of the hole, the conventional mechanical, i.e., traction free condition, and the impermeable electrical, i.e., $\boldsymbol{D} \cdot \boldsymbol{n} = 0$, boundary conditions are implemented. Figure 4a presents the distribution of the remanent polarization vectors in a non-fully poled state ($E_0 = 0.76E_c$). It can be clearly seen that during the poling process, in which the applied electric filed is less than the coercive field, the material near the cavity is poled but not that of the other region. The highest concentration of the polarization appears at the apex of the major axis of the cavity.

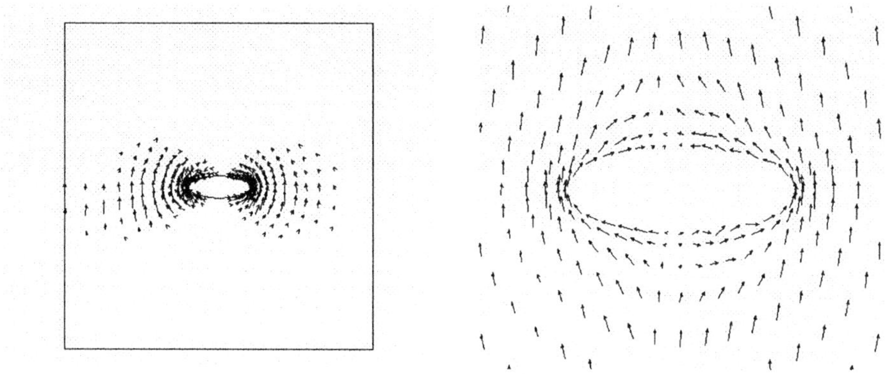

(a) (b)

Fig. 4 Distribution of remanent polarization vector

Figures 4b show the vector-field distributions of remanent polarization in a state where the applied electric field is higher than the coercive field. It can be seen that the directions of the remanent polarization vectors near the boundary of the elliptic hole are along the tangential direction of the said boundary. This implies that domain switching is easier to occur along the tangential direction near the boundary of the elliptic hole. It is obvious that concentrations of the electric field, strain and remanent polarization exist near the said boundary. The density and magnitude of the vector indicate that such concentrations decrease with increasing distance away from the boundary of the hole. The larger vector of the remanent polarization at the apex of the major axis of the

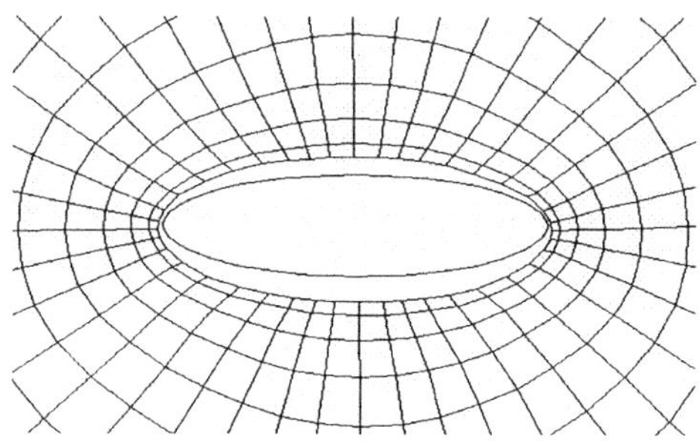

Fig. 5 the deformed mesh

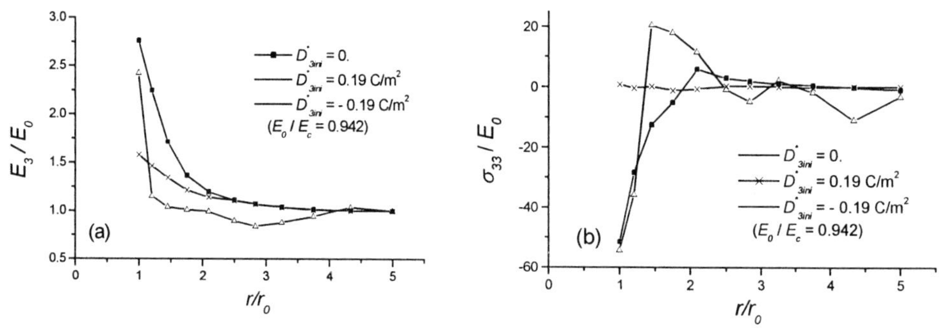

Fig. 6 Effects of initial remanent polarization

elliptic hole reveals that the concentrations near the apex are stronger than that in other places. Figure 5 shows the deformed mesh of the model studied. Note that inside the deformed mesh is an original elliptic hole before the loading is applied. It is obvious that the tensile deformation along the said boundary makes the hole more open. Therefore, it may be concluded that an applied positive electric field enlarges the hole. This is in agreement with the experimental observation made by Fang and Li [11], in which they concluded that an applied positive electric field might lead to opening of a crack.

The effects of the initial remanent polarization on the field distributions near the circular hole are of great interest to researchers. Figure 6 shows the variation of the normalized electric field, E_3/E_0, and the normalized stress, σ_{33}/E_0, with the initial remanent polarization. The ratio of the applied electric field to the coercive field is set equal to 0.942 in the plots. Note that D^*_{3ini} refers to the overall initial remanent polarization in the x_3-direction. A minus value of D^*_{3ini} means that it is in the negative x_3-direction, and $D^*_{3ini} = 0$ means that the material is unpoled. Figure 6 shows that the electric field and stress concentrations for the unpoled case are close to those of the case of negative initial remanent polarization. However, a positive initial remanent polarization significantly reduces the concentrations. It is worth noting that the observation of significant electric field and stress concentrations arising from a negative remanent polarization is consistent with the conclusion made by Lu et. al. [12], in which a negative remanent polarization is equivalent to a positive electric field.

4.3 EXAMPLE 3: A POLED PZT PLATE WITH A SINGLE EDGE CRACK SUBJECTED TO PRESCRIBED ELECTRIC LOADS

Due to symmetry, only one half of a poled PZT plate with a single edge crack that is

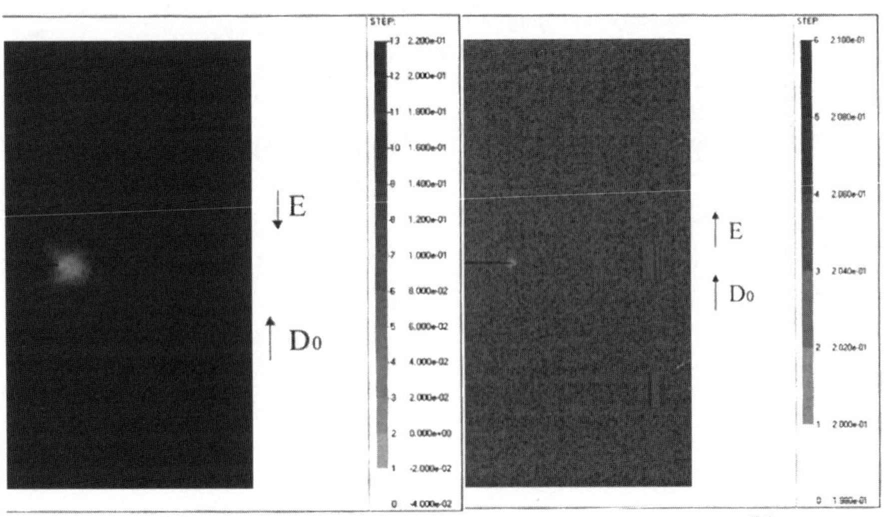

Fig. 7 Distributions of electric displacement D_3 under: (a) negative E_0; (b) positive E_0

subjected to prescribed electric loads needs to be modeled. Either a negative or a positive electric field E_0 ($=0.72E_c$) is applied in the x_3-direction. On the crack surfaces, the conventional mechanical, i.e., traction free condition, and the impermeable electrical, i.e., $\boldsymbol{D} \cdot \boldsymbol{n} = 0$, boundary conditions are implemented. The distributions of the electric displacement, D_3, under either an applied negative or a positive electric field are shown in Figure 7, which show that the negative electric field makes more domain switch than the positive electric field. It is obvious that the domain switching zone near the crack tip in Figure 7(a) is larger than that in Figure 7(b). The distributions of the stress, σ_{33}, under either an applied negative or a positive electric field are shown in Figure 8. It is important to note that although the applied electric field E_0 is less than the coercive field E_c, it still lead to significant electric field and stress concentrations, which are sufficient to induce domain switching at the crack tip. However, these electric field and stress concentrations decrease with increasing distance away from the crack tip. Figure 8(a) shows that a compressive stress appears at the crack tip when subjected to a negative electric stress. It is interesting to note from Figure 8(b) that under an applied electric field the tensile stress appears not only at the crack tip but also in the trail region of the crack. Thus, it may be concluded that a positive electric field enhances crack growth while a negative electric field impedes crack propagation since it produces a compressive stress near crack tip, which is consistent with both the experimental observations [9,11] and the theoretical results [4,6,9,12].

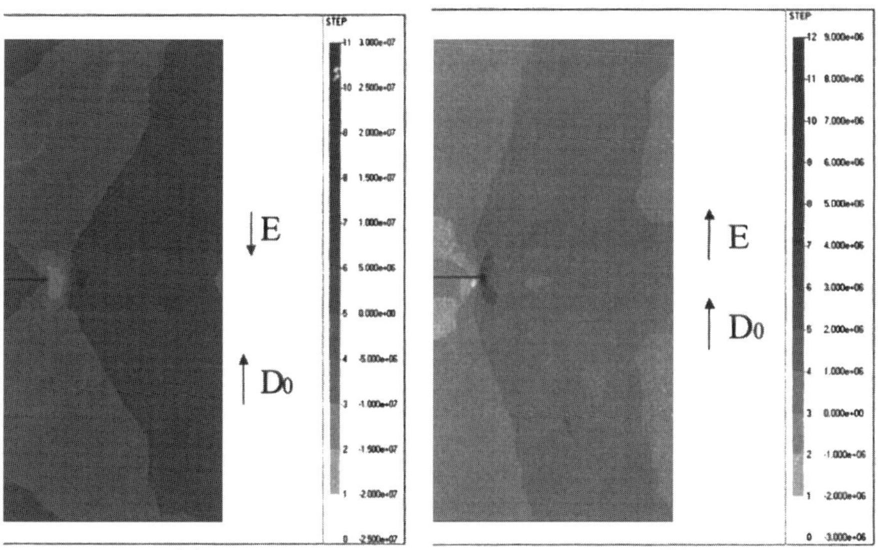

(a) (b)

Fig. 8 Distributions of stress σ_{33} under: (a) negative E_0; (b) positive E_0

5. Concluding Remarks

(1) The distribution of the remanent polarization near a defect, such as a circular or an

elliptic hole or a crack, in a fully poled ceramic plate subjected to a large applied load, shows that the hysteresis nonlinear behavior is dominant. This behavior is due to electric domain switching.

(2) Even when the remote applied electric field is lower than the coercive field, the remanent polarization is still concentrated near the boundary of a circular or an elliptic hole or the tip of a crack. This is also true for the electric and stress-strain fields, which are sufficiently large to induce domain switching, near the said boundary or the crack tip. However, the electric field and stress concentrations decrease with increasing distance away from the said boundary or the crack tip.

(3) A positive applied electric field has the tensioning effect on the specimen consisting of either a circular or an elliptic hole. In other words, a positive applied electric field enlarges the cavity.

(4) Tensile stress appears not only at the crack tip but also in the trail region of the crack under an applied electric field while a negative electric field produces a compressive stress at crack tip. In fact, the stress-concentration near crack tip increases with increasing the applied electric load. Thus, it can be concluded that a positive electric field enhances crack growth while a negative electric field impedes crack propagation, which is consistent with both the experimental and theoretical results.

Acknowledgement

Support from the National Science Foundation of China is acknowledged.

References

[1] M. E. Lines and A. M. Glass, *Principles and Applications of Ferroelectrics and Related Materials*, Oxford, Oxford University Press, (1977).

[2] Ikeda, T., *Fundamentals of Piezoelectricity*, Oxford: Oxford University Press, (1990).

[3] C. S. Lynch, The effect of uniaxial stress on the electro-mechanical response of 8/6/35 PLZT, *Acta Metall. Mater.* **44**, 4137, (1996).

[4] H. J. Gao, T. Y. Zhang and P. Tong, Local and global energy release rates for an electrically yielded crack in a piezoelectric ceramic, *J. Mech. Phys. Solids,* **45,** 491, (1997).

[5] Hwang S. C., Lynch C. S. and McMeeking R. M., Ferroelectric/ ferroelastic interactions and a polarization switching model. *Acta metall.mater.*, **43**, 2073, (1995).

[6] Sosa, H. Plane problems in piezoelectric media with defects, *Int. J. Solids Structures* **28**, 491, (1991).

[7] Y. E. Pak, Crack extension force in a piezoelectric material, *J. Appl. Mech.* **57**, 647, (1990).

[8] Z. Suo, C. M. Kuo, D. M. Barnet and J. R. Willis, Fracture Mechanics for Piezoelectric Ceramics. *J. Mech. Phys. Solids,* **40,** 739, (1992).

[9] S. B. Park and C. T. Sun, Fracture criteria for piezoelectric ceramics, *J. Am. Ceram. Soc.* **78**, 1475, (1995).

[10] T. Zhu and W. Yang, Toughness variation of ferroelectrics by polarization switch under non-uniform electric field, *Acta Mater.* **45**, 4695, (1997).

[11] D.N. Fang and C. Q. Li, Nonlinear electric-mechanical behavior of a soft PZT-51 ferroelectric ceramic, *Journal of Materials Science*, **34**, 4001, (1999).

[12] W. Lu, D.N. Fang and K.C. Hwang, Nonlinear Electric-mechanical Behavior and Micormechanics Modeling of Ferroelectric Domain Evolution, *Acta Mater.,* **47**(10), 2913, (1999).

Two-way Shape Memory Effect Obtained by Training of Combined Loading

M. Tokuda*, S. Sugino*, T.Inaba* and P.Sittner**

* Department of Mechanical Engineering, Mie University,
Kamihama 1515, Tsu 514-8507, Japan.
** Institute of Physics, Czech Academy of Sciences,
Na Slovance 2, 18040 Prague 8, Czech Pepublic

ABSTRACT

Two-way shape memory effect obtained by using the combined loading training were experimentally investigated by applying the combined load of axial force and torque to the thin-walled tubular specimen of Cu-based shape memory alloy (SMA). Some fundamental and unique behaviors are shown and discussed in this paper.

1. INTRODUCTION

The development of two-way shape memory effect [1-3] has been expected for the engineering application of shape memory alloy (abbreviated as SMA、 hereafter). When the SMA (shape memory alloy) is applied to the actuator (or thermal engine), the bias-stress has to be normally added in some way (e.g., by using the bias-spring). That is, SMA can deform cyclically during the temperature cycles, under the (bias-, or externally-added) load. The two-way shape memory effect is here defined as the actuating behavior without any external bias stress (bias loading). In this case, the actuating (engine) system using SMA becomes to be quite simple, and thus the development of two-way memory effect has been expected from the engineering application point of view, especially from the components of smart structures or intelligent materials systems. One of the technique to get the two-way shape memory effect without any bias (external) load is to produce the internal (residual) stress in the shape memory material, instead of the external bias stress. This kind of internal (residual) stress can be induced by the formation of stable dislocation structure. The structure can be obtained by the cyclic deformation called as "training". It is quite interesting that the dislocation plays an important role in the two-way shape memory effect, though the appearance of dislocation is rather harmful to the functions of shape memory alloy, that is, one-way shape memory effect, pseudo-elastic effect and so on. As well-known, various kinds of dislocation structures can be obtained by using the combined (multi-axial, complex) cyclic loading. That is, various kinds of training are possible and thus, various kinds of two-way shape memory behavior may be obtained. In this idea, some fundamental experiments related with the combined loading training were performed in the author's laboratory. Some interesting experimental results are shown in this report.

2. EXPERIMENTAL PROCEDURE

The specimen selected in this work is the thin-walled tube made of Cu-based shape memory alloy whose chemical constituent is Cu-8wt%Al-10wt%Zn-5wt%Mn. The geometry of

Q.P. Sun (ed.), IUTAM Symposium on Mechanics of Martensitic Phase Transformation in Solids, 55–62.

specimen is shown in Fig. 1. As shown in the figure, the outer and inner diameters are 8mm and 5mm, respectively, and thus the thickness is 1.5mm. The heat treatment is 650℃ for 2 hours followed by quenching into the iced–water, and after that, the aging : 100℃ for 20min. is conduced. The characteristic temperatures measured by the DSC technique under the stress-free condition are as follows:

$$M_S = -30℃,\quad M_f = -55℃,\quad A_S = -35℃,\quad A_f = -15℃.$$

The experiments were performed by applying the combined loads of axial force and torque to the thin-walled tube specimen mentioned above. The experimental apparatus is the Shimadzu Auto-graph Combined Loading Testing Machine [4-6] with the electric furnace which controls the surrounding temperature. Strain is measured by electric strain gauge.

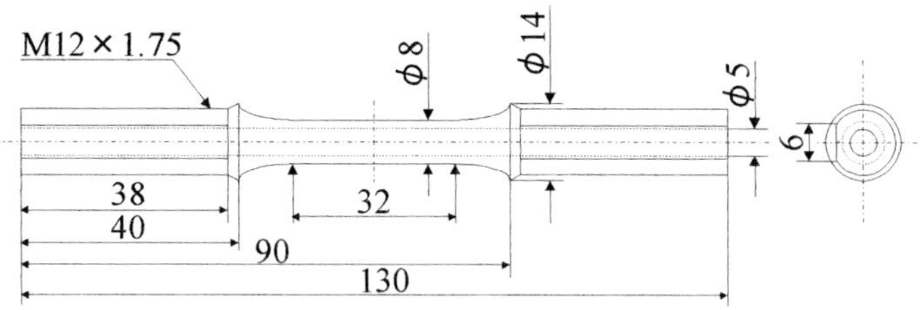

Fig. 1 Geometry of specimen

3. EXPERIMENTAL RESULTS AND DISCUSSIONS

3.1 Simple loading training

Figure 2(a) shows the two-way shape memory behavior of Cu-based shape memory alloy tube before training. As found from this figure, some small cyclic changes of axial strain ε and shear strain γ are observed during the temperature cycle between 20℃ and –80℃. That is, the material shows small two-way shape memory effect before training. This effect may be caused by the residual stress which exists in the as-received state (before the training). Figure 2(b) shows the two way shape memory behavior obtained by the simple torsion training of tube, whose stress amplitude is $C_1\tau$ =340MPa (between 0MPa and 340MPa) and the number of cycles is 5. As found from this figure, a large cyclic change of shear strain γ can be obtained by the torsion training. Also, the cyclic change of axial strain becomes much smaller than that observed in Fig.2(a). This is a typical evidence of the path dependency of inelastic behavior of shape memory alloy. The coefficients C_1 and C_2 are the material constants related with the equivalent stress σ_{EQ} and equivalent strain ε_{EQ}, respectively, defined as follows:

$$\sigma_{EQ}=\sqrt{[\sigma^2+(C_1\tau)^2]},\qquad \varepsilon_{EQ}=\sqrt{[\varepsilon^2+(C_2\gamma)^2]}.$$

The stress-strain relations for the uniaxial tension, simple torsion, and proportional loading of axial stress σ and torsion stress τ coincide very well with each other by using the above defined equivalent stress σ_{EQ} and strain ε_{EQ}. If the material is of so-called Mises type, $C_1=\sqrt{(3)}$ and $C_2=1/\sqrt{(3)}$. In this material, C_1=1.2 and C_2=0.5 [4-6].

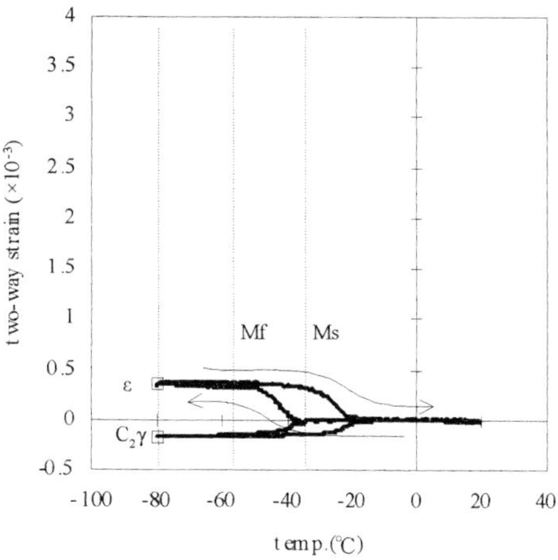

Fig. 2(a) Two-way shape memory strain without any training

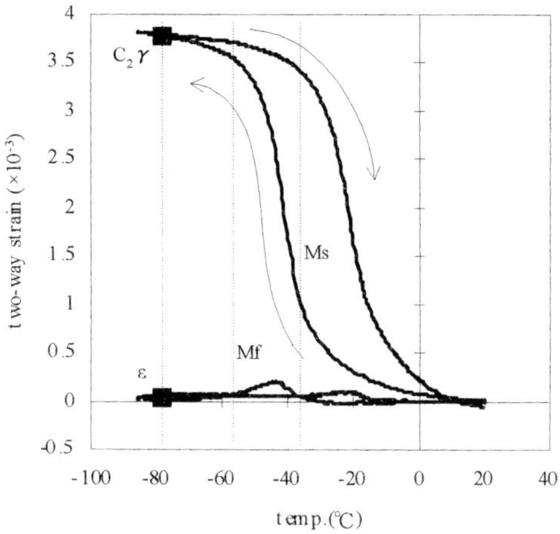

Fig. 2(b) Two-way shape memory strain obtained by simple torsion training

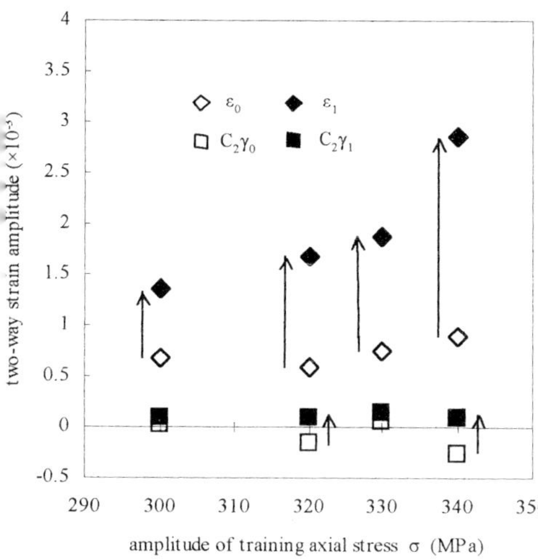

Fig. 3 Relation between two-way memory amplitude and the training stress amplitude in the case of uniaxial tension training

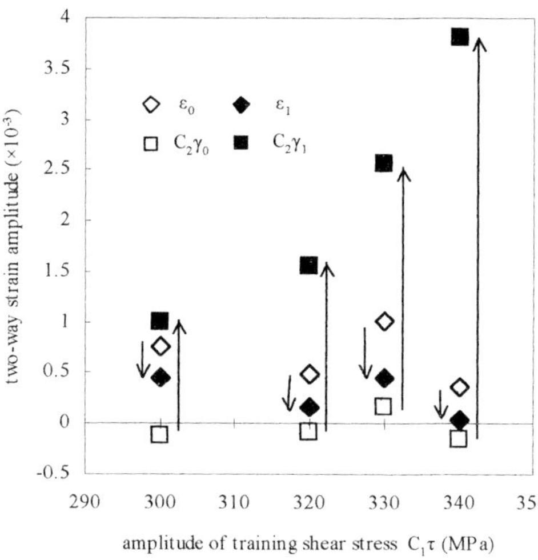

Fig. 4 Relation between two-way memory amplitude and the training stress amplitude in the case of simple torsion training

Figures 3 and 4 show the two-way strain amplitude (the difference of two strains at 20℃ and –80℃) obtained by four different stress amplitudes. Figure 3 shows the results obtained by the uniaxial tension training, and Fig. 4 shows the results obtained by simple torsion training. In these figures, ◆ denotes the axial strain amplitude (ε_1) obtained after training, ◇ denotes the axial strain amplitude (ε_0) obtained without training (before training), ■ denotes the shear strain amplitude ($C_2\gamma_1$) obtained after the training, and □ denotes the shear strain amplitude ($C_2\gamma_0$) without the training. The amplitude (◇ and □) obtained before the training are not related with the stress amplitude. These results show the difference of as-received states. As found from these figures, the two way shape memory effects (two-way strain amplitudes) shown by ◆ and ■ are increased with the increase of the training stress amplitude. The path-dependency can be clearly observed again in Fig.4. That is, the existed two-way memory effect of axial strain (◇) before the training decreases with the simple torsion training (◇ ⇒ ◆). This is the so-called fading-memory phenomenon, which has been well-observed in the (normal, or dislocation-slip) plastic behavior of metallic materials. The common behavior noticed in these figures is that the two-way memory strain amplitude increases with the increase of training stress amplitude. Additionally it seems that the simple torsion training is more effective than the uniaxial tension training from the view point of the obtained two-way shape memory strain amplitude. The reason (mechanism) is uncertain, present now. Further discussion may be necessary.

3.2 Combined loading training (I)

Figures 5-7 show the examples of experimental results obtained by combined loading training. Figure 5 shows the two-way shape memory phenomenon obtained by the following training process (process A): 5 cyclic simple torsion loading-unloading whose amplitude is $C_1\tau$ =320MPa after 5 cyclic uniaxial loading-unloading whose amplitude is σ=320MPa. As found from this figure, the tube specimen deforms cyclically in a two-dimensional way, that is, the cyclic two-way shape memory deformation composed of simultaneous axial strain and torsion strain is obtained. It is quite interesting that the amplitudes of axial strain and shear strain are different with each other, though those two-way strains were obtained by the uniaxial tension and simple torsion whose loading amplitudes are the same: 320MPa. It may be related with the residual stress state induced by the combined training process. Figure 6 shows the two-way shape memory phenomenon obtained by the exchange of loading order of uniaxial loading and torsion loading (process B: 5 cyclic uniaxial loading whose amplitude is σ=320MPa after 5 cyclic torsion loading whose amplitude is $C_1\tau$=320MPa. The two-dimensional deformation during the change of temperature without any external applied stress is also observed. The difference between the results shown in Figs. 5 and 6 may be related with that the torsion training produces larger two-way memory strain than the uniaxial tensile training. Figure 7 shows the two-way shape memory phenomenon obtained by using the following training (process C): 5 cycles of torsion training whose amplitude is $C_1\tau$ =310MPa after 5 cycles of uniaxial tensile training whose amplitude of σ=330MPa. Figure 8 shows the two-way memory strain paths obtained by three experiments (processes A, B, and C) mentioned above. Here, it should be noted that there exists some differences between the forward deformation (with the decrease of temperature) and the reverse deformation (with the increase of temperature) in the case of Process C (hysteretic performance), and all obtained two-way behavior are slightly different from the proportional deformation.

3.2 Combined loading training (II) [7]

Figure 9 shows the experimental results obtained by the proportional cyclic load training. Figure 9(a) shows the stress-strain relation during the training. For the training, 5 repeated loading–unloading between stress states (σ, $C_1\tau$)=(305MPa, 258MPa) and (0, 0) was employed. Figure 9(b) shows the two-way behavior obtained by the training. As found from the figure, the thin-walled tubular specimen shows the cyclic and proportional deformation only by the change of temperature without any bias (external) stress, in the case of the proportional load training. Figure 10 shows an example of experimental results

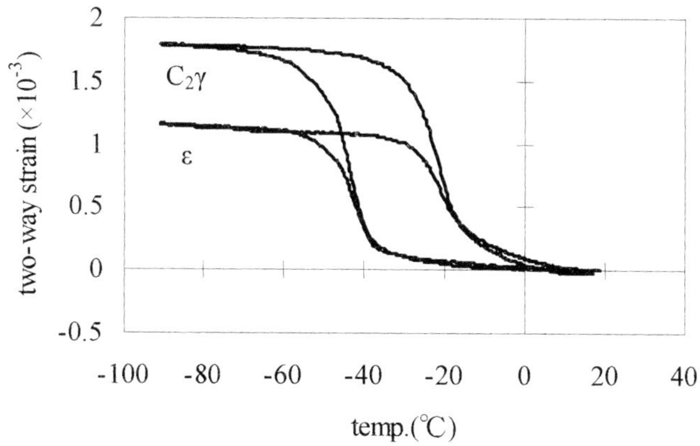

Fig. 5 Two-way memory strain obtained by combined loading training (A)

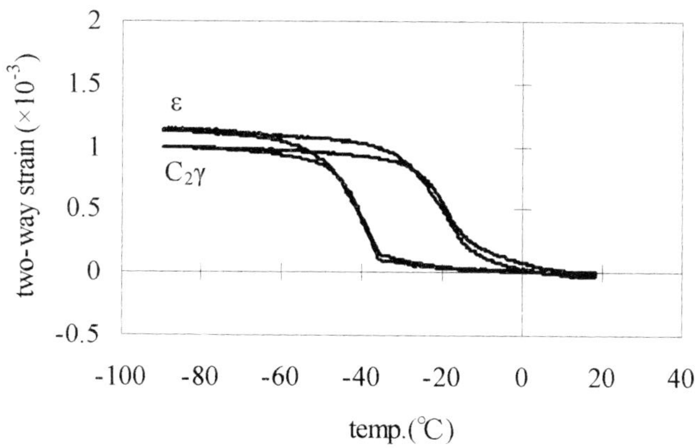

Fig. 6 Two-way memory strain obtained by combined loading training (B)

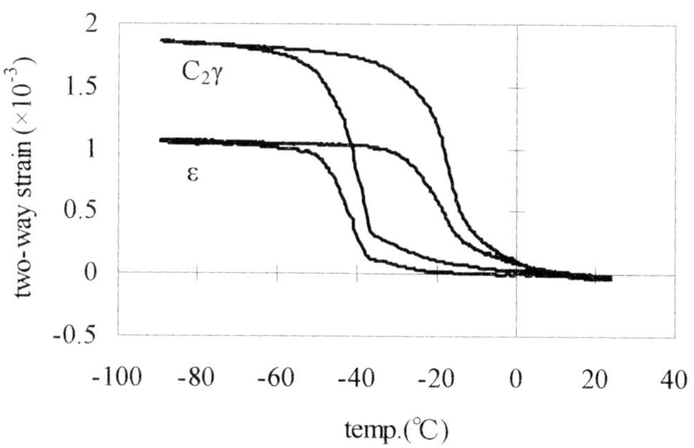

Fig. 7 Two-way memory strain obtained by combined loading training (C)

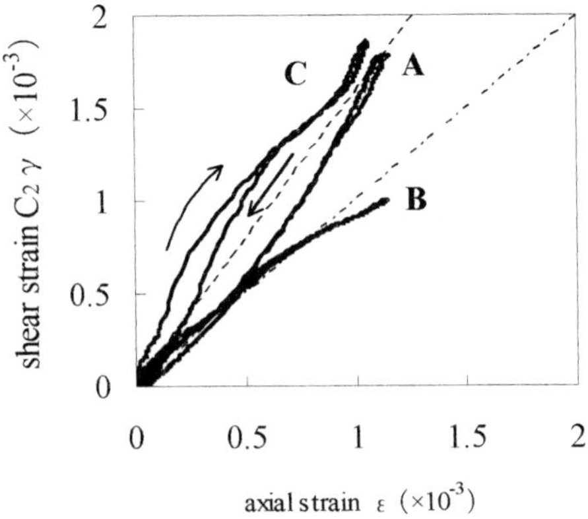

Fig. 8 Comparison of two-way memory strain paths obtained by the training processes A, B and C

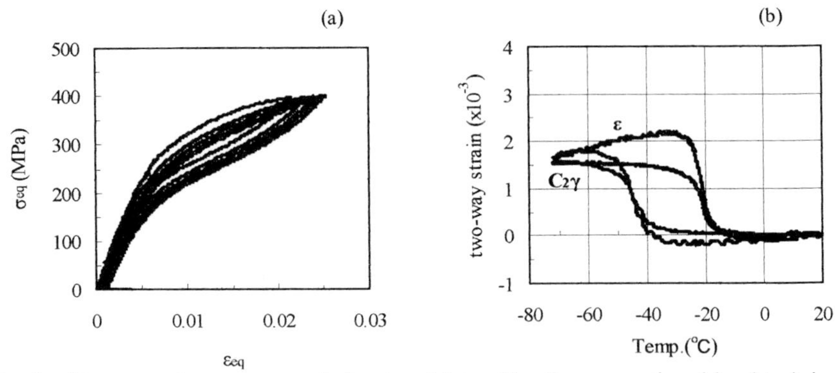

Fig. 9 Two-way shape memory behavior obtained by the proportional load training

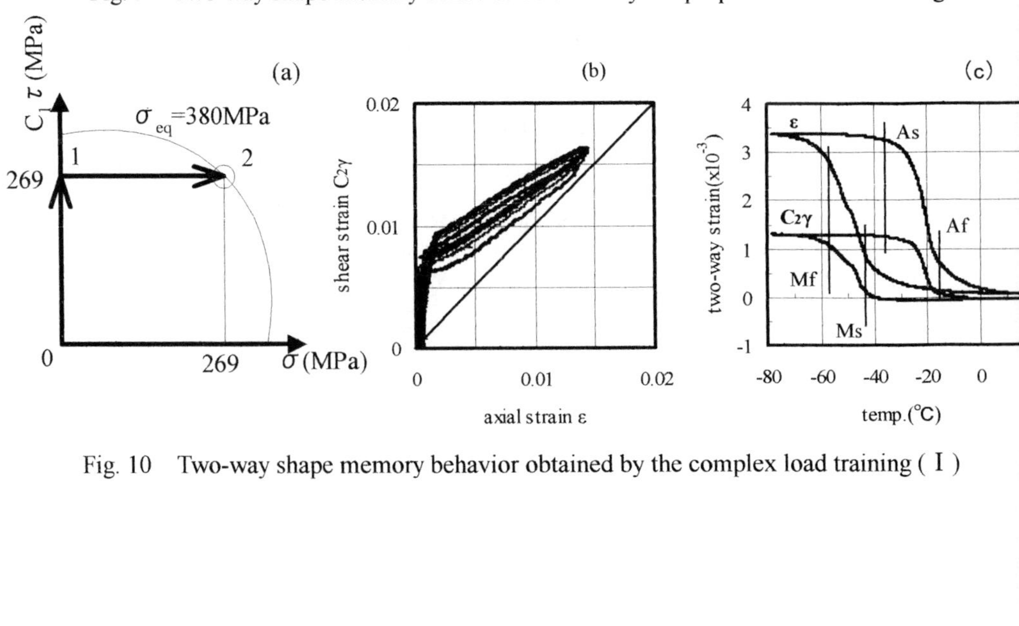

Fig. 10 Two-way shape memory behavior obtained by the complex load training (I)

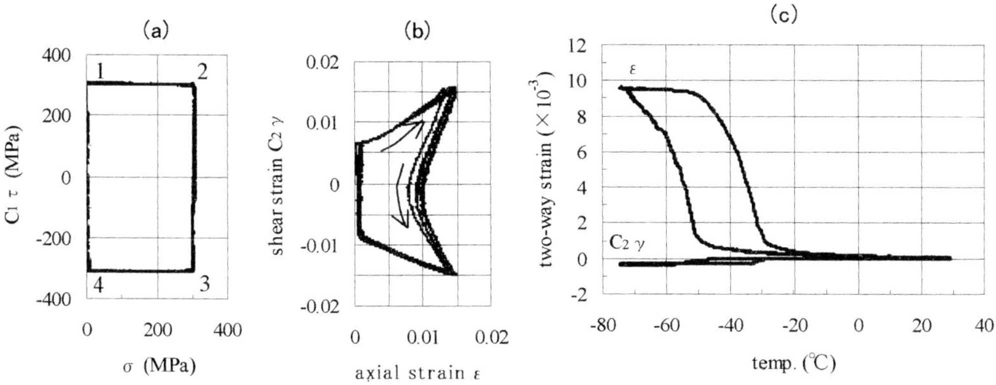

Fig. 11 Two-way shape memory behavior obtained by the complex load training (Ⅱ)

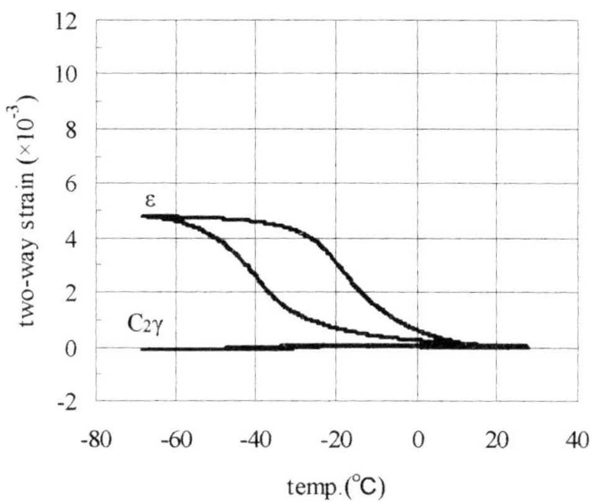

Fig. 12 Two-way shape memory behavior obtained by the simple tension

obtained by the complex loading training. Figure 10(a) shows the complex load training path: that is, 0-1 is the torsion loading, 1-2 is the axial stress loading while keeping the shear stress constant, 2-1 is the unloading of axial stress, and 1-0 is the unloading of shear stress. The number of training cycles is 5. Figure 10(b) shows the strain response during the 5 times training. Figure 10(c) shows the obtained 2-way behavior. As found from the figure, there is a slight phase difference between the axial strain and shear strain during the change of temperature. That is, the axial strain starts first at (around)-15 ℃, and the shear strain starts at (around) –40℃. That is, this two-way memory deformation is not of proportional type : at first, the specimen shows the elongation, and after some elongation, the tube specimen shows the simultaneous deformation of elongation and torsion. Also, the axial strain is much larger than the shear strain. This difference may be explained by the path dependency of shape memory alloy, especially related with the Fading Memory of Plasticity, again. Figure 11 shows another examples of complex loading training. Figure 11(a) shows the training path: 0-1 is shear stress loading, 1-2 shows the axial loading while keeping the shear stress constant,

2-3 is the unloading and reverse loading of shear stress while keeping the axial stress, 3-4 is the unloading of axial stress constant and 4-0 is the unloading of shear stress. Fig. 11(b) shows the strain response during the training, and Fig. 11(c) shows the two-way memory phenomenon. As found from the figure, the shear strain is quite small. Figure 12 shows the two-way axial strain obtained by the simple tensile training whose amplitude is 300MPa. As found from the comparison between Fig. 11(c) and Fig. 12, the two-way axial strain obtained by the complex loading training is much larger than the two-way strain obtained by the uniaxial tensile training, whose cyclic amplitude is the same:300MPa. That is, the training of shear stress component increases the two-way shape memory axial strain. Why it happens is now open problem. We need further experiments and discussions.

4. CONCLUSIONS

In this paper, the following tentative conclusions were achieved:

(1) The larger training stress amplitude gives the larger 2-way shape memory strain.

(2) The torsion deformation training produces two-way shape memory larger than the uni-axial tensile training. This difference is quite interesting from the mechanics of two-way shape memory effect, but this question is not solved yet.

(3) Various kinds of 2-way shape memory behavior may be obtained by the training by using the combined loading training. It is quite interesting to investigate the two-way shape memory effect obtained by various training process including the change of temperature and the change of 6 independent stress components.

(4) It may be very important to investigate the fundamental mechanism to produce the two-way shape memory phenomenon from the view point of the relationship between residual stress and variant changes induced by phase transformation.

(5) Further systematic investigation may be necessary in the training of complex loading and change of temperature. Also, the prediction of residual stress is quite important and interesting from the micro-mechanics point of view.

REFERENCES

1. Stalmans, R., Humbeek, J.V. and Felaey, L., (1991), "Training and the two-way memory effect in copper based shape memory alloy", Journal de Physique IV, Colloque C4, supplement au Journal de Physik Ⅲ, Vol.1, pp.C4-403-408.
2. Bo, Z. and Lagoudas, D.C., (1999), "Thermo-mechanical modeling of SMAs under cyclic loading Part Ⅲ, evolutions of plastic strain and two-way shape memory effect", International Journal of Engineering Sciences 37, pp.1175-1203.
3. Perkins, J. and Hodgson, D., (1990), "Engineering Aspects of Shape Memory Alloys", (eds. Duering, T.W., et al.), pp.195-201, Butterworth-Heimeman.
4. Tokuda, M., Sittner, P., Takakura, M. and Ye, M., (1995), "Experimental Study of Performances in Cu-based Shape Memory Alloy under Multi-axial Loading Conditions", Materials Science Research International, 1(4), pp.260-265.
5. Sittner, P., Hara, Y. and Tokuda, M., (1995), "Experimental Study on the Thermoelastic Martensitic Transformation in Shape Memory Alloy Polycrystal induced by Combined External Force", Metallurgical and Materials Transactions A, 26A, pp.2923-2935.
6. Tokuda, M., Ye, M., Takakura, M. and Sittner, P., (1999), "Thermomechanical behavior of shape memory alloy under complex loading conditions", International Journal of Plasticity, 15, pp.223-239.
7. Tokuda, M., Sugino, S. and Inaba, T., (2001), "Two-way shape memory behavior obtained by combined loading training", Proceedings of 11th International Conference on Adaptive Structures and Technologies (Nagoya, Japan), pp.382-389, Technomic Publishing Co., INC, Lancaster/Basel.

OBSERVATION OF DOMAIN MOTION IN SINGLE-CRYSTAL BARIUM TITANATE UNDER COMBINED ELECTROMECHANICAL LOADING CONDITIONS

E. BURCSU, G. RAVICHANDRAN, AND K. BHATTACHARYA
Division of Engineering and Applied Science
California Institute of Technology
Pasadena, CA 91125, U.S.A.

Abstract

The nonlinear electromechanical behavior of ferroelectric materials is governed by the motion of domains. Since many common ferroelectric materials, such as barium titanate and PZT, are also ferroelastic, the domain motion is highly affected by stress as well as electric field. Experiments are performed on (001) and (100) oriented single crystals of barium titanate under combined electromechanical loading conditions. The crystal is subjected to a constant compressive stress (dead load) and an oscillating electric field along the [001] direction. Global deformation and polarization are measured as a function of electric field at different values of compressive stress. The use of semi-transparent electrodes and transmitted illumination allows in situ, real-time microscopic observations of domain patterns using a long working-distance, polarizing microscope. The combined electromechanical loading results in a cycle of stress and electric field induced 90° domain switching. This is an electrostrictive behavior with measured strains of up to 0.9%.

1. Introduction

Ferroelectric materials have found wide application due to their excellent dielectric and electromechanical properties. Such applications include sensors and actuators for use in ultrasonics, micro-positioning and active damping. The properties of interest for these applications are piezoelectricity and electrostriction. Piezoelectricity, as expressed by the converse piezoelectric effect, is a linear relationship between strain and electric field, while electrostriction is a quadratic relationship [8]. For the purposes of this paper, electrostriction will be considered in a more general sense as electric field induced deformation that is independent of field polarity.

Barium titanate is a ferroelectric material of the perovskite class and has been extensively studied [2]. At high temperature it has the cubic perovskite structure, as shown in Figure 1. When cooled below 120°C, it transforms to a tetragonal phase. In addition to the strain induced by the lattice distortion, there is a spontaneous polarization along the c-axis of the unit cell. Thus, at phase transition the unit cell can take any of six crystallographically equivalent combinations of strain and polarization. Furthermore, different regions

Q.P. Sun (ed.), IUTAM Symposium on Mechanics of Martensitic Phase Transformation in Solids, 63–70.

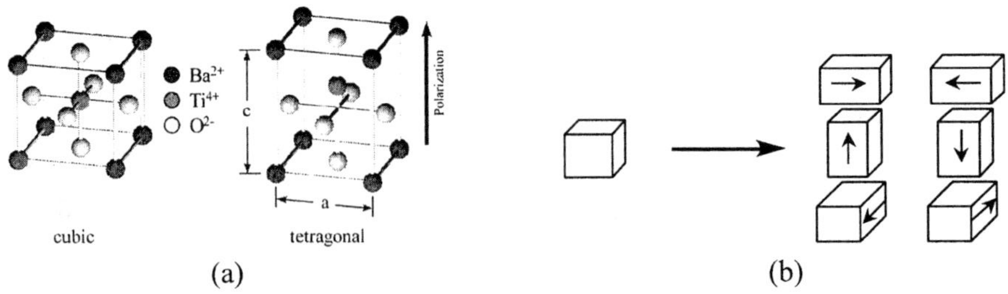

Figure 1. (a) The cubic (high temperature) and tetragonal (room temperature) structures of barium titanate. (b) Upon the cubic to tetragonal phase transition, the unit cell can take any of six equivalent combinations of strain and polarization. The arrow indicates the direction of polarization.

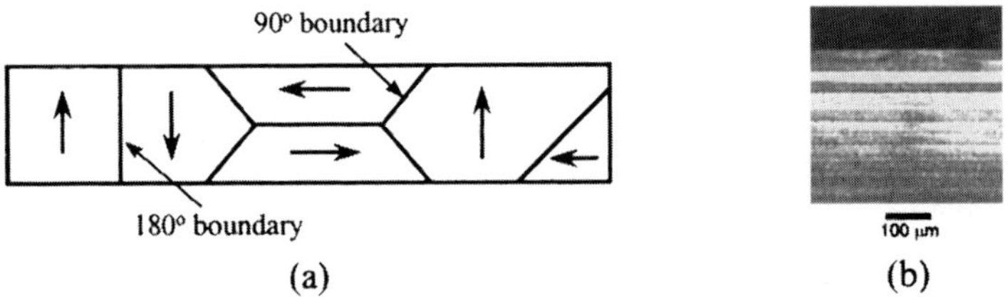

Figure 2. (a) Schematic diagram of the subgranular structure of domains separated by 90° and 180° boundaries and (b) a domain pattern photographed using a polarizing microscope.

of a single crystal or single grain in a polycrystal can take on different directions of polarization. The regions of constant polarization are known as a ferroelectric domains and are separated by 90° or 180° boundaries, as shown in Figure 2. The domain boundaries can be nucleated and moved by electric field or stress. This work is a study of these processes under combined electrical and mechanical loading conditions. The importance of these processes are their relationship to 'poling' of piezoelectric ceramics and the enhanced piezoelectric response of some materials. Poling is the process of generating a domain texture to result in a global average polarization and is required to make a ferroelectric ceramic piezoelectrically active. The details of the mechanism of the poling and depoling process in polycrystals is very complex due to the presence of residual stresses and grain boundary mismatch, and is the subject of a number of experimental and theoretical studies [3, 4, 5].

2. Theory

While detailed theories of ferroelectrics have been developed, most recently by Shu and Bhattacharya using the framework of finite deformation continuum mechanics [7], for the purposes of this paper, a simplified approach is sufficient. Consider a flat plate ferroelectric with electrodes on two faces, exposed to a constant compressive stress. The equilibrium condition can be analyzed by minimization of the following energy density

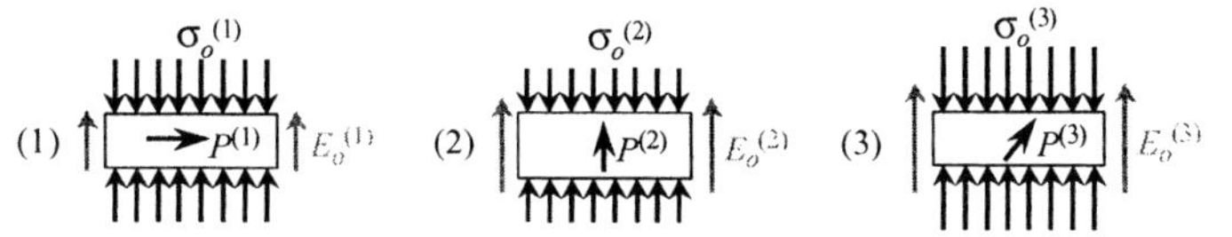

Figure 3. Possible flat plate equilibrium configurations of barium titanate under compressive stress and electric field loading.

function:

$$G(\epsilon, P, \theta) = W(\epsilon, P, \theta) - E_o \cdot P - \sigma_o \cdot \epsilon \quad , \tag{1}$$

where the first term is the free energy density, the second term describes the interaction with the applied electric field and the third the interaction with the applied mechanical load. W depends on the strain, ϵ, the polarization, P, and temperature, θ, and has multiple ground states corresponding to the equivalent polarized states of the crystal. For barium titanate at room temperature, these states correspond to any of the six $\langle 100 \rangle$ directions.

The possible equilibrium configurations for a crystal of barium titanate cut in a flat plate configuration with a $\{100\}$ orientation under combined electric field and compressive load, are shown in Figure 3. Any configuration with the polarization oriented in-plane will be energetically equivalent. This is shown as state (1) with an appropriate combination of stress and electric field. At a low stress and high electric field, the stable equilibrium configuration will be state (2), the out-of-plane configuration. For very high values of compressive stress and electric field, the orthorhombic phase will be the equilibrium configuration, shown as state (3). Thus, the domain state and, therefore, the global strain and polarization can be altered by application of stress or electric field. In fact, combined electromechanical loading, as described, can allow switching between states (1) and (2) resulting in a strain of up to 1%.

3. Experiment

An experimental setup was designed to reproduce the conditions outlined above. The system was designed to apply a constant compressive load and variable electric field to a ferroelectric crystal. It was further designed to allow accurate measurement of strain and polarization, as well as allow *in situ* observation of the crystal during the experiment. Experiments were performed on single crystals of barium titanate of 5×5×1 mm dimensions with (001) and (100) orientation (Superconix, Lake Elmo, MN). The crystals were poled prior to polishing, so were nearly single domain in most cases. Semi-transparent, sputtered platinum electrodes were deposited on the major surfaces of the crystal.

The experimental setup consists of a dead weight loading mechanism, displacement measurement transducer for strain measurement, high-voltage power supply, Sawyer-Tower circuit for polarization measurement, and video microscope. The mechanical and electrical components are described elsewhere [1]. Real time, *in situ* observations are made of the crystal using a long working-distance video microscope. The microscope consists of a 10× long working distance objective, extension tube and CCD video camera (Sony XC-75) connected to a digital frame grabber. This system has a field of view of

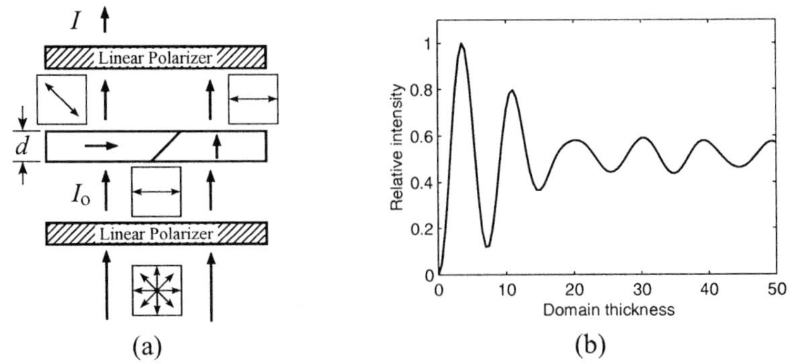

Figure 4. (a) Generation of birefringence contrast in tetragonal barium titanate and (b) predicted relative intensity vs. in-plane domain thickness, *d*.

about 0.65×0.5 mm with a lateral resolution of approximately 1 μm at a working distance of 50 mm. The crystal is imaged through the electrodes along the loading axis. Light is transmitted from a high-intensity halogen lamp through crossed polarizers, one below the specimen and one above. Images are captured at 30 frames per second, allowing observation of the evolution of the domain pattern during the course of the experiment.

The domain structure is visible due to the natural birefringence of the material. The principles of birefringence contrast in ferroelectrics are covered very thoroughly in the work of Schmid [6]. When a crystal is placed between crossed polarizers as illustrated in Figure 4(a), if a region of the crystal has an out-of-plane state, the polarization of the incident light is unchanged and the transmitted light is filtered out. For a region with an in-plane state, the polarization angle of the light is rotated and the transmitted intensity is dependent on the wavelength, λ, and thickness, d. The predicted intensity detected by the CCD as a function of d for the current system is shown in Figure 4(b).

4. Results

During the experiment, a slow voltage signal is generated and the strain and polarization are measured over ten cycles. Figure 5 shows strain and polarization data from a typical experiment with an input signal frequency of 0.05 Hz. The electric field input signal is shown as the solid curve in Figure 5(a) and the strain signal is shown as the dashed curve. The strain responds equivalently to both positive and negative electric field, as is evident when the strain is plotted as a function of the electric field in Figure 5(b). Polarization vs. electric field, shown in Figure 5(c), exhibits a more typical hysteresis. The difference in the character of the strain and polarization hysteresis is explained by the fact that there is no strain change associated with a 180° reversal of polarization.

The strain vs. electric field behavior is highly affected by stress as is seen in Figure 6. At zero stress, the magnitude of the actuation strain is very low, appearing as a nearly horizontal line in the plot. As stress is increased, the magnitude of the actuation strain increases to a maximum of about 0.9% at a stress of 2.14 MPa. As the stress is further increased, the actuation strain decreases and the hysteresis loops broaden as the applied

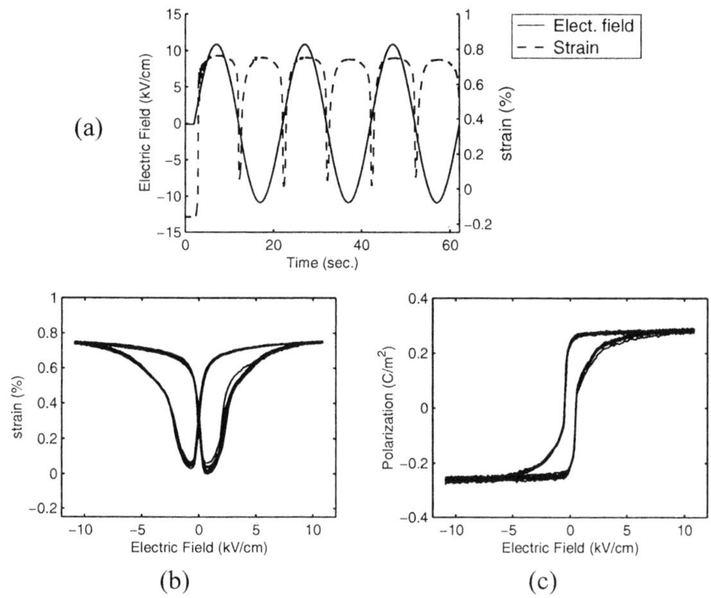

Figure 5. (a) Electric field and strain vs. time. (b) Strain and (c) polarization vs. electric field for a (100) oriented crystal at 1.07 MPa compressive stress.

field is unable to overcome the applied stress.

As the strain and polarization are recorded, a series of images of the domain pattern are captured during the fifth cycle of the experiment at a rate of 30 frames per second. Figure 7 shows five selected images from a test at a compressive stress of 1.07 MPa. The images in the figure are numbered in sequence corresponding to the labeled plots of strain and polarization. The initial image is taken at an electric field of -6,400 V/cm. At this stage the crystal is at nearly its maximum level of strain and minimum level of polarization. The image is nearly black, indicating that the crystal is completely $[00\bar{1}]$ polarized. In the second image (E = -180 V/cm) a bright band of in-plane oriented domains has entered the field of view. The strain has decreased by about 0.4% and polarization has begun to increase. In the third image, as the electric field further increases to the level of the coercive field (E = 500 V/cm) the image is completely bright with in-plane oriented domains throughout the field of view. At this point the strain has reduced to its minimum level and the polarization is zero. In the fourth image (E = 1170 V/cm) the 90° domains have begun to disappear from the field of view as the strain begins to increase and the polarization is begins to level off. By the time the electric field has reached 5,600 V/cm in the fifth image, the strain and polarization have nearly reached their maximum levels and the image is again completely black.

The video images are useful for getting qualitative information of local behavior. To attempt to make a more quantitative comparison between the image data and the crystal averaged values of strain and polarization, an average image intensity is calculated. This

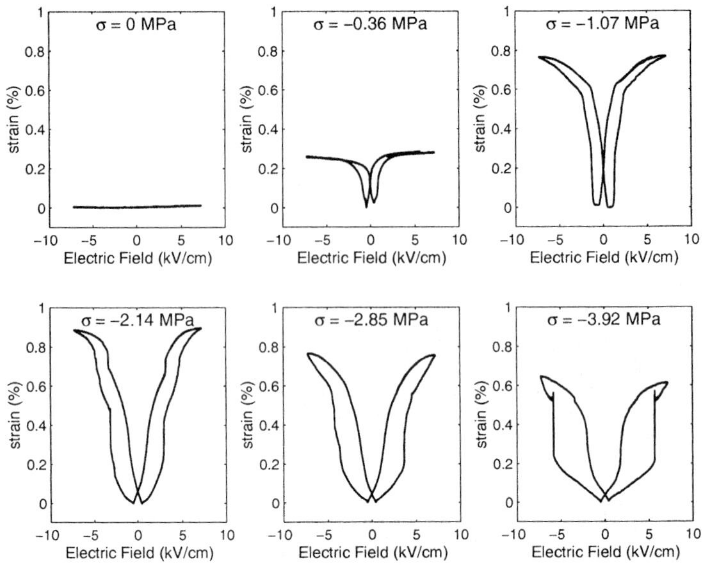

Figure 6. Strain vs. electric field for different values of compressive stress with an initially (001) oriented crystal with input amplitude of ±7.5 kV/cm.

intensity is calculated by summing the gray level (or brightness) of all pixels in an image (644×480 pixels) and normalizing by the maximum possible value (644×480×255). This calculation generates an average quantity over the entire image area which can be compared with other data. The intensity as a function of electric field is plotted along with the strain in Figure 8. For clarity, the intensity curves have been shifted and the strain axis has been inverted. For the 0.72 MPa case the intensity smoothly rises, peaks and then abruptly falls. This trend is closely followed by the strain. For higher levels of stress the intensity slowly begins to rise and then drops before abruptly rising to a plateau. This plateau in the intensity level corresponds to the leveling off of the predicted intensity seen in Figure 4(b). The intensity then suddenly drops, recovers a little bit and then smoothly drops again. The strain signal in each of these cases has a much more smooth rise and fall and a somewhat broader range that envelops the intensity peak. The apparent lag in the intensity history in some cases gives an indication of the inhomogeneity of the deformation. Since the field of view is only about 1% of the crystal area, even the average intensity can be considered to be a local measurement.

5. Conclusions

The electrostrictive response of single-crystal barium titanate under combined electrical and mechanical loading has been shown to be highly dependent on the level of applied stress. A maximum strain of 0.9% has been measured at a stress of approximately 2 MPa. Polarized light microscopy has been used to observe the evolution of the domain pattern during the test. An average image intensity has been calculated which correlates well

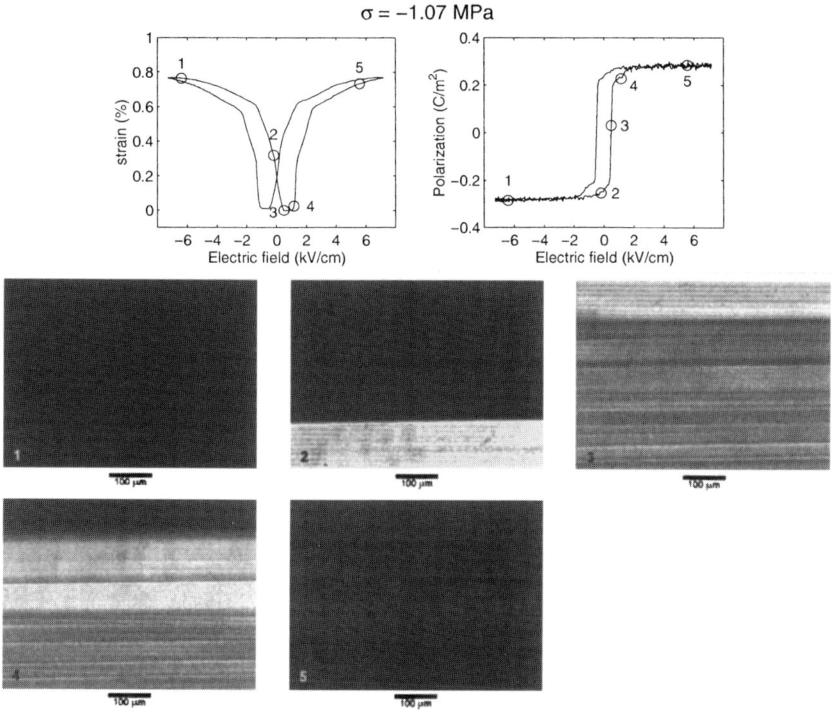

Figure 7. Sample images from experiment at 1.07 MPa compressive stress with corresponding strain and polarization.

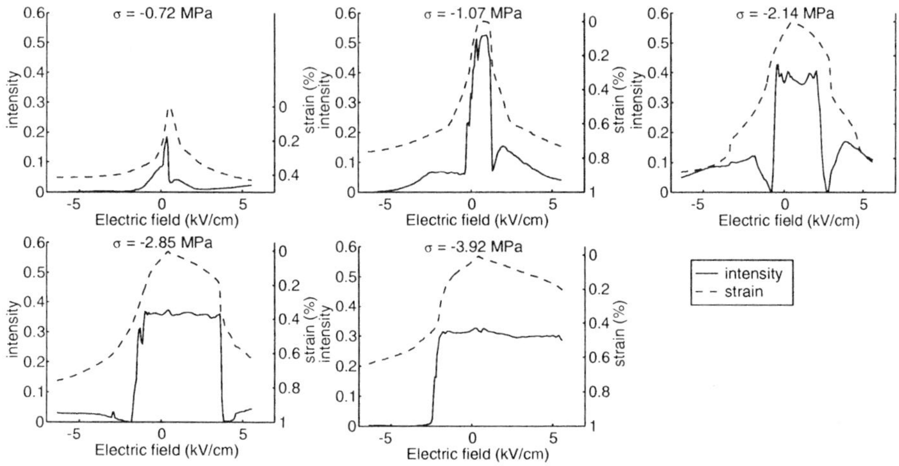

Figure 8. Comparison of average image intensity and strain vs. electric field.

with the measured strain signal and gives insight into the level of inhomogeneity of the crystal deformation. Future work on barium titanate will attempt to determine whether the domain microstructure evolves differently with each cycle or whether the same microstructure appears over and over again. Additional experiments will also focus on other materials such as bulk single-crystal lead titanate and thin film PZT.

6. Acknowledgments

We gratefully acknowledge the support of the Army Research Office (M.A. Zikry, Program Manager) under grant DAAD 19-99-1-0319. This work was supported in part by the MRSEC Program of the National Science Foundation under Award Number DMR-0080065.

7. References

1. Burcsu, E., Ravichandran, G. and Bhattacharya, K.: 2000, 'Large strain electrostrictive actuation in barium titanate'. *Appl. Phys. Lett.* **77**, 1698–1700.
2. Jona, F. and Shirane, G.: 1962, *Ferroelectric Crystals*. New York: Pergamon. Reprint, Dover, New York, 1993.
3. Li, S. P., Bhalla, A. S., Newnham, R. E., Cross, L. E. and Huang, C. Y.: 1994, '90° domain reversal in $Pb(Zr_xTi_{1-x})O_3$ ceramics'. *J. Mater. Sci.* **29**, 1290–1294.
4. Li, Z., Foster, C. M., Dai, X.-H., Xu, X.-Z., Chan, S.-K. and Lam, D. J.: 1992, 'Piezoelectrically-induced switching of 90° domains in tetragonal $BaTiO_3$ and $PbTiO_3$ investigated by micro-Raman spectroscopy'. *J. Appl. Phys.* **71**, 4481–4486.
5. Lynch, C. S.: 1996, 'The effect of uniaxial stress on the electro-mechanical response of 8/65/35 PLZT'. *Acta mater.* **44**, 4137–4148.
6. Schmid, H.: 1993, 'Polarized Light Microscopy (PLM) of Ferroelectric and Ferroelastic domains in transmitted and reflected light'. In: N. Setter and E. L. Colla (eds.): *Ferroelectric Ceramics: Tutorial reviews, theory, processing, and applications*. Zurich: Monte Verità, pp. 107–126.
7. Shu, Y. C. and Bhattacharya, K., 'Domain patterns and macroscopic behavior of ferroelectric materials'. *to appear in Phil. Mag. B*.
8. Uchino, K.: 2000, *Ferroelectric Devices*. New York: Marcel Decker.

TESTING AND MODELING A CU-AL-BE SHAPE MEMORY ALLOY BEHAVIOR UNDER COMPLEX STRESS LOADING

C. BOUVET[1], S. CALLOCH[2], C. LEXCELLENT[1] and A. VIVET[3]

[1] *Laboratoire de Mécanique Appliquée R. Chaléat, CNRS UMR 6604, Université de Franche-Comté, 24 chemin de l'Epitaphe, 25000 BESANCON, France*
[2] *Laboratoire de Mécanique et Technologie, Ecole Normale Supérieure de Cachan, CNRS, Université Paris 6, 61 avenue du Président Wilson, 94235 CACHAN Cedex, France*
[3] *Laboratoire Universitaire de Recherche Scientifique d'Alençon IUT Alençon, Université de Caen Basse-Normandie 61250 DAMIGNY, France*

Abstract

On one hand, bi-compression tests and tension (compression)-internal pressure tests on a Cu-Al-Be shape memory alloy (SMA) polycrystals permit the determination of the yield surface of phase transformation initiation (Austenite → Martensite). On the second hand, the lattice measurements of the $L2_1$ austenite and the monoclinic martensite cells for this alloy permit to determine the "nature" of the phase transformation i.e. an exact interface between the parent phase and an untwinned martensite variant. Hence, a micro-macro integration allows to predict the yield curve of the same polycrystalline alloys. At last, some tools are given for a general phenomenological modeling at the macroscopic scale, in the frame of the thermodynamic of irreversible process of isotropic pseudoelasticity in SMA. These two different investigations fit well the experimental results.

1. Introduction

The present paper is concerned with the "pseudoelastic" behavior of SMAs, i.e. the stress induced phase transformation (from austenite A to martensite M) under isothermal external conditions. The high-temperature phase (A) transforms in the martensite (M) phase diminishing generally its crystalline symmetry. The lower crystalline symmetry of M gives rise to several variants. Each variant has an associated stress-free transformation strain.

In order to use SMAs as efficient elements in a smart structure as actuators or sensors, accurate and reliable predictions of their tridimensional behavior including their hysteresis response are required. Until now, thermomechanical models of the pseudoelastic behavior are validated only with uniaxial experimental observations. But, some recent experimental results in tension-torsion (Rogueda et al., 1996, Raniecki et al., 1999) or under triaxial stress states (Gall et al., 1998) show that the mechanical behavior of SMAs is affected by the multiaxial state of stress. A few comparisons between simulations with classical models and experimental results under non-proportional loadings show the need for more extensive experimental and

Q.P. Sun (ed.), IUTAM Symposium on Mechanics of Martensitic Phase Transformation in Solids, 71–78.

theoretical studies to understand and model the formation and reorientation mechanisms of martensitic platelets under non-proportional loading conditions (Lim and Mc Dowell, 1999).

The present paper consists of three main sections. In the first section, a series of proportional and non-proportional biaxial loadings (tension-internal pressure and bi-compression) is presented and analyzed. In the second section, starting from the lattice measurements of the austenite and the martensite cells, a yield phase transformation of this Cu-Al-Be polycrystal is predicted.

In the third section, some tools for a general macroscopic modeling of the SMAs pseudoelastic behavior are given.

2. Experimental investigations under biaxial loadings

The studied materials are Cu 87.75–Al 11.33–Be 0.49 (wt %) polycrystals. The heat treatment consists in soaking at 650°C for 20 mn followed by a water quench at 100°C for one hour. The Differential Scanning Calorimetry measurements permit to obtain the four classical phase transformation (p. t.) temperatures :

for the cube alloys n° 1 : $M_s^0 = 21°C$, $A_s^0 = 19°C$, $M_f^0 = -7°C$, $A_f^0 = 32°C$

for the tube alloys n° 2 : $M_s^0 = 14°C$, $A_s^0 = 12°C$, $M_f^0 = -13°C$, $A_f^0 = 26°C$

The testing equipment and the validation of boundaries loading conditions for the samples are described in the paper of Bouvet and co-authors (Bouvet et al., 2000). At first, the tension-compression tests revealed that the asymmetric behavior is obvious. Secondly, a series of radial paths (σ_1/σ_2 = Cte) for bi-compression (•) and tension(compression)-internal pressure tests (♦) permits to obtain a yield surface for the initiation of p. t. with a p. t. rate vector normal to this surface (fig. 1).

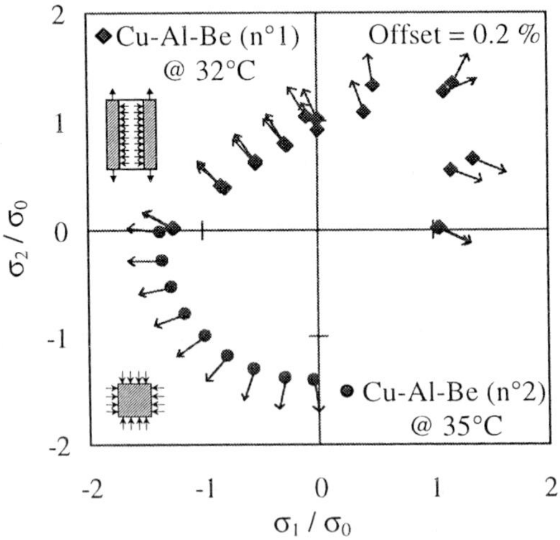

Figure 1 : Yield surface of phase transformation initiation for bi-compression (•) and tension(compression)-internal pressure tests (♦)

Moreover, some non-proportional loadings are performed : a square path ($\sigma_{zz}, \sigma_{\theta\theta}$) for tension-internal pressure loading (fig. 2) and a quarter-circle path for bi-compression loading (fig. 3) are shown.

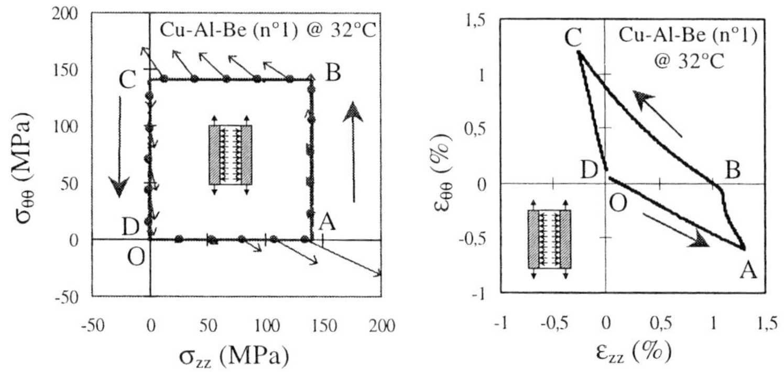

(a) **(b)**

Figure 2 : Square path for tension-internal pressure loading : (a) loading path with transformation strain rate and (b) strain response

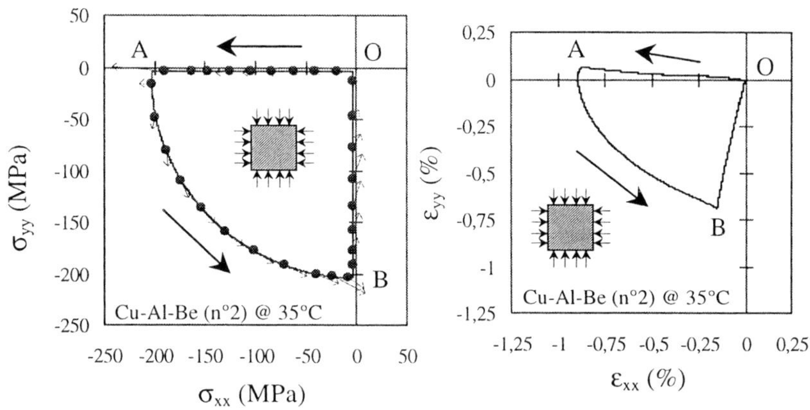

(a) **(b)**

Figure 3 : Quarter-circle path for bi-compression loading loading : (a) loading path with transformation strain rate and (b) strain response

These non-proportional paths correspond to a coupling between a creation-annihilation and a reorientation of martensite platelets during the loading history.

3. "Micro-macro" modeling (Lexcellent et al., 2001)

3.1. BASIC THEORY

Lattice parameters measured by X Ray on a slightly different composition (Cu-12.3 Al-0.5 Be (wt %)) of the present alloy have been used. The austenite cell is $L2_1$ with $a_o = 5.82$ Å and the martensite cell is monoclinic (M18R) with $a = 4.46$ Å, $b = 5.22$ Å, $c = 38.25$ Å, $\theta = 83.3°$) (Moreau et al., 1995).

The strain gradient tensor $\underline{F}$ between the cubic cell $L2_1$ and the product monoclinic cell M18R can be written as

$$\underline{F} = \begin{pmatrix} \alpha & \gamma\cos\theta & 0 \\ 0 & \gamma\sin\theta & 0 \\ 0 & 0 & \beta \end{pmatrix}_{\text{(frame of M18R cell)}} \tag{1}$$

where the transformation stretches α, β, γ are :

$$\alpha = \frac{\sqrt{2}\,a}{a_o} \quad , \quad \beta = \frac{b}{a_o} \quad , \quad \gamma = \frac{\sqrt{2}\,c}{9a_o} \tag{2}$$

The tensor $\underline{E}^t$ associated with the p. t. strain can be written in a classical way

$$\underline{E}^t = \frac{1}{2}(\underline{U}^2 - 1) \text{ with } \underline{U}^2 = {}^T\underline{F}.\underline{F} = \begin{pmatrix} \alpha^2 & \alpha\gamma\cos\theta & 0 \\ \alpha\gamma\cos\theta & \gamma^2 & 0 \\ 0 & 0 & \beta^2 \end{pmatrix} \tag{3}$$

In order to find the interface between the phases, a compatibility equation must be satisfied. In particular, the compatibility equation between austenite and "untwinned" martensite can be written as :

$$\tilde{\underline{R}}\,\underline{U} - \underline{1} = \vec{b} \otimes \vec{m} \tag{4}$$

where the unknowns are the rotation $\tilde{\underline{R}}$, the shape strain vector $\vec{b}$ and the habit plane normal $\vec{m}$.

Equation (4) is called the habit plane equation. The necessary and sufficient conditions of the existence of solutions of this equation are that the matrix $\underline{U}^2$ has ordered eigenvalues $0 < \lambda_1 \le \lambda_2 = 1 \le \lambda_3$.

For austenite-untwinned microstructure, an eigenvalue of the symmetric matrix $\underline{U}^2$ is equal to one if and only if α, γ and $\tilde{\theta}$ are related by

$$\cos^2\tilde{\theta} = \frac{(1-\alpha^2)(1-\gamma^2)}{\alpha^2\gamma^2} \tag{5}$$

In the present case, $\tilde{\theta} = 84.8°$ is closed to $\theta = 83.3°$.

Finally, for completeness, it can be shown that there are twelve distinct variants U_1 – U_{12} which the matrix expressions can be found in the paper of Hanes (Hanes, 1999).

Moreover, the components of $\vec{b}$ and $\vec{m}$ can be obtained.

3.2. MICRO-MACRO INTEGRATION

The procedure used to calculate yield surface of polycrystal is purely phenomenological and is adapted from a paper by W. Huang (Huang, 1999) : i) A polycrystalline material is represented by n grains which constitutes an isotropic texture e.g. a random distribution of the grain crystallographic orientation in the space of the Euler angles. The number m of possible variants $\underline{E}_i^t$ is equal to 12. Interactions between the grains are not taken into account because the yield surface represents the apparition of the first martensite platelets (ii) Under a given stress condition $\underline{\sigma}^o$, for each grain k et among the m possible variants the one representing the higher factor K in the equation

$$\underline{\sigma} : \underline{\varepsilon}^t = K(T) \tag{6}$$

($\underline{\varepsilon}^t$ is the p. t. tensor in the geometrical sample frame) is selected. A set of n factor K_k^{max} can be calculated under a different stress condition. $K_{tension,k}^{max}$ stands for the results under uniaxial tension. 4i) The ratio λ and the p. t. start stress $\underline{\sigma}^t$ are obtained :

$$\lambda = \frac{\text{Average}\left(\sum_{k=1}^{n} K_k^{max}\right)}{\text{Average}\left(\sum_{k=1}^{n} K_{tension,k}^{max}\right)}, \quad \underline{\sigma}^t = \frac{1}{\lambda}\underline{\sigma}^o \tag{7}$$

5i) A new stress condition is applied and the corresponding p. t. stress is determined and so on. Hence, a convex yield surface is determined for the polycrystalline Cu Al Be alloy.

4. Some tools for a general macroscopic modeling

This work constitutes an extension of a thermodynamical study of isotropic pseudoelasticity in SMAs considering the tension-compression asymmetry performed by Raniecki and Lexcellent (Raniecki, Lexcellent, 1998).

The Gibbs function g can be represented in the form (Raniecki et al., 2001)

$$\begin{aligned} \rho g(\underline{\sigma}, z, T) &= \rho(u_o^{*(1)} - T s_o^{*(1)}) + \rho Cp\left[\Delta T - T Ln(T/T_o)\right] \\ &- \rho z\, \pi_o^f(T) - \underline{\sigma} : \underline{\underline{M}} : \underline{\sigma}/2 - \Delta T(\alpha \underline{1} : \underline{\sigma}) + W^{(in)} - z\underline{k} : \underline{\sigma} \end{aligned} \tag{8}$$

where $\pi_o^f = \Delta u^* - T\Delta s^* (\Delta u^* = u_o^{*1} - u_o^{*2}, \Delta s^* = s_o^{*1} - s_o^{*2})$, $\Delta T = T - T_o$ (T_o the reference temperature), $0 \le z \le 1$ is the variable volume fraction of the product phase M ;

ρ, M, Cp, α are respectively the mass density, the compliance, the specific heat, the thermal expansion (the superscript 1 is for A and 2 for M phases). W^{in} is the strain energy of eigenstrain incompatibilities caused by the internal stress field and can be presented in the form

$$W^{(in)} = z(1-z)\,\phi_{it}(T)\ ,\ \phi_{it}(T) = \overline{u}_o - T\overline{s}_o \tag{9}$$

The term $\underline{k}$ present in the definition of macroscopic phase distortion $\underline{\varepsilon}^T = z\,\underline{k}$ is, in general function of the unspecified number of orientation variables, say h_K (K = 1,2...). The calculation of g extremes with respect to h_k gives the equilibrium values of h_K and $\underline{k}$, which under assumption that extremal $\underline{k}$ possesses the potential g* is such that

$$\underline{k} = \rho\frac{\partial g^*}{\partial \underline{\sigma}}\ ,\ \underline{k}:\underline{\sigma} = \rho g^* \tag{10}$$

The incremental energy dissipation is

$$\pi^f\,dz \geq 0 \tag{11}$$

Here, π^f represents the energy-conjugate force of z

$$\pi^f = -\frac{\partial g}{\partial z} = g^*(\underline{\sigma}) - (1-2z)\phi_{it} + \pi^f_o \tag{12}$$

The unstable states where $\pi^f = 0$, $z = 0$ and $z = 1$ are the critical states at which the forward (e.g. A → M) and reverse (e.g. M → A) p.t. are respectively initiated

$$\pi^f(\underline{\sigma}^{AM}, z=0, T) = g^*(\underline{\sigma}^{AM}) - \phi_{it} + \pi^f_o = 0 \tag{13a}$$

$$\pi^f(\underline{\sigma}^{MA}, z=1, T) = g^*(\underline{\sigma}^{MA}) + \phi_{it} + \pi^f_o = 0 \tag{13b}$$

For instance, the equation (13a) represents the yield surface of p.t. initiation (A → M) and g*($\underline{\sigma}$) can be chosen to fit the experimental points shown in figure 1.

For instance, when volumetric macroscopic phase distortion is negligible (tr $\underline{\varepsilon}^T$ = tr $\underline{k}$ = 0) for isotropic solids, one assumes that

$$\rho g^* = \gamma\,\overline{\sigma}\,f(y) \tag{14}$$

with

$$\overline{\sigma} = \left(\frac{3}{2}\,\mathrm{dev}\,\underline{\sigma} : \mathrm{dev}\,\underline{\sigma}\right)^{1/2}\ ,\ J_3' = \det(\mathrm{dev}\,\underline{\sigma}) \tag{15}$$

and

$$y = \frac{27}{2}\frac{J_3'}{\overline{\sigma}^3} \qquad -1 \leq y \leq 1 \tag{16}$$

γ is a constant and $f \geq 0$ is the "shape" function of the limit surface.

Provided that (14) holds, the equivalent macroscopic phase distortion is

$$\overline{\varepsilon}^{T} = \left(\frac{2}{3} \underline{\varepsilon}^{T} : \underline{\varepsilon}^{T} \right)^{1/2} = z\, \overline{\gamma}(y)$$
$$\overline{\gamma}(y) = \gamma \left(f^{2} + 9(1-y^{2}) \left(\frac{df}{dy} \right)^{2} \right)^{1/2} \tag{17}$$

The choice of the f(y) function corresponds to the necessity to have a convex yield surface and we choose

$$f(y) = \cos\left(\frac{\arccos(1-a(1-y))}{3} \right) \qquad 0 < a < 1 \tag{18}$$

As a first important result, the value a = 0.7 fits well the experimental results and the micro-macro prediction as exhibited on figure 4.

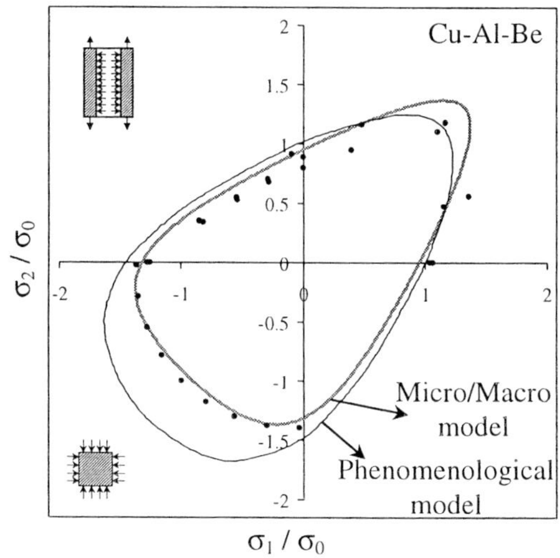

Figure 4 : Yield surface of phase transformation initiation for bi-compression (•) and tension-internal pressure tests (♦) and phnenomological modeling

The derivation of $\underline{\varepsilon}^{T}$ gives

$$\dot{\underline{\varepsilon}}^{T} = \underline{k}\, \dot{z} + \dot{\underline{k}}\, z$$
$$\dot{\underline{\varepsilon}}^{T} = \dot{\underline{\varepsilon}}^{Pe} + \dot{\varepsilon}^{Re} \tag{19}$$

$$\underline{k} = \rho \frac{\partial g^*}{\partial \underline{\sigma}} = \frac{2}{3} f(y)\, \underline{N} + \frac{9}{2} \frac{df}{dy} (3\underline{N}^2 - y\underline{N} - \frac{2}{3} \underline{1}) \tag{20}$$

$$\text{with} \quad \underline{N} = \frac{dev(\underline{\sigma})}{\bar{\sigma}}$$

For the initiation of the P.T. z = 0 and the expression of $\underline{\dot{\varepsilon}}^T(z=0) = \underline{k}\,\dot{z}$ takes into account the experimental observations (normality of the P.T. rate vector to the yield surface).

For modeling the non-proportional loadings, the first right term of equation (19) can be set for the formation (annihilation) of martensite platelets $(\dot{\varepsilon}^{Pe})$ and the second for the reorientation $(\dot{\varepsilon}^{Re})$.

An extension can also be performed by the introduction of an internal stress $\underline{X}$ where the applied stress $\underline{\sigma}$ becomes ($\underline{\sigma}$ - $\underline{X}$) an effective one.

5. Conclusion

For proportional loadings, the theoretical determination of the shape of the yield p.t. surface, by a homogenization process or by a macroscopic approach doesn't lay any problem. But for non-proportional loadings, the question is opened, because a coupling between the formation (annihilation) and the reorientation of martensite platelets must be considered for the model. In this purpose, some tools are given such as the introduction of an internal kinematical stress tensor in the model.

References

[Gall et al., 1999]: Gall K., Sehitoglu H., Maier H.J. and Jacobus K., "Stress Induced Martensitic Phase transformation in Polycrystalline Cu-Zn-Al shape memory alloy under different Stress States", In : Metal. Mat. Trans, 29A, pp. 765-773

[Rogueda et al., 1996]: Rogueda C., Lexcellent C. and Bocher L., "Experimental Study of Pseudoelastic Behavior of a Cu-Zn-Al polycrystalline shape memory alloy under tension-torsion proportional and non proportional loading tests", Arch. Mech., 48, pp. 1025-1045

[Lim and McDowell, 1999]: Lim J.T. and Mc Dowell D.L., "Mechanical behavior of an Ni-Ti shape memory alloy under axial-torsional proportional and nonproportional loadings", J. Eng. Mat. Tech., 121, pp. 9-18

[Bouvet et al., 2000]: Bouvet C., Calloch S. and Lexcellent C., "Experimental investigations under biaxial loadings on Cu-Al-Be shape memory alloy", Proc. of EUROMAT 2000, Advances in Mechanical Behaviour Plasticity and Damage (Ed. D. Miannay, P. Costa, D. François, A. Pineau), 2, pp. 793-798

[Moreau et al., 1995]: Moreau F., Tidu A., Barbe A., Eberhadt A., Heizman J.J., "Study of Cu Al Be shape memory alloy by X-Ray diffraction (Ed. by A. Planes, J. Ortin, L. Manosa), J. Phys. IV, 5(C2), pp. 269-274

[Lexcellent et al., 2001]: Lexcellent C., Vivet A., Bouvet A., Blanc P., Satto C. and Schryvers D., "From the lattice measurements of the austenitic and the martensitic cells to the macroscopic mechanical behavior of shape memory alloys", 5th European Mechanics of Material Conference, Delft, The Netherlands, March 5-9

[Hanes, 1999]: Hanes K.F., "Bulk and thin films microstructures in untwinned martensite", J. Mech. Phys. Solids, 47(9), pp. 1917-1939

[Huang, 1999]: Huang W., "Yield surface of shape memory alloys and their applications", Acta Mater., 47(9), pp. 2769-2776

[Raniecki and Lexcellent, 1998]: Raniecki B., Lexcellent C., "Thermodynamics of isotropic pseudoelasticity in shape memory alloys", Eur. J. Mech. A/Solids, 17(2), pp. 185-205

[Raniecki et al., 2001]: Raniecki B., Tanaka K., Ziolkowski A., "Testing and modeling of Ni Ti SMA at complex stress state", Selected results of Polish-Japanese Research Cooperation, to appear in Mat. Sci. Res. Int.

Thermomechanical Properties of Polyurethane-Shape Memory Polymer

H.Tobushi, K.Okumura, M.Endo,
Aichi Institute of Technology, 1247 Yachigusa, Yagusa-cho Toyota, 470-0392 Japan

S.Hayashi,
Mitsubishi Heavy Industries, Ltd., Nagoya Research & Development Center, 1 Takamichi, Iwatsuka-cho, Nakamura-ku, Nagoya, 453-8515 Japan

1. Introduction

The development of applications of shape memory polymers (SMPs) is currently of great interest [1-4]. Because SMPs are light, high in shape recovery, easy to manipulate and economical compared with shape memory alloy, their development is being actively promoted. In many SMPs, the phase transition temperature is around room temperature. Characteristics such as shape recovery and shape fixity exist due to the difference in mechanical properties of materials above and below the phase transition temperature. SMPs possessing these characteristics are being used as temperature sensors and actuators.

The SMP of polyurethane series can be moulded by such methods as injection, extrusion and blowing, similar to conventional plastics, and the shape recovery temperature can be set anywhere in the range of room temperature $\pm$50K, making a wide range of applications possible. In the SMP of polyurethane series, both mechanical and optical properties are different above and below the glass transition temperature T_g. With respect to the thin film, moisture permeability varies significantly above and below T_g. The application of this property to clothing is of interest.

In the present study, the thermomechanical deformation properties of the SMP thin film and foam of polyurethane series are examined. The cyclic properties of shape fixity and shape recovery are investigated experimentally. The thermomechanical constitutive model is proposed.

2. Shape fixity and shape recovery

2.1 MATERIAL AND SPECIMEN

The material used in the experiment was an SMP thin film of polyurethane series. The glass transition temperature T_g of the material was about 328K. The casting film was

Q.P. Sun (ed.), IUTAM Symposium on Mechanics of Martensitic Phase Transformation in Solids, 79–86.

made from solution (Diary MS5510; produced by Mitsubishi Heavy Industries Ltd) under drying conditions. The weight ratio of polymer to dimethyl formamide (DMF) is 3:7. The viscosity of solution at 298K is 100Pa·s. The specimen was a thin film of uniform shape. The dimensions of the testing area were length 25mm, thickness 50μm and width 5mm.

2.2 EXPERIMENTAL PROCEDURE

The three-dimensional stress-strain-temperature relationship during the thermomechanical cycling test is shown schematically in Figure 1. Firstly ① at temperature T_h, the maximum strain ε_m was applied at a constant strain rate of 50%/min; secondly ② maintaining ε_m, the specimen was cooled to T_l. After holding at T_l for 5 min, the specimen was unloaded ③. Under the no-load condition, it was heated from T_l to T_h and held for 5 min at T_h ④. The thermomechanical paths ① to ④ were repeated 10 times. The heating and cooling rate was about 8K/min.

2.3 STRESS-STRAIN-TEMPERATURE RELATIONSHIP

The stress-strain curves, strain-temperature curves and stress-temperature curves obtained by the thermomechanical cycling test for maximum strain ε_m=100% are shown in Figure 2(a), (b) and (c), respectively. Labels ① to ④ correspond to paths ① to ④ in Figure 1, and N represents number of cycles. As seen in Figure 2(a), strain is recovered by unloading and heating under no stress. The yield stress decreases markedly in the early cycles but decreases gradually thereafter. The loading curves gradually shift to the right and downward. The cyclic deformation properties of the material change depending on entanglement decoupling of molecular chains, orientation of molecular chains and hardening due to crystallization. The stress that increases when the specimen is cooled to temperature T_l under constant ε_m, ②, is thermal stress due to thermal contraction. Because the modulus of elasticity at temperature T_l is large, strain is recovered elastically with a steep slope during the unloading process at T_l, ③. The strain ε_u at the termination point of unloading is nearly equal to ε_m and does not vary with cycling. This means that excellent shape fixity of the material can be obtained.

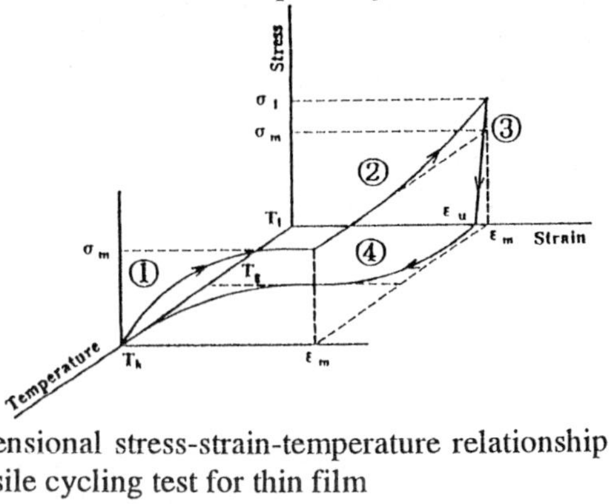

Figure 1. Three-dimensional stress-strain-temperature relationship showing the loading path in tensile cycling test for thin film

As seen in Figure 2(b), during the heating process in the first cycle, ④, strain starts to be recovered at temperatures near T_g-5K, is recovered markedly in the vicinity of T_g and is recovered gradually at temperatures above T_g. This phenomenon may be caused as follows. At low temperature, strain is constant at ε_u under no stress because the micro-Brownian motion of soft segments is frozen. If it is heated to high temperatures above T_g without restraint, the original shape is recovered because micro-Brownian motion becomes active. The strain-temperature curves during the heating process, ④, shift to the right with an increase in N. This means that recovery temperature rises with cycling. This phenomenon may be caused as follows. Because hardening due to orientation of molecular chains and crystallization progress with cycling, a larger thermal energy is required for strain to be recovered, leading to a rise in the recovery temperature. This trend is the same as a rise in the transformation temperature with an increase in N for shape memory alloys subjected to cyclic deformation.

As seen in Figure 2(c), stress decreases at T_g+20K with ε_m constant at T_h. The decrease in stress is due to stress relaxation that occurs to keep ε_m constant at T_h. Stress increases gradually below T_g during the cooling process, ②, and increases significantly in the vicinity of T_l. The increase in stress $\Delta\sigma$ is the thermal stress for thermal contraction due to the variation in temperature ΔT.

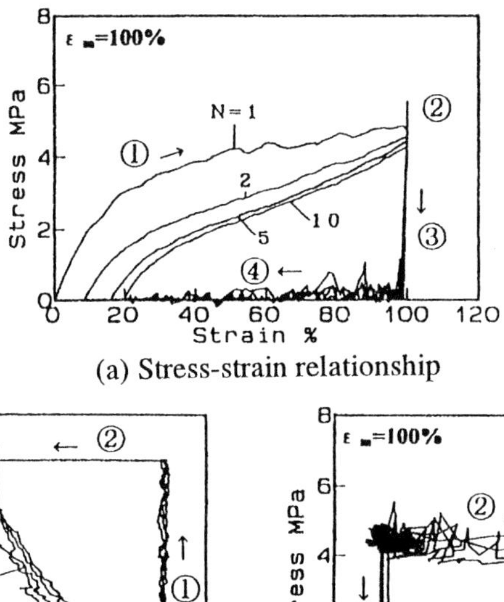

(a) Stress-strain relationship

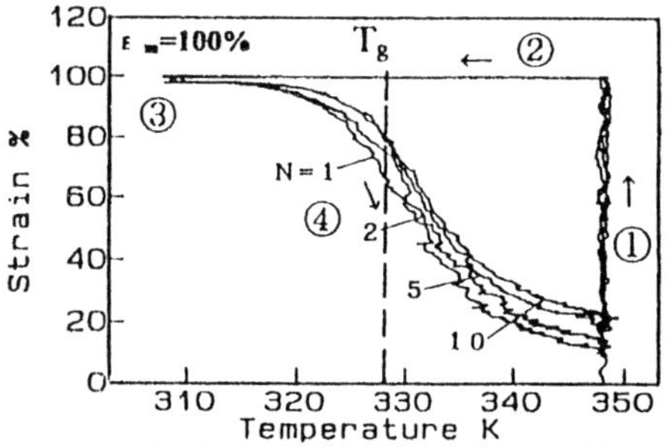

(b) Strain-temperature relationship

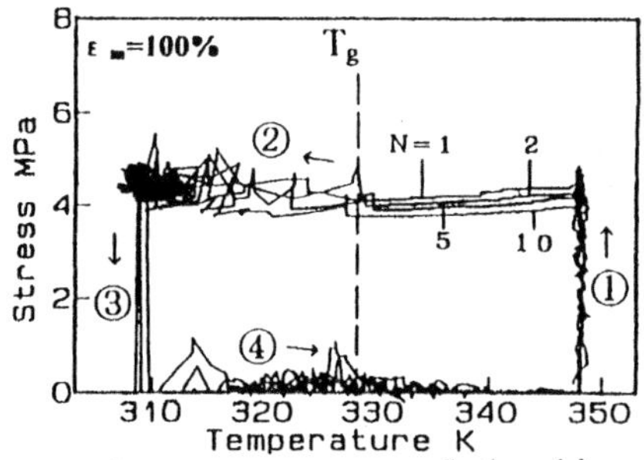

(c) Stress-temperature relationship

Figure 2. Experimental results

3. Constitutive model

3.1 CONSTITUTIVE EQUATION

The authors proposed a linear constitutive model that was developed by modifying a linear viscoelastic theory [5]. The modified linear constitutive model involves a slip element due to internal friction in a three-element standard linear viscoelastic model. The model expresses the fact that, in the process of creep recovery under no-load which follows loading with creep strain ε_c, a certain part of ε_c remains as irrecoverable strain ε_s. The relationship in the modified linear constitutive equation is expressed as follows

$$\dot{\varepsilon} = \frac{\dot{\sigma}}{E} + \frac{\sigma}{\mu} - \frac{\varepsilon - \varepsilon_s}{\lambda} + \alpha \dot{T} \tag{1}$$

$$\varepsilon_s = S\varepsilon_c \tag{2}$$

where σ, ε and T denote stress, strain and temperature, respectively. The dot denotes time derivative. E, μ, λ and α represent modulus of elasticity, viscosity, retardation time and coefficient of thermal expansion. E, μ and λ depend on temperature. As expressed by Eq. (2), a certain ratio S of creep strain ε_c appears as irrecoverable strain ε_s. In order to express nonlinear strain, let us consider a nonlinear term expressed by a power function of stress. By considering nonlinear terms, Eq. (1) becomes

$$\dot{\varepsilon} = \frac{\dot{\sigma}}{E} + m\left(\frac{\sigma - \sigma_y}{k}\right)^{m-1}\frac{\dot{\sigma}}{k} + \frac{\sigma}{\mu} + \frac{1}{b}\left(\frac{\sigma}{\sigma_c} - 1\right)^n - \frac{\varepsilon - \varepsilon_s}{\lambda} + \alpha \dot{T} \tag{3}$$

where σ_y and σ_c denote proportional limits of stress in the time-independent term and the viscous term, respectively, and corresponded to yield stress and creep limit.

If large strain is applied at a constant strain rate in the loading process, seemingly time-independent plastic strain ε_p remains after unloading. Let us consider ε_p corresponding to time-dependent ε_c in the right-hand side of Eq. (2). In order to express irrecoverable strain ε_s, Eq (2) becomes

$$\varepsilon_s = S(\varepsilon_c + \varepsilon_p) \tag{4}$$

where S is proportional coefficient.

3.2 DEPENDENCE OF COEFFICIENTS ON TEMPERATURE

The relationship between modulus of elasticity E and temperature T obtained by the dynamic mechanical tests on SMP is shown in Figure. 3(a). In Figure 3(a), E is expressed in logarithmic scale and T in a reciprocal normalized by T_g. As seen in Figure 3(a), E varies markedly in the glass transition region but is almost constant above and below the glass transition region ($T = T_g \pm 15$K). As shown in Figure 3(b), the relationship between log$E$ and T_g/T is approximately represented by a straight line with a slope of a. Thus E in the glass transition region $T_g - T_w \leq T \leq T_g + T_w$ becomes

$$E = E_g \exp\left[a\left(\frac{T_g}{T} - 1\right)\right] \tag{5}$$

where E_g is the value of E at $T = T_g$. As seen in Figure 3(b), E_h and E_l denote the values of E above and below the glass transition region, respectively. The exponential function

(5) is similar to the equation for viscosity that was derived theoretically by Eyring and empirically by Arrhenius.

In the same manner as E, in order to express the deformation properties in the glass transition region, the material parameters k, σ_y, μ, σ_c, λ and S are expressed by an exponential function of temperature T. These parameters are denoted by x and are expressed as follows

$$x=x_g\exp\left[a\left(\frac{T_g}{T}-1\right)\right] \tag{6}$$

where x_g is the value of x at $T=T_g$. In the present study, for simplicity of the theory, α is assumed to be constant for each coefficient. Therefore, by taking into a constant a, each coefficient x can be evaluated by the characteristic value x_g at T_g.

3.3 MATERIAL PARAMETERS

From the results of the tests, the temperature width T_w of the glass transition region was determined to be 15K and a in Eqs. (5) and (6) was decided to be 38. The values of x_g at T_g for the coefficients in Eqs. (3) and (4) are shown in Table 1. The parameters m, n, b and α were determined to be 2, 1.01, 10^4s and 11.6×10^{-5}/K, respectively.

3.4 STRESS-STRAIN-TEMPERATURE RELATIONSHIP

The thermomechanical test with $T_h=T_g+20$K, $T_l=T_g-20$K, $\varepsilon_m=20\%$, strain rate of 50%/min, average heating rate of 4K/min and average cooling rate of 8K/min was preformed. The stress-strain curves, stress-temperature curves and strain-temperature curves obtained from the test are shown in Figure 4(a), (b) and (c), respectively. In these figures, symbols ①~④ denote the loading process. The solid curves denote the calculated results. As seen in Figure 4, the calculated results express well the inclination of the experimental results such as shape fixity, shape recovery and recovery stress.

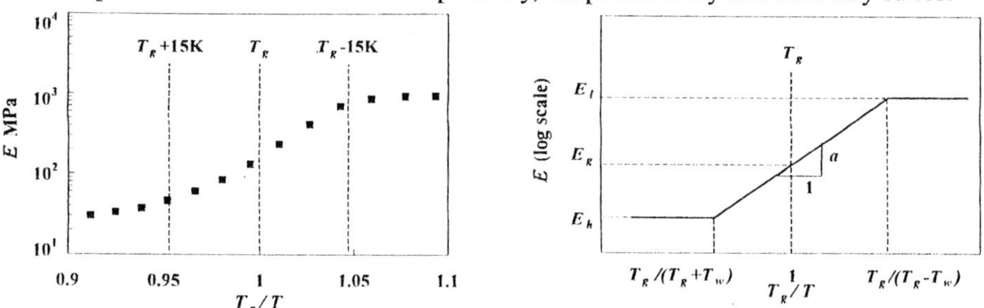

(a) Experimental result (b) Approximate curve

Figure 3. Relationship between elastic modulus and temperature

Table 1. Values of coefficients at T_g

E_g MPa	k_g MPa	σ_{yg} MPa	μ_g GPa·s	σ_{cg} MPa	λ_g s	S_g —
146	40	4.0	35.0	0.5	390	0.1

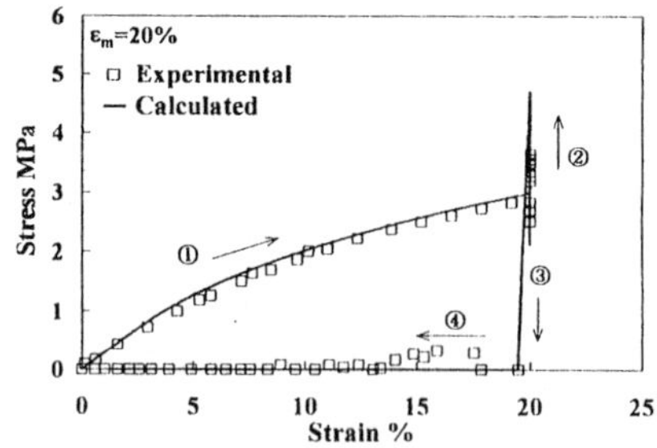

(a) Stress-strain relationship

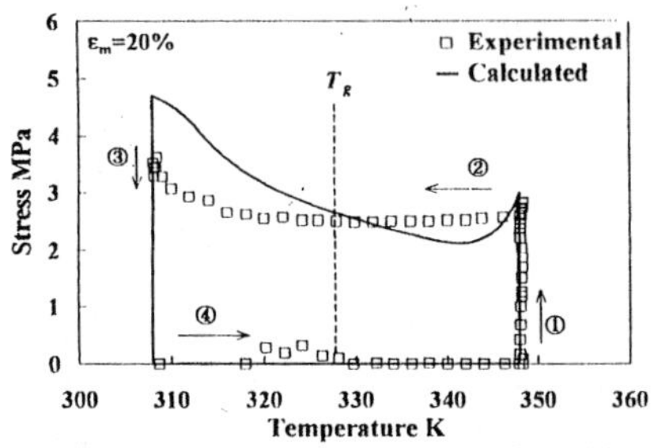

(b) Stress-temperature relationship

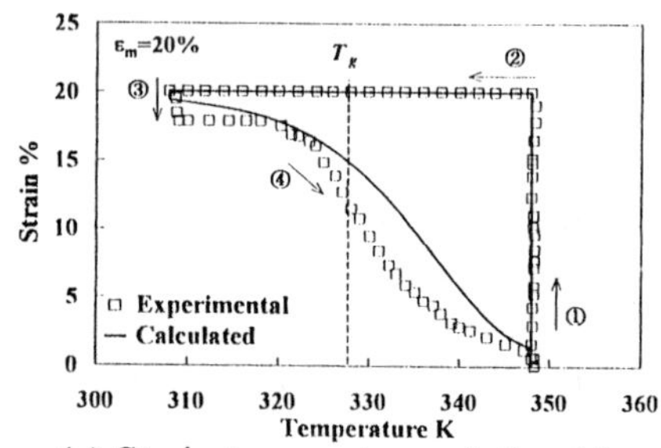

(c) Strain-temperature relationship

Figure 4. Experimental results and calculated results

4. Thermomechanical properties of foam

4.1 MATERIAL AND SPECIMEN

The material used in the experiment was SMP foam of polyurethane of the polyether polyole series (Diary MFNo.21; produced by Mitsubishi Heavy Industries, Ltd.). The foams were made by chemical foaming. The expansion ratio was about 30. The foams were produced as slabstock. The specimen was cut off into a cube with length of the side 20mm. The glass transition temperature T_g was about 328K.

4.2 EXPERIMENTAL PROCEDURE

The three-dimensional stress-strain-temperature diagram showing the loading path in the thermomechanical cycling test is shown in Figure 5. First ① maximum strain $-\varepsilon_m$ was applied at high temperature T_h above T_g. Second ② maintaining $-\varepsilon_m$, the specimen was cooled to low temperature T_l below T_g. Last ③ it was heated from T_l to T_h under no-load condition. The thermomechanical paths ① to ③ were repeated 10 times. The experimental conditions were as follows: T_h=T_g+30K, T_l=T_g-30K, strain rate $\dot{\varepsilon}$ =25%/min, ε_m=78%.

4.3 SHAPE FIXITY AND SHAPE RECOVERY

The stress-strain curves, stress-temperature curves and strain-temperature curves obtained by the thermomechanical cycling tests are shown in Figure 6(a), (b) and (c), respectively. N denotes the number of cycles. As seen in Figure 6(a), maximum stress in the loading process ① is a little large in N=1 and changes little thereafter. The cyclic deformation properties may be caused as follows. In the initial loading, incomplete cells may be destroyed but their variations may be small thereafter. Strain increases under stress plateau during the strain region between 10% and 60%. The increase in strain during this region may correspond to buckling of cells. If strain increases above 60%, stress increases markedly. In the final region, deformation resistance becomes very large because post-buckling cells are deformed. In the cooling process ②, stress diminishes. Therefore maximum strain is fixed in its state and the rate of strain fixity is 100%.

As seen in Figure 6(b), stress decreases during the cooling process ②. In the initial stage of the process ②, stress decreases due to stress relaxation at high temperature. In the middle stage of the process ②, stress decreases gradually due to thermal contraction caused by decrease in temperature. Because the crosshead of the testing machine is held during the process in the tests and the compressed SMP foam contracted furthermore due to decrease in temperature, stress decreases. In the latter stage of the process ②, stress decreases markedly. Modulus of elasticity is very large at temperatures below T_g. Therefore the variation in thermal stress due to thermal contraction caused by decrease in temperature becomes large, resulting in large decrease in stress. The stress-temperature curves do not change under cyclic loading.

As seen in Figure 6(c), strain decreases at temperatures in the vicinity of T_g during the heating process ③. In the initial stage of the process ③ at low temperature in the vicinity of T_l, strain is constant at ε_m because the micro-Brownian motion of soft segments of SMP is frozen. If it is heated to high temperature, strain is recovered because the micro-Brownian motion becomes active above T_g. The strain-temperature curves do not change under cyclic deformation. The rate of strain recovery is 99%.

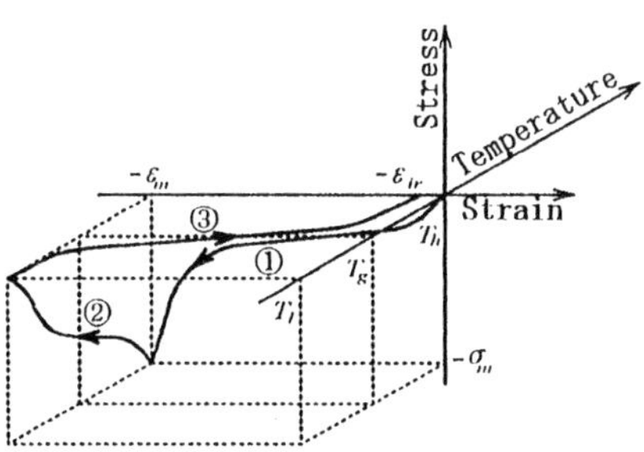

Figure 5. Three-dimensional stress-strain-temperature relationship showing the loading path in compressive cycling test for foam

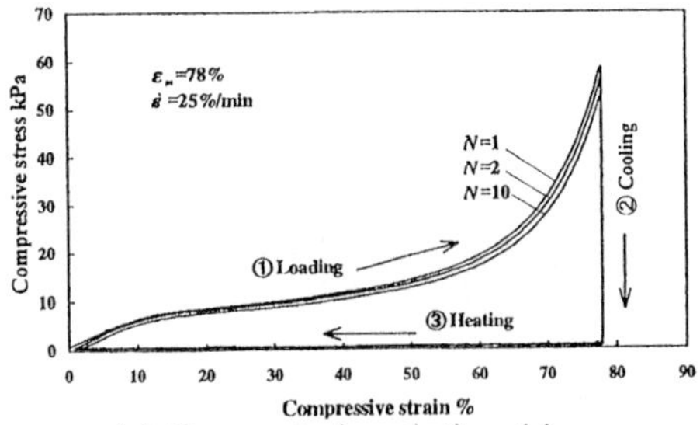

(a) Stress-strain relationship

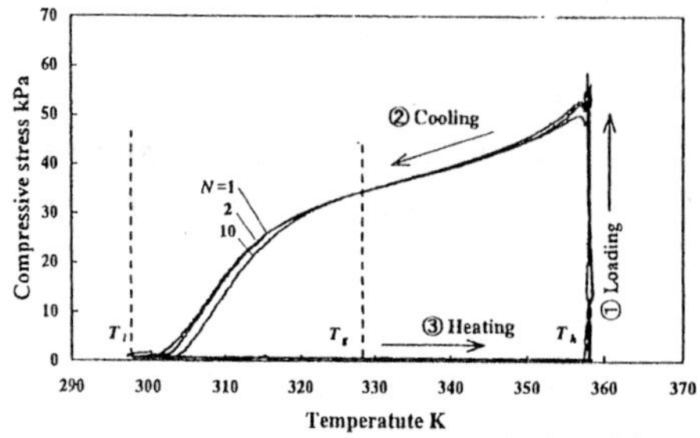

(b) Stress-temperature relationship

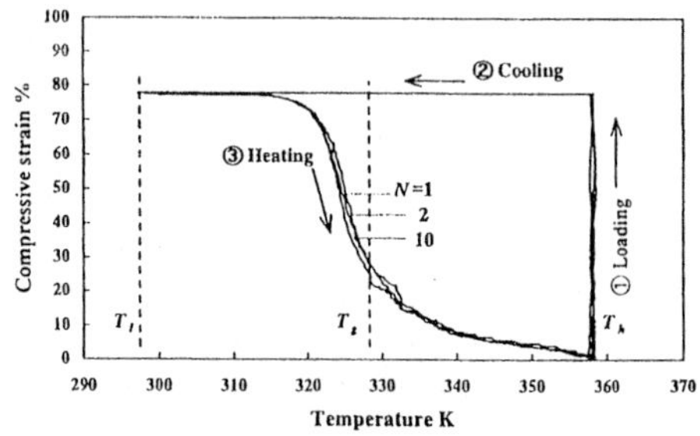

(c) Strain-temperature relationship

Figure 6. Experimental results

5. Conclusions

The thermomechanical properties of the polyurethane-SMP thin film and foam were investigated. The results are summarized as follows.

(1) Strain is recovered at temperatures in the vicinity of T_g during the heating process under no load.

(2) In the case of small maximum strain, both the rate of strain fixity and the rate of strain recovery are larger than 98%, showing high performance as an intelligent material.

(3) In the case of large maximum strain, the stress-strain curves vary significantly in the early cycles but only slightly thereafter.

(4) The proposed constitutive model expresses well the thermomechanical behaviors such as shape recovery, shape fixity and recovery stress.

References

1. S.Hayashi, Inter. Progress in Urethanes, 6, 90-115, 1993.
2. H.Tobushi, H.Hara, E.Yamada and S.Hayashi, Smart Mater. Struct. 5, 483-491, 1996.
3. H.Tobushi, T.Hashimoto, N.Ito, S.Hayashi and E.Yamada, J. Intell. Mater. Syst. Struct., 9, 127-136, 1998.
4. H.Tobushi, K.Okumura, M.Endo and S.Hayashi, Trans. MRS-J, 26(1), 351-354, 2001.
5. H.Tobushi, T.Hashimoto, S.Hayashi and E.Yamada, J. Intell. Mater. Syst. Struct., 8, 711-718, 1997.

TRANSFORMATION INDUCED PLASTICITY (TRIP) IN A MARAGING STEEL

The TRIP behavior for General Loading Paths

T. ANTRETTER, F.D. FISCHER AND U. CIHAK
Institute of Mechanics, Montanuniversität Leoben,
Franz-Josef-Straße 18, A-8700 Leoben, Austria
K. TANAKA
Dept. of Aerospace Engineering, Tokyo Metropolitan Institute of Technology, Asahigaoka 6-6, J-191-0065 Hino/Tokyo, Japan
G. CAILLETAUD
Centre des Matériaux, Ecole des Mines,
B.P. 87, F-91003 Evry Cedex, France
AND
P. FRATZL AND B. ORTNER
Institute of Metal Physics, Montanuniversität Leoben,
Jahnstraße 12, A-8700 Leoben, Austria

1. Introduction

An extensive experimental program on a 9 % Ni, 12% Cr, 2% Mo steel maraging steel is introduced. For the specific purpose of studying the physical phenomenon of transformation induced plasticity this material exhibits some very desirable features: The martensite start temperature (M_s) can be found around 150°C; the martensitic finish temperature (M_f) is at ca. 80°C. The material has a yield stress of about 200 MPa at ca. 200°C (fully austenitic state) and of ca. 900 MPa at room temperature (fully martensitic state). Plastification is accompanied by rather small hardening up to a strain of roughly 8 %. Since the martensitic transformation takes place at a low temperature level upon cooling on air only, this material can be considered as an ideal testing material with practically no creep effects during and after transformation. First of all dilatation tests are demonstrated for two types of specimens: longitudinal, also called axial specimens with a fiber texture and radial specimens, as shown in Figure 1b. A material characterization using M_s and M_f lines, i.e. martensite start and finish temperature

Q.P. Sun (ed.), IUTAM Symposium on Mechanics of Martensitic Phase Transformation in Solids, 87–94.

monitored for different loading conditions and stress levels reveals the influence of the type of loading (tension, shear, compression). Consequently two different loading histories are tested:

- loading above M_s and then cooling, accompanied by partial or full unloading,
- loading at a constant temperature in the transformation temperature interval.

As already pointed out in some earlier papers ([1, 2]), aside from what is considered as the classical TRIP effect (see e.g. [3]), the orientation of the martensitic variants becomes the dominant feature being responsible for the different deformation behavior at the same global equivalent load (and unloaded) stress state. An extensive micromechanical modeling is performed on the basis of a numerical implementation of a transformation condition ([4]) into a finite element algorithm (see [5]). Due to a reasonable correlation between experimental and numerical results a quantification of the orientation effect depending on the loading type has been established. For the issue of an improved constitutive law for the TRIP strain rate the reader is referred to a recent publication by the group of authors [6] where a recommendation, referring to a proposal by the French group [7] and [2], is given for a new evolution law including the backstress and its development. A comprehensive report on experimental work and micromechanical modeling is presented therein showing a new interpretation of the TRIP effect as well as new constitutive TRIP equations for steels.

2. Experiments

2.1. TEXTURE EFFECT

In previous dilatation studies it has been observed that for axial specimens a complete heating - cooling cycle under zero applied stress leaves some residual strain in the material eventually resulting in the notorious ratchetting phenomenon for repeated test cycles. This gap in the dilatation loop can typically be explained by a pre-existing backstress introduced during the rolling and forging processes. This backstress does not appear to the same extent if the specimen has been machined perpendicular to the rolling direction (i.e., the radial specimen) out of the original rod. Such an anisotropic response is commonly referred to as texture. Annealing far beyond the austenite start temperature is usually employed to erase such a texture and thus homogenize the material. Figure 1a shows the effect of different annealing temperatures for an axial specimen. Evidently, the specimens have to be heated up to about 1200°C to entirely eliminate the backstress and hence close the loop. For radial specimens the tendency does not turn out to be as clear, so an alternative way of judging the efficiency of annealing is devised. To this end the so called transformation jump $\Delta\varepsilon$ being the difference between the total strain at martensite start (M_s) and at room temperature is introduced (see Figure 2a). $\Delta\varepsilon$ plotted vs. the annealing temperature for both radial and axial specimens is displayed in Figure 2b. The transformation jump for both kinds of specimes approaches a limit value of about 0.55% suggesting that perfect isotropy has successfully been achieved.

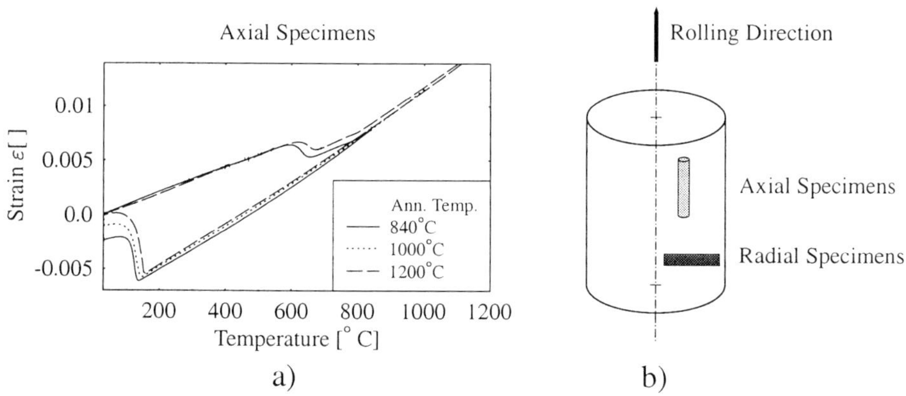

Figure 1. a) Dilatation curve of an axial specimen for different annealing temperatures. b) Axial and radial specimens.

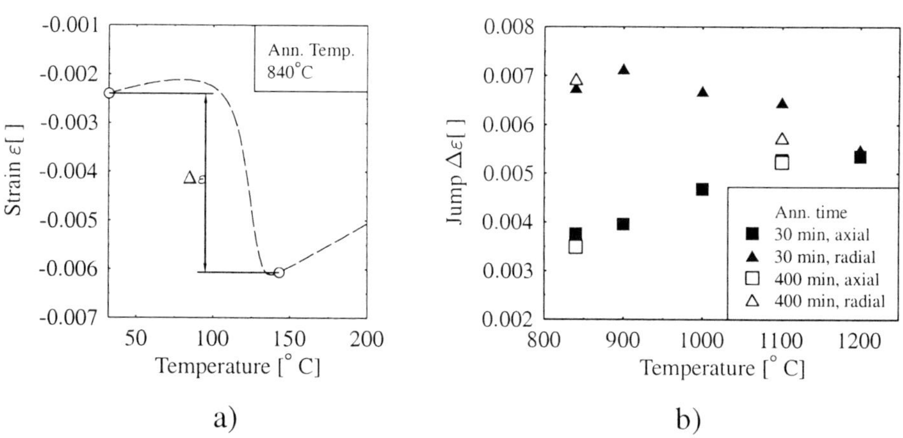

Figure 2. a) Definition of the transformation jump. b) $\Delta\varepsilon$ as a function of annealing temperature and time.

The reverse martensite to austenite transformation starting during heating at around 600°C seems to be more complex than is commonly believed. The details of the dilatation curve in this particular temperature range indicate that a significant amount of martensite transforms back into austenite in a displacive way before diffusional processes commence. Furthermore, the slope of the dilatation curve representing the continuing expansion cannot be attributed to the thermal expansion of any α' - γ mixture suggesting that another yet unknown intermediate phase may be involved. The research on this topic is in its very early stages and will be a major issue of future efforts.

2.2. TRANSFORMATION BEHAVIOR UNDER LOAD

A sophisticated multiaxial tension torsion machine has been used to determine the strain response of the material to a heating-cooling cycle under load. Details and discussions on the experimental procedure are given in [8]. In the following an initially backstress-free specimen will always serve as the starting point for all further discussions. For the textured specimen this requires applying a constant hold stress σ_h of 28 MPa to obtain a closed dilatation loop. In subsequent experiments the transformation characteristics depending on different hold stress levels, in tension, torsion and compression have been measured. From the raw data obtained in these experiments the martensite start and finish temperatures, M_s and M_f, respectively, can be determined. The results are depicted in Figure 3. In good agreement with preliminary theoretical estimations ([9]) M_s shows a linear dependence on σ_h where the slope of the fit line is steepest for compression, followed by shear and then tension. This descending order is also predicted by the theory. While M_s can be determined fairly accurately the values for M_f can only be given with some uncertainty. No dependence on the load is observed, i.e., the M_f lines turn out to be almost vertical.

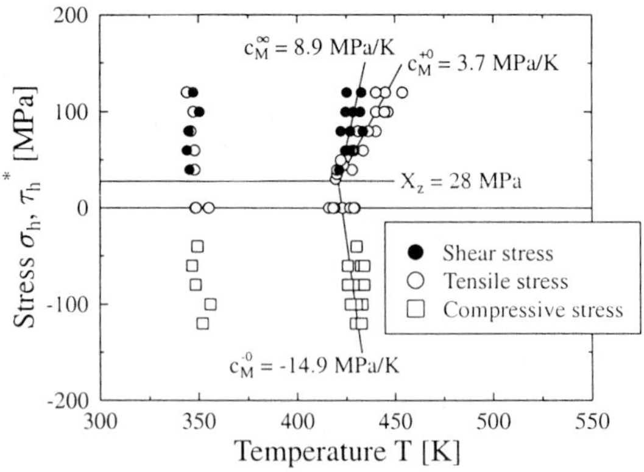

Figure 3. Martensite start and finish lines for different load cases.

Structural engineers are chiefly interested in the mechanical performance of the material expressed in terms of a yield limit as a function of the operating temperature. Using the same experimental setup as described above such a strength characteristics can be obtained by cooling down a specimen to a certain temperature level and then carrying out a conventional tensile test under isothermal conditions. This procedure can be repeated as often as necessary in order to determine the yield stress for various temperature levels. The results are shown in Figure 4 where both the 0.05% (hollow squares) and the 0.005% (filled squares) proof stress are displayed. At temperatures around 460 K the yield limit suddenly drops due to the onset of stress induced martensitic transformation (SIMT). At lower temperatures the already existing fraction of martensite is large enough, so that the hardening effect of martensite prevails over the softening effect of SIMT. Thus at lower testing temperatures the overall behavior of the material is essentially governed by the laws of composite mechanics rather than the influence of the transformation process itself. Figure 4 also contains the martensite start line for tension (hollow circles) whose slope has an impact on the shape of the dip observed in the strength characteristics of the material.

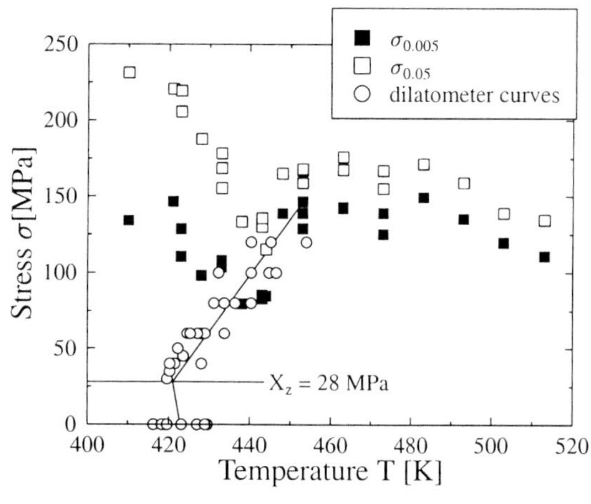

Figure 4. Yield stress defined as 0.05% and 0.005% proof stress, respectively, as a function of temperature.

The experimental equipment also allows to carry out unloading tests where the load is instantaneously reduced or even removed at a given temperature or martensite fraction, respectively. In contrast to the predictions of the classical theory the measured TRIP strain does not remain constant after unloading but a sudden drop can be observed which henceforth in this paper will be termed "backflow" (see Figure 5). From a macroscopic point of view this phenomenon can be blamed on the evolution of a backstress during transformation, similar to what is known in standard plasticity for kinematically hardening materials. Substituting the applied stress σ by an effective stress σ_{eff} defined as σ-X_z with X_z being the backstress enables to explain the backflow behavior. If σ_{eff} becomes negative upon full or partial removal of the applied load then it keeps driving the TRIP strain evolution, however in the opposite direction. Alternatively a microscopic perspective may be chosen: Typically a predominant external applied stress state only selects limited sets of martensitic variants as the ones that generate the maximum mechanical driving forces for martensitic transformation. If this stress state changes its sign or is replaced by disordered internal stresses then new variants may appear that may arrange in self-accommodating groups in a way which macroscopically entails a negative TRIP strain rate.

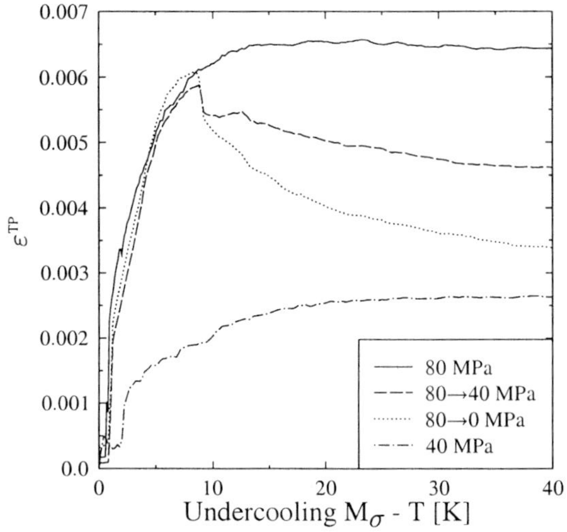

Figure 5. TRIP strain response to partial and full removal of a tensile load.

3. Numerical Simulations

Starting from the physical principles of thermo-mechanical processes a transformation condition has been established as described in detail in [4]. Based on this condition an elastic predictor - radial return algorithm [5] can be employed and coded as a subroutine that can be linked to the Finite Element Package ABAQUS ([10]). The finite element model itself consists of a regular cubic array of 8 node brick elements where each element represents one grain with the integration points as nucleation sites of new martensitic variants. Periodic boundary conditions are assumed in order to rule out any unrealistic boundary artifacts. In the first step the model response to various levels of a constant load and a cooling path has been studied resulting in the transformation topology displayed in Figure 6 for the case of tension. In addition lines of equal martensite fraction, so called iso-ξ lines, can be constructed, thus essentially defining the transformation kinetics under constant hold stress. Note that the iso-ξ lines are almost parallel except for a narrow band at very low martensite fractions where the slope changes noticeably thus accounting for differently shaped kinetics at different load levels. There is, however, a small quantitative discrepancy between the slope of the M_s line found experimentally and the one determined numerically due to inaccurate assumptions for two essential material properties, i.e. the volume dilatation δ and the transformation shear γ. Recalibration of the model with recently obtained data for the lattice constants should eliminate these errors.

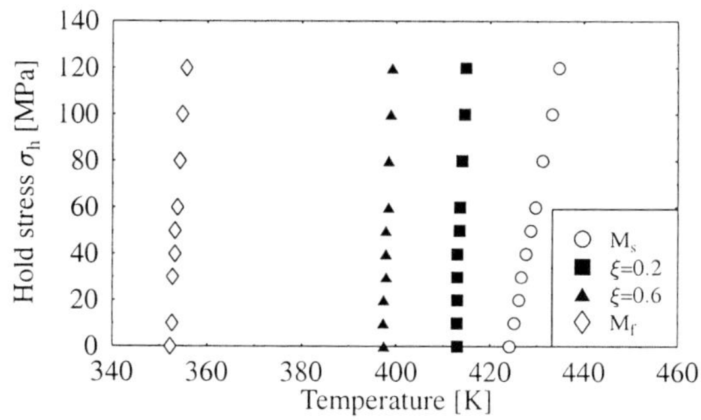

Figure 6. Numerically determined iso-ξ lines for tension.

In analogy to the experiments described in Figure 4 the same temperature loading paths have been simulated using the present finite element model. Four representative flow curves corresponding to different temperature levels within and around the transformation interval are depicted in Figure 7. Again the same tendency as observed in the experiments is obtained purely numerically, i.e. the transformation has a significant impact on the overall material behavior. Especially in the area of interest, the numerical data agree very well with the yield limits found in the tensile tests.

4. Conclusions

The numerical model as well as the experiments show a sudden drop of the TRIP strain ε^{TP}, termed "backflow" upon partial or full unloading, which is not predicted by the classical theory. Analogously to the standard plasticity theory of kinematically hardening materials a backstress concept has been introduced capable of accounting for the material resonse under non-proportional loading paths. From a microscopic point of view the backflow

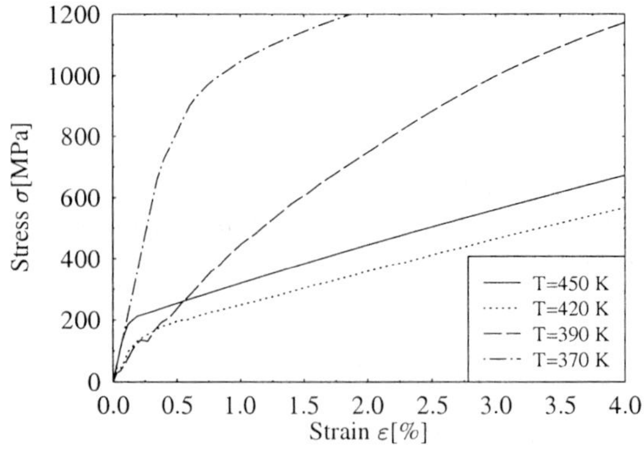

Figure 7. Simulated flow curves evaluated at various temperature levels within and around the transformation range.

behavior can be explained by the formation of self-accommodating groups of martensitic variants also referred to as orientation effect. The reverse transformation from martensite to austenite is essentially unexplored and will be an issue of future research.

References

1. Marketz, F., Fischer, F.D., Tanaka, K.: Micromechanics of Transformation-induced Plasticity and Variant Coalescence, *Journal de Physique IV, Colloque C1, Supplément au Journal de Physique III* **6** (1996), C1-445 – C1-454.
2. Fischer, F.D., Reisner, G., Werner, E., Tanaka, K., Cailletaud, G., Antretter, T.: A New View on Transformation Induced Plasticity (TRIP), *Int. J. Plasticity* **16** (2000), 723-748.
3. Fischer, F.D., Sun, Q.-P., Tanaka, K.: Transformation-Induced Plasticity (TRIP), *Appl. Mech. Rev.* **49** (1996), 317-364.
4. Fischer, F.D., Reisner, G.: A Criterion for the Martensitic Transformation of a Microregion in an Elastic-Plastic Material, *Acta mater.* **46** (1998), 2095-2102.
5. Reisner, G., Werner, E.A., Fischer, F.D.: Micromechanical Modelling of Martensitic Transformation in Random Microstructures, *Int. J. Solids & Structures* **35** (1998), 2457-2473.
6. Fischer, F.D., Antretter, T., Azzouz, F., Cailletaud, G., Pineau, A., Tanaka, K., Nagayama K.: The Role of Backstress in Phase Transforming Steels, *Archives of Mechanics* **52** (2000), 569-588.
7. Videau, J.C., Cailletaud, G., Pineau, A.: Experimental Study of the Transformation Induced Plasticity in a Cr-Ni-Mo-Al-Ti Steel, *Journal de Physique IV, Colloque C1, Supplément au Journal de Physique III* **6** (1996), C1-465 – C1-474.
8. Nagayama, K., Terasaki, T., Tanaka, K., Fischer, F.D., Antretter, T., Cailletaud, G., Azzouz, F.: Mechanical Properties of a Cr-Ni-Mo-Al-Ti Maraging Steel in the Process of Martensitic Transformation, *Materials Science and Engineering* **A308** (2000), 25-37.
9. Marketz, F., Fischer, F.D.: A Mesoscale-Study on the Thermodynamic Effect of Stress on Martensitic Transformation, *Met.Trans A* **26A** (1995), 267-278.
10. http://www.hks.com.

FREE ENERGY OF MARTENSITIC TRANSFORMATION - FROM DISCRETE TO HOMOGENEOUS MICROSTRUCTURE

C. MESSNER, E.A. WERNER
Lehrstuhl für Mechanik
CD Laboratorium für Moderne Mehrphasenstähle
TU München, Germany
Q.P. SUN [1]
Dept. of Mechanical Engineering
Hong Kong University of Science and Technology, PR China
G. REISNER
Andritz AG, Graz, Austria

Abstract

In this paper we present a macroscopic continuum model for inclusions (i.e. martensite bands) distributed periodically along the axis of a cylindrical rod such as an SMA specimen, where each inclusion carries a uniform axisymmetric eigenstrain. The effects of microstructure's periodicity and the growth of the inclusions on the elastic strain energy and subsequently on the mechanical response of the cylindrical rod under tensile loading are evaluated. In the limiting case of a length of periodicity approaching zero the discrete nature of the process changes to a homogeneous one. The effect of heterogeneous elastic material properties on the free energy and the overall response is emphasized and the solutions obtained are discussed and compared with analytical results available.

Key Words: localized inelastic strain, periodic inclusions, continuum model, non-convex free energy.

1 Introduction

The deformation of materials and structures under thermomechanical loading is sometimes realized by the formation and evolution of distributed defects like the nucleation and growth of inclusions or cracks. The elastic stress and strain fields and the strain energy of such systems (with distributed defects) are of great importance for understanding and modeling their mechanical behavior. This paper focuses on periodically distributed growing inclusions in a long cylindrical rod under tension. Macroscopically this may represent the single-band (large length of periodicity) or multiple-band (small length of periodicity) propagation phenomenon of localized martensitic transformation observed in tensile tests of NiTi shape memory alloy (SMA) specimens under different loading rates [1, 2]. Microscopically this may symbolize the multi-site martensite nucleation and growth inside grains of polycrystalline SMAs [3].

Even though in general the response of SMAs during transformation is a coupled thermo-mechanical process, the pure mechanical aspects of the martensitic transformation are still very important and become dominant in the isothermal, quasi-static case. For the

[1] corresponding author: Tel.: +852-2358-8655; Fax: +852-2358-1543; E-mail: meqpsun@ust.hk

Q.P. Sun (ed.), IUTAM Symposium on Mechanics of Martensitic Phase Transformation in Solids, 95–102.

mechanical interaction, we formulate a consistent macroscopic continuum model for the considered matrix-inclusion system where the thermal effect can be incorporated in the future. Although a great amount of work has been done on the traditional matrix-inclusion problems such as those in mechanics of composite materials and polycrystals, the micro-macro transition in the case of localized transformation phenomenon via growing inclusions so far is only available for homogeneous material properties [4]. In this paper heterogeneous material properties are addressed, which is significant for NiTi SMAs where Young's modulus of martensite is roughly half that of austenite. By reducing the length of periodicity the transformation process changes from a discrete to a homogeneous one. The influence of the length of periodicity and the heterogeneous material properties on the macroscopic response of the system are discussed.

2 Continuum model for periodic microstructure

For the evaluation of the structural (i.e. macroscopic) response of a specimen with growing inclusions we assume a periodic arrangement of inclusions in an infinite circular rod with constant cross section $A_0 = R^2\pi$ (Fig. 1). The two phases (martensite and parent austenite) are separated by planar interfaces which are perpendicular to the axis of the wire (direction 3). Both phases are assumed to be elastic isotropic with different Young's moduli E_1 and E_2 for phase 1 and 2, respectively. In the following we consider phase 1 as the parent phase transforming to phase 2 during forward transformation. The transformation strain is assumed as homogeneous within phase 2 and is defined by the transformation strain $\epsilon_{33}^{\mathrm{tr}}$ in tensile direction. The axial symmetric transversal transformation strain components are $\epsilon_{11}^{\mathrm{tr}} = \epsilon_{22}^{\mathrm{tr}} = -\frac{1}{2}\epsilon_{33}^{\mathrm{tr}}$.

A representative volume element (RVE) of length L_0 can be introduced. The nominal strain $\mathbf{E}$ in tensile direction is the external state variable defining the system under displacement controlled loading. The volume fraction of inclusions $\xi = \frac{l_{\mathrm{m}}}{L_0}$ describes the transformation process as an internal variable, where l_{m} denotes the length of the inclusion. We further introduce the dimensionless length of periodicity λ given by $\lambda = \frac{L_0}{R}$. Since we consider only the growth of the inclusions, their volume fraction is prescribed as monotonically increasing from 0 to 1 during forward transformation.

The total Helmholtz free energy $\Phi(\mathbf{E}, \xi, T)$ for the RVE can be formulated as [4]:

$$\begin{aligned}\Phi(\mathbf{E}, T, \xi) = {} & \frac{1}{2}\frac{A_0 L_0 E_1}{1+\xi\left(\frac{E_1}{E_2}-1\right)}(\mathbf{E} - \xi\epsilon_{33}^{\mathrm{tr}})^2 + W^{\mathrm{int}}(\xi) + \\ & + \left(W_0^{\mathrm{chem}}(T) + \xi k(T - T_0)\right) A_0 L_0. \end{aligned} \tag{1}$$

where the term $W^{\mathrm{int}}(\xi)$ expresses the internal elastic strain energy due to the transformation strain. With $\epsilon_{ij}^{\mathrm{el}}$ denoting the elastic strain the internal elastic strain energy can be obtained as [5]

$$W^{\mathrm{int}}(\xi) = \frac{1}{2}\int_{\mathrm{A_0 L_0}} \sigma_{ij}(r)\epsilon_{ij}^{\mathrm{el}}(r)dV. \tag{2}$$

Due to the periodic arrangement of inclusions the internal elastic strain energy implicitly depends on λ. The change of the chemical free energy with the temperature T is given by

$(W_0^{\rm chem}(T) + \xi k(T - T_0))A_0 L_0$ with $W_0^{\rm chem}(T)$ and $k(T - T_0)$ denoting the chemical free energy density of the initial phase and the change of free energy density due to the phase transformation (here linearly approximated near the equilibrium temperature T_0), respectively.

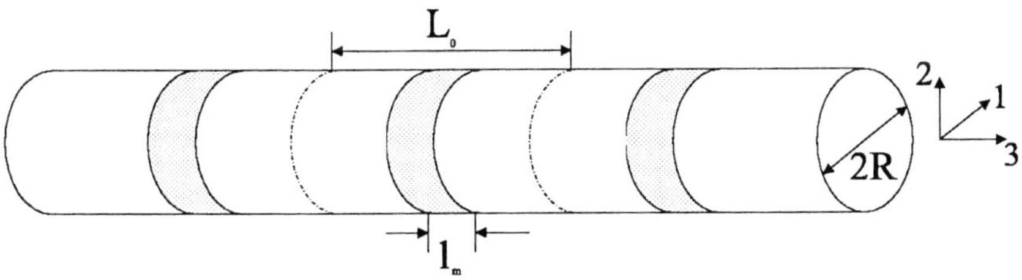

Figure 1: Schematic illustration of inclusions (dark) in a cylindrical rod (bright).

The transformation condition requires equality of the energy release rate and the dissipation rate [4]. In the simpliest case the energy dissipation rate can be approximated to be proportional to the increase of the band length: $\dot{W}^d = \pm d_0 A_0 L_0 \dot{\xi}$, (+) for forward and (-) for backward transformation, respectively. The specific dissipation d_0 can be determined experimentally, e.g., from the stress hysteresis in the quasi-static isothermal stress-strain curve [1, 2]. With the use of the constitutive relation

$$\Sigma_{33} = \frac{E_1}{1 + \xi(\frac{E_1}{E_2} - 1)} \left(\mathbf{E} - \xi\epsilon_{33}^{\rm tr}\right), \tag{3}$$

the nominal macroscopic tensile stress can be calculated as

$$\Sigma_{33}(\xi) = \eta \left(-\epsilon_{33}^{\rm tr} + \sqrt{(\epsilon_{33}^{\rm tr})^2 + \frac{2}{\eta}\left(\frac{1}{A_0 L_0}\frac{\partial W^{\rm int}(\xi)}{\partial \xi} + q_0\right)}\right), \tag{4}$$

where $\eta = E_1 E_2/(E_1 - E_2)$ and $q_0 = k(T - T_0) \pm d_0$.

The internal elastic strain energy is gained numerically by solving the corresponding boundary value problem by means of FEM [6]. Without loss of generality the radius R of the wire is set to unity. The length of periodicity is varied between $\lambda = 0.1$ to $\lambda = 4$. The smallest martensite band length is chosen as $0.01R$ which is close to the limit of convergency of the numerical calculation. The material properties entering the numerical evaluation of the $W^{\rm int}(\xi)$ are the Young's moduli of the two phases, their Poisson's ratios and the macroscopic transformation strain. Published material data for polycrystalline equiatomic NiTi are used [1, 2]: $E_1 = 67000$ MPa; $\nu_1 = \nu_2 = \nu = 0.4$; $\epsilon_{33}^{\rm tr} = 0.043$ and $\kappa = \frac{E_1}{E_2} = 1$ and 2 for homogeneous and inhomogeneous material properties, respectively. Interested only in the purely mechanical response of the considered matrix-inclusion systems no thermal effects and no hysteresis are considered in our simulations, thus $d_0 = 0$ and $T = T_0$.

3 Results

Internal elastic strain energy Fig. 2-a depicts the change of the normalized internal elastic strain energy W^* as a function of the volume fraction of martensite ξ for homogeneous material properties ($\kappa = 1$). The actual values of $W^{\text{int}}(\xi)$ can be obtained by multiplying the given values with $\frac{\pi R^2 L_0 E_1 (\epsilon_{33}^{\text{tr}})^2}{4(1-\nu)}$. Depending on λ the shape of the curves changes significantly. In the case of large values of λ, e.g. for $\lambda = 4$ (lower curve), the

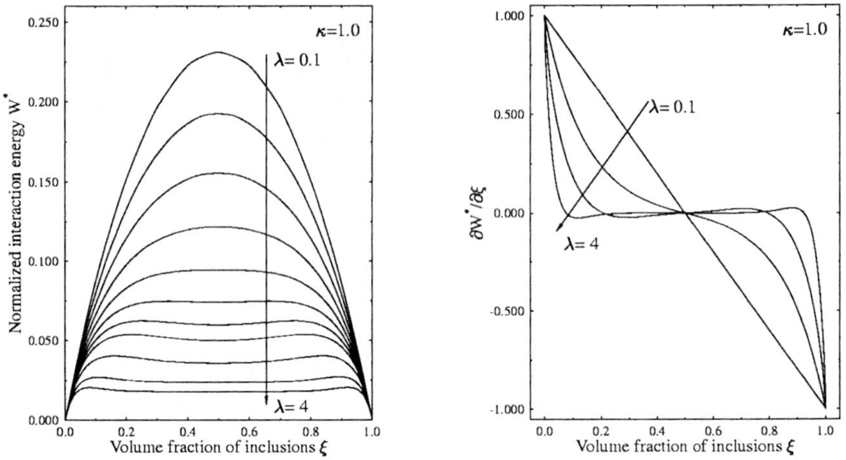

(a) (b)

Figure 2: Normalized internal elastic strain energy (a) and first derivative of W^* with respect to ξ (b) as a function of the volume fraction ξ and different lengths of periodicity λ for homogeneous material properties ($\kappa = 1$).

internal elastic strain energy approaches the values belonging to a single martensitic band propagating in an infinite rod. W^* increases and quickly reaches a peak value with increasing ξ. Further growth of the inclusion causes a small decrease in W^* and soon a steady state value is obtained. Since the stress state at the matrix-inclusion interface is only due to the transversal transformation strains the elastic strain energy is stored in finite regions adjacent to the interfaces. In the initial stage of the transformation the stress fields of two neighboring interfaces interact strongly because of the small band length which causes the steep increase of W^*. In the steady-state regime the stress states in the vicinity of the interfaces do not interact and W^* remains constant even upon further increase of ξ.

For $\lambda = 0.1$, which represents a very dense distribution of inclusions, the normalized internal elastic strain energy exhibits a parabolic variation with ξ approaching asymptotically the theoretically derived expression for the homogeneous material transformation [7], $W^* = \xi(1 - \xi)$. A similar parabolic relationship [8] for the strain energy is found in the treatment of spherical inclusions with shear and dilatation transformation strains in a three-dimensional matrix as well as for the interfacial energy of randomly distributed domains of martensite and austenite [9].

According to Eq. 4 the first derivative of the internal elastic strain energy with respect to the volume fraction of inclusions can be seen as the dominant term determining the macroscopic response of the matrix inclusion system. Fig. 2-b depicts $\frac{\partial W^*}{\partial \xi}$ as a function of the volume fraction ξ for different λ. The normalization is the same as in Fig. 2-a.

For $\lambda = 4$ a very steep monotonic decrease can be observed after nucleation of the inclusion. The function becomes negative upon further growth and finally reaches its steady state value zero which corresponds to the plateau in Fig. 2-a. At the end of the transformation process another steep decrease can be observed. For $\lambda = 0.1$ the first derivative of the internal elastic strain energy is almost linearly decreasing and approaches the limiting function $1-2\xi$ characteristic for the homogeneous transformation. Since we prescribe the volume fraction of inclusions as linearly increasing from 0 to 1 starting with infinitesimal small ξ, the value of the first derivative is 1 irrespective of λ. When ξ reaches 1 similar considerations yield the value -1 that is independent of the length of periodicity.

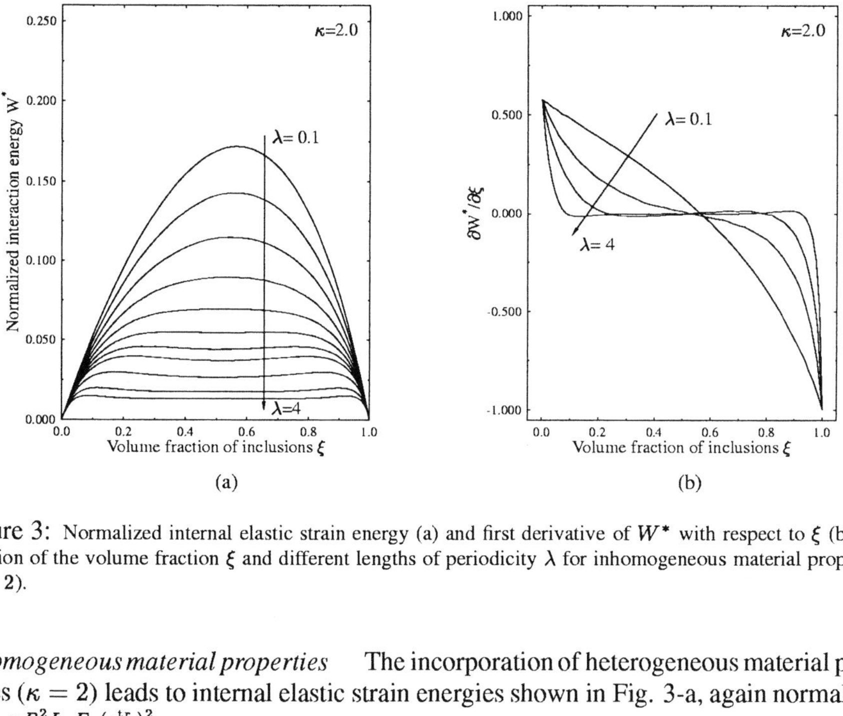

Figure 3: Normalized internal elastic strain energy (a) and first derivative of W^* with respect to ξ (b) as a function of the volume fraction ξ and different lengths of periodicity λ for inhomogeneous material properties ($\kappa = 2$).

Inhomogeneous material properties The incorporation of heterogeneous material properties ($\kappa = 2$) leads to internal elastic strain energies shown in Fig. 3-a, again normalized with $\frac{\pi R^2 L_0 E_1 (\epsilon_{33}^{tr})^2}{4(1-\nu)}$. The difference in elastic properties shifts the curves towards the less stiffer phase (i.e. phase 2 - right hand side) and the symmetry with respect to $\xi = 0.5$ is lost. Estimates of the macroscopic free energy based on Voigt and Hashin-Shtrikman bounds confirm the significant influence of heterogeneity and the loss of symmetry of W^{int} [10]. A comparison of the different estimates shows that the calculated values of W^* are best approximated by the Voigt bound. Due to the special geometry of the considered matrix inclusion system the transformation strain within the inclusions is constrained by the surrounding matrix in transverse direction only. Thus, for small lengths of periodicity the difference in the total strain between the narrow matrix and inclusion domains approaches zero which corresponds to Voigt's assumption of uniform strain.

In the case of inhomogeneous material properties the peak value of $\frac{\partial W^*}{\partial \xi}$ at beginning of the transformation is approximately half that of the corresponding value for $\kappa = 1$, since the elastic modulus of the inclusion is only half that of the matrix (Fig. 3-b). The minimum of $\frac{\partial W^*}{\partial \xi}$ at coalescence of the inclusions is comparable to that for homogeneous properties. Analysis with inhomogeneous material properties clearly shows the strong impact of the elastic properties on the free energy during the transformation process.

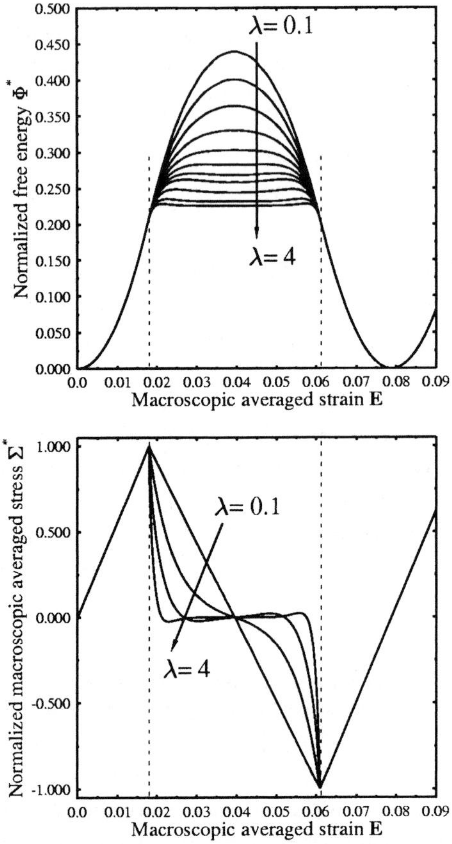

Figure 4: Helmholtz free energy and average stress strain relation for homogeneous material properties.

Helmholtz free energy - Average stress-strain relation With the values for $\frac{\partial W^{\text{int}}(\xi)}{\partial \xi}$ we can calculate the Helmholtz free energy as a function of the external macroscopic strain in tensile direction $\mathbf{E}$ (Eq. 1). The upper plot of Fig. 4 shows the normalized Helmholtz free energy Φ^* for homogeneous material properties (normalization factor: $\frac{\pi R^2 L_0 E_1 (\epsilon_{33}^{\text{tr}})^2}{4(1-\nu)}$). Starting from $\mathbf{E} = 0$ the function is parabolic during the purely elastic deformation until $\mathbf{E}$ reaches the macroscopic critical strain $\mathbf{E}^{\text{crit}}$: $\Phi(\mathbf{E}, \xi) = \frac{1}{2} E_1 \mathbf{E}^2 \, \pi R^2 L_0$. Since we do not consider any nucleation mechanism in the framework of our model, $\mathbf{E}^{\text{crit}}$ has to be determined by experiments. In Fig. 4 the critical strain (see left vertical dashed line) is chosen so that the tangent of the Helmholtz free energy is equal before and after nucleation of the inclusions. Due to the evolution of the internal elastic strain energy the Helmholtz free energy becomes non-convex during the transformation process:

$$\frac{\partial^2 \Phi}{\partial \mathbf{E}^2} < 0, \tag{5}$$

which leads to an instability of the material response in tensile loading [11]. For $\lambda = 0.1$ the Helmholtz free energy is non-convex during the whole transformation process. For more dilute bands ($\lambda = 4$) the Helmholtz free energy is only non-convex at the beginning and the end of transformation. For intermediate values of ξ there is a region of stable band propagation. After the whole specimen has transformed, i.e. $\xi = 1$ and $\mathbf{E} = \mathbf{E}^{\mathrm{crit}} + \epsilon_{33}^{\mathrm{tr}}$ (right vertical dashed line), the function becomes convex again due to the further elastic deformation: $\Phi(\mathbf{E}, \xi) = \frac{1}{2} E_1 (\mathbf{E} - \epsilon_{33}^{\mathrm{tr}})^2 \, \pi R^2 L_0$.

Finally, from the Helmholtz free energy the macroscopic averaged stress-strain relations can be obtained. The lower plot of Fig. 4 depicts the macroscopic stress Σ^* normalized with $\frac{E_1 \epsilon_{33}^{\mathrm{tr}}}{4(1-\nu)}$. In the case of isolated bands a sharp load drop can be observed after nucleation and shortly before the transformation fronts merge. During steady state growth of the inclusions the macroscopic stress becomes zero. This corresponds to the stable transformation from phase 1 to 2 [11, 9]. For homogeneous transformation the macroscopic averaged stress decreases linearly and the whole transformation is accompanied by a strain softening response. It has to be noted that only in the case of the purely mechanical model ($q_0 = 0$) the propagation stress becomes zero. Otherwise, Σ^* increases proportional to the temperature rise and dissipation rate during the transformation process.

4 Discussion and conclusions

One of the challenging aspects of modeling localized martensitic transformation in SMA is the fact that the constitutive relations cannot be gained trivially from experiments as in the case for stable material behavior. As soon as localization of the deformation occurs, each experiment becomes dependent on the material as well as on the structure of the specimen even for the simple case of a tensile loading. While it is rather straight forward to model localization via an appropriate, assumed up-down-up type of constitutive behavior, it is very difficult to gain the constitutive behavior experimentally.

In our approach we avoid prescribing the up-down-up behavior and use stable, linear elastic deformational behavior of both phases instead . The transformation is modeled via growth of periodically arranged inclusions, which bear the transformation eigenstrain. For large lengths of periodicity the discrete transformation process is obtained exhibiting a pronounced stress drop before the steady state propagation stress is reached. For the limiting case of vanishing length of periodicity the transformation becomes homogeneous and the macroscopic response is linearly decreasing during transformation. Conceptually, our model bridges the gap between a discrete transformation process and a homogeneous one.

The predicted nominal stress-strain relationship for the homogeneous transformation process can be applied in turn as the homogeneous one-dimensional constitutive law for further investigations on material instabilities. This will lead to an infinite number of solutions since the intrinsic strain softening of the material behavior leads to a loss of uniqueness of the solution [11]. However, experiments and thermodynamic considerations already have shown, that only a single band will nucleate and propagate in reality, except there are other mechanisms operating such as non-isothermal conditions. To overcome this contradiction between experimental observation and theoretical prediction of the structural response with material instabilities a promising way seems to be the incorporation of the strain gradient effect in the simulation.

Acknowledgements The work is supported by the Research Grant Council of Hong Kong (Project no. HKUST 6038/99E), the National Science Foundation of China (Project no. 19825107) and the German-Hong Kong joint research scheme (Project no. D/9800728) of the Deutscher Akademischer Austauschdienst. We acknowledge that Hibbit, Karlsson and Sorensen, Inc. made ABAQUS available under an academic license.

References

[1] J.A. Shaw and S. Kyriakides. Thermomechanical aspects of NiTi. *J. Mech. Phys. Solids*, 43(8), 1243–1281, 1995.

[2] J.A. Shaw and S. Kyriakides. On the nucleation and propagation of phase transformation fronts in a NiTi alloy. *Acta Mater.*, 45(2), 683–700, 1997.

[3] K. Otsuka and C.M. Wayman. *Shape memory materials.* Cambridge University Press, Cambridge, 1998.

[4] Q.P. Sun and Z. Zhong. An inclusion theory for the propagation of martensite band in NiTi shape memory alloy wires under tension. *Int. J. Plasticity*, 16, 1169–1187, 2000.

[5] T. Mura. *Micromechanics of defects in solids.* Martinus Nijhoff, Dordrecht, 1987.

[6] C. Messner, G. Reisner, Q.P. Sun and E. Werner. On instabilities and autocatalytic effects associated with the nucleation of martensite bands in polycrystalline SMA flat tensile specimens. *Comp. Mat. Sci.*, 19, 313–319, 2000.

[7] Z. Zhong and Q.P. Sun. Elastic solution of a cylindrical rod containing periodically distributed inclusions with axisymmetric eigenstrains. 2001. not published.

[8] Q.P. Sun, K.C. Hwang and S.W. Yu. A micromechanics constitutive model of transformation plasticity with shear and dilatation effects. *J. Mech. Phys. Solids*, 39, 507–524, 1991.

[9] I. Müller and H. Xu. On the pseudoelastic hysteresis. *Acta Metall. Mater.*, 39, 263–271, 1991.

[10] D. Bernardini. On the macroscopic free energy functions for shape memory alloys. *J. Mech. Phys. Solids*, 49, 813–837, 2001.

[11] J.L. Ericksen. Equilibrium on bars. *J. Elasticity*, 5, 191, 1975.

OBSERVATIONS ON STRESS-INDUCED TRANSFORMATIONS IN NITI ALLOYS

[1]HUSEYIN SEHITOGLU, [1]X. Y. ZHANG, [2]Y.I. CHUMLYAKOV, [3]I. KARAMAN, [4]K. GALL, [5]H. J. MAIER

[1]University of Illinois, 1206 West Green Street, Urbana, IL 61801, USA, [2] Tomsk State University, 1, Novo-Sobornay, 634050 Tomsk, Russia, [3] Texas A&M University, Mail Stop 3123, College Station, TX 77843, USA, [4]University of Colorado at Boulder, ECME 279, Campus Box 427, Boulder, CO 80309-0427 USA, ,[5]University Paderborn, Lehrstul fur Werkstoffkunde, FB10, 33098 Paderborn, Germany

ABSTRACT

The work focuses on stress-induced transformation in single crystal NiTi alloys. Many of the basic features of these alloys are known but there are still a number of concepts that have not been fully understood and require further clarification. In this short paper, we focus on four issues, namely tension-compression asymmetry, crystal orientation effects on stress-strain response, the role of precipitates on transformation and slip resistance and elastic moduli measurements on austenite, martensite and the two-phase microstructure. We show that coherent precipitates lower the transformation stress at small strains while increasing the slip resistance of austenite and martensite domains at higher strains. Based on the resolved shear stress factors (RSSF), we identify hard and soft orientations explaining tension-compression asymmetry and crystal orientation dependence. Also, we note that the martensite modulus is significantly higher than the austenite modulus for the single crystal orientations studied.

1. INTRODUCTION

Shape memory alloys have unique properties such as recoverability of deformation upon heating and pseudoelastic stress-strain behavior strains near 10% [1]. These are promising alloys finding applications ranging from bioengineering and medicine to mechanical and civil engineering. The materials science background has been documented for these alloys over the years in a variety of materials journals, but a number of unexplained issues remain including tension-compression asymmetry [2-5], crystal orientation dependence of stress-strain response [2-5], the role of precipitates [1,5], and the differences in austenite and martensite elastic moduli. We will demonstrate these key effects to provide a better physical understanding of the response of shape memory alloys.

2. DEMONSTRATION OF TENSION-COMPRESSION ASYMMETRY

There is increasing interest in understanding the compression behavior of NiTi alloys [1]. Aside from practical reasons, such as understanding bending behavior, the martensitic transformation, and slip behavior can be studied without premature fracture of the specimens. It has been known that compressive stresses in polycrystalline NiTi are higher than the tensile stresses for the same values of applied strain [2,5]. We illustrate in this section the differences

Q.P. Sun (ed.), IUTAM Symposium on Mechanics of Martensitic Phase Transformation in Solids, 103–109.

in critical transformation stress between tension and compression for the [111] orientation which is a dominant fiber texture in cold-drawn polycrystalline NiTi. In Figure 1, the stress-strain response is shown at room temperature for both tension and compression cases. We note that theoretical transformation strains are near 2.98% in [111] compression. In Figure 1, the transformation to martensite is nearly complete in compression while in tension the transformation strains are much higher theoretically (5.98%). The transformation stresses in tension are lower than compression. These experimental results (Figure 1) can be simply explained based on the calculation of RSSF for the austenite to martensite transformation. Using the habit planes, and the transformation directions the RSSF's for all possible twin types have been calculated for tension and compression in Figures 2(a) and 2(b) [6-7]. Among the four twin types, the Type II-1 has been observed most commonly in experiments and also used in mechanics simulations [2,8, 9].

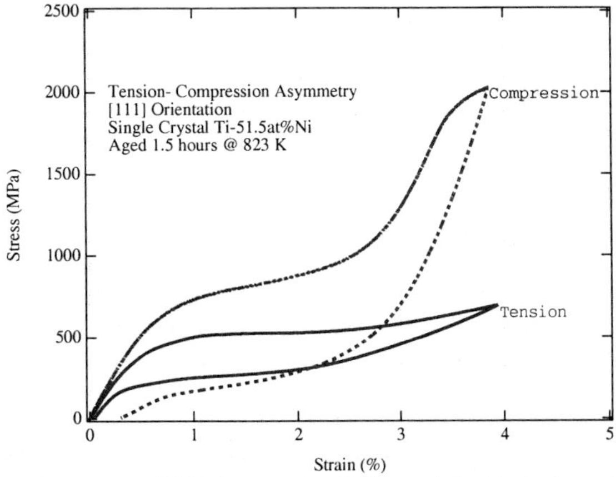

Figure 1 The stress-strain response of NiTi single crystals ([111] orientation) in tension and compression.

The resolved shear stress factor, RSSF, is simply defined as, $\mathrm{RSSF} = (\mathbf{b}\cdot\mathbf{e})(\mathbf{m}\cdot\mathbf{e})/|\mathbf{b}|$ where **e** denotes the crystallographic direction of interest, **b** is the transformation direction and **m** is the habit plane normal. For the [111] orientation the RSSF is 0.253 for compression and 0.439 in tension. We note that the location of the maximum resolved shear stress factor changes depending on the type of twinning. A total of 192 habit planes (with 96 correspondent variant pairs (CVPs)) have been identified for the NiTi system. A CVP , correspondent variant pair, is a formal term for two twin related martensite variants. The four twin types exclude the compound twinning solutions, which have been addressed in a recent paper [10]. Typically, 24 variants (corresponding to Type II-1) are used in the calculations. To our knowledge, this is the first time where such a complete representation of theoretical results is provided for the NiTi alloy system (also see [1]).

3. CRYSTAL ORIENTATION DEPENDENCE

To illustrate the crystal orientation effects, we consider the experimental results on single crystals under compression in Figure 3. The stress-strain response for [001], [112], [111] and [148] are shown. According to Figure 2(a) the [148] orientation has the highest RSSF (0.512) while the [111] corner of the stereographic triangle has the lowest RSSFs under compression (0.253). This is consistent with the experimental findings in Figure 3 showing "hard' response for [111] and 'soft' response for the [148] case. It is also clear from Figure 3 that the transformation behavior and the stress-strain response as a function of strain differ considerably among these four orientations. For example, in the [148] orientation the transformation proceeds over a nearly constant stress due to the presence of a single CVP (correspondent variant pair) while multiple CVPs are activated in all other cases. Upon completion of the transformation at high strains, martensite first deforms elastically, then slip

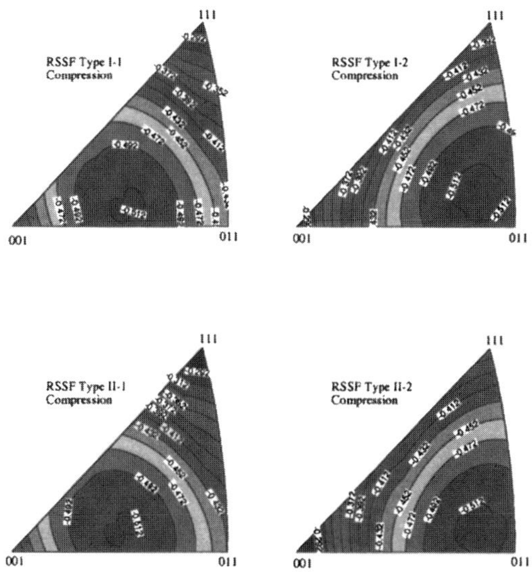

The Maximum Resolved Shear Stress Factors (RSSF)

(a)

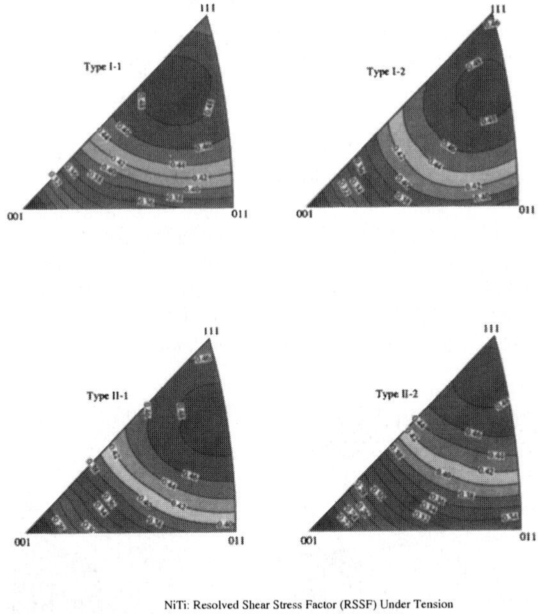

(b)

Figure 2 (a) Resolved shear stress factors in compression [1] (b) Resolved shear stress factors in tension

deformation occurs. The [001] orientation is not favorable to slip and the response remains elastic even at very high stress levels [1]. We note that the unloading modulus at high strains is higher compared to the austenite elastic modulus, and this point will be further examined in Section 5.

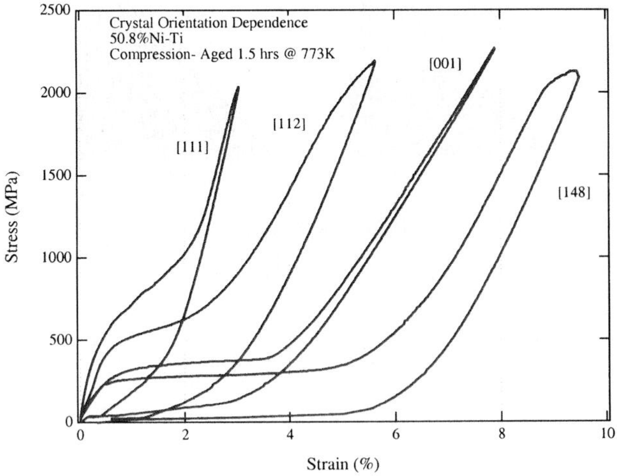

Figure 3 Compressive stress-strain response of NiTi displaying crystal orientation dependence.

4. THE ROLE OF PRECIPITATES

A number of early works on NiTi alloys have been conducted in the solutionized state. In the solutionized state, shape memory behavior has been observed in most cases. With the introduction of precipitates pseudoelasticity can occur over a broad range of temperatures. The presence of the precipitates provides several differences in mechanical response, including an increase in martensite start temperature, a lower transformation stress, and the pseudoleasticity phenomenon. TEM micrographs of precipitates for peakaged (723K, 1.5 h) and overaged(823K, 1.5h) 50.8%Ni Ti are shown in Figures 4(a) and 4(b) respectively. To understand the role of precipitates, we consider the deformation behavior of the [148] orientation in compression loadings in Figure 5. Two different heat treatments have been undertaken. In the peak aged case the precipitates are nearly coherent while they are incoherent for the overaged case. We note that in the peak aged case the transformation stress is lower while the slip resistance is higher (exceeding 2000 MPa). Therefore, the presence of precipitates leads to optimum materials design conditions for shape memory alloys.

(a)

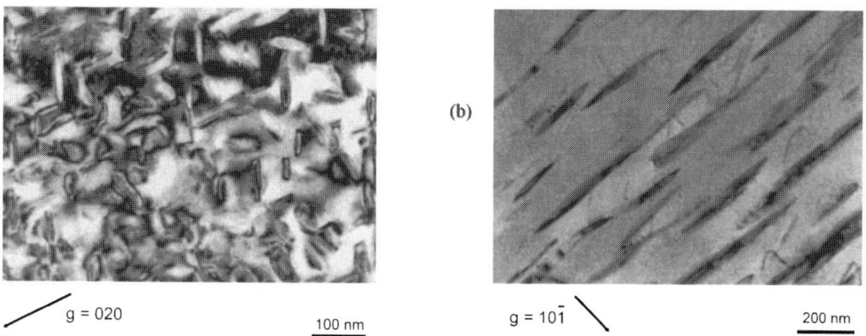

Figure 4 Two-beam bright-field TEM micrographs of NiTi specimen in the peakaged (a) and overaged (b) condition[1]

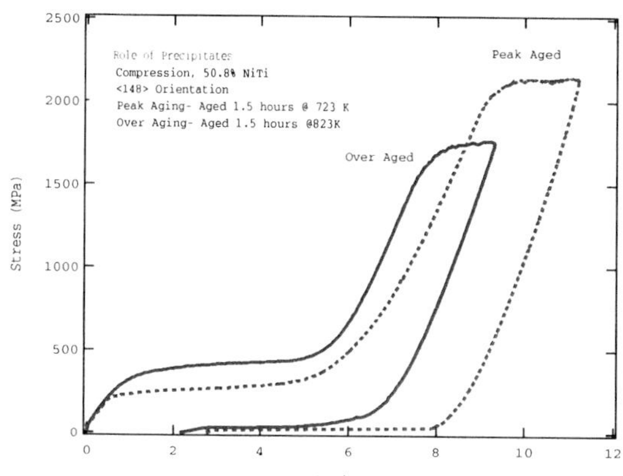

Figure 5 Stress-strain response of [148] crystals in compression showing the role of precipitates on the stress-strain response (50.8%Ni)

5. ELASTIC MODULUS OF AUSTENITE AND MARTENSITE PHASES

In a number of publications on NiTi alloys [11], it has been reported that the martensite modulus is substantially lower than the austenite modulus. We note that either these experiments were conducted under test temperatures below the martensite start temperature, or the elastic modulus was measured from cases where upon unloading, reverse transformation occurs without a distinct elastic region. In the first case, the 'apparent' modulus of a thermally induced martensite is measured where deformation proceeds due to martensite reorientation and detwinning (the detwinning is the conversion of two twin related martensite variants within a CVP to a single variant). In the second case, high internal stresses produce a spontaneous reverse transformation. In either case, these previous measurements report lower martensite modulus than in reality.

When we examine the elastic modulus of the specimens that have undergone austenite to martensite transformation under compression (such as Figures 1,3,5) and deformed in the plastic regime further a distinct elastic region is observed upon unloading. The elastic modulus upon a series of unloadings at different strain levels is given in Figure 6 as a function of applied strain. Results from both the peakaged and the overaged cases are included in Figure 6. We note that for the case of [001] the elastic modulus increases gradually with increasing strain and reaches an asymptotic value. This orientation is particularly interesting because upon austenite to martensite transformation the [001] direction in the austenite lattice remains as the [001] direction in the martensite lattice [1]. The increase in elastic modulus is approximately from 35GPa to 70GPa. Experiments on other crystal orientations produce similar findings. For example, the [112] austenite orientation upon transformation corresponds to the [110] orientation in martensite [1,10]. We have measured the [110] austenite modulus as 75 GPa and the martensite [110] modulus as 125GPa confirming the higher elastic modulus of martensite relative to austenite.

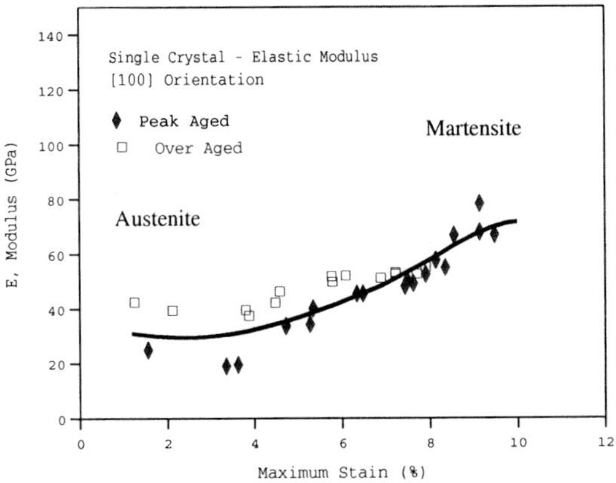

Figure 6 The elastic modulus of NiTi as a function of strain. At intermediate strains the modulus reflects the two phase microstructure; at high strains the modulus of martensite is reached. All these measurements are obtained from unloading at specified strain levels in compression.

6. CONCLUSIONS

The work supports the following conclusions:

(1) Both the tension-compression asymmetry and the crystal orientation dependence are linked to the difference in resolved shear stress factors in tension and compression.
(2) The role of precipitates on stress-strain response is significant, with coherent precipitates lowering transformation stress and increasing slip resistance at the same time.
(3) The elastic moduli of martensite phase is higher than the austenite phase as demonstrated from repeated unloading experiments in compression.

7. REFERENCES

1. Sehitoglu, H., I. Karaman, R. Anderson, X. Zhang, K. Gall, H. J. Maier, and Y. Chumlyakov, *Acta Materialia*, (2000), **48, 13**, 3311-3326, see also ibid (2001), **49,4**, 747.
2. Gall, K. and H. Sehitoglu, *International Journal of Plasticity*, (1999), **15**, 69-92.
3. Gall, K., H. Sehitoglu, Y. I. Chumlyakov, I. V. Kireeva, and H. J. Maier, (1999), *Journal of Engineering Materials and Technology*, **12**,19-27
4. Gall, K., H. Sehitoglu, Y. I. Chumlyakov, I. V. Kireeva, and H. J. Maier, (1999), *Journal of Engineering Materials and Technology*, **12,** 28-37,
5. Gall, K., H. Sehitoglu, Y. Chumlyakov, and I. Kireeva,, *Acta Materialia*, (1999), **47:4**, 1203-1217
6. Ball, J.M., R.D. James, (1987), Arch. Rat. Mech. Anal., **100,** 13-52.
7. Wechsler, M.S., Lieberman, D. S, Read T. A., (1953), Trans. AIME, **197**, 1503-1515
8. Patoor, E. N. Siredey, A. Eberhardt, and M. Berveiller, (1995), *J. De Phys III*, **5,8** 227.
9. Nishida, M., H. Ohgi, I. Itai, A. Chiba, K. Yamauchi, (1995) Acta Materialia,, **43,** 1219-1227.
10. Sehitoglu, H, J. Jun, X.Zhang, I. Karaman, Y. Chumlyakov, H. J. Maier, K. Gall, (2001), to appear in Acta Materialia
11. Hodgson, D. E., M. H.Wu, R. J. Biermann (1990) ASM Handbook, , **2** , 897.

8. ACKNOWLEDGEMENTS

Portions of the research is supported by a grant from the National Science Foundation contract CMS 99-00090, Mechanics and Materials Program, Arlington, Virginia, and Air Force Office of Scientific Research, Directorate of Aerospace and Materials Sciences, Arlington, Virginia. Professor Chumlyakov received support from the Russian Fund for Basic Researches, Grant Nos. 02-95-00350, 99-03-32579. The facilities at Microanalysis of Materials, Materials Research Laboratory were used. This laboratory is funded by DOE-DMS grant DEFG02-96ER45439.

DEPENDENCE OF THE ENERGY RELEASE RATE ON THE PROPAGATION SPEED OF MARTENSITIC TRANSFORMATION IN MATERIALS

Biao Wang[1], Zhongmin Xiao[2] and Xiaoxia Wu[1]
[1]Research Center for Composite Materials and Electro-Optics Research Center, Harbin Institute of Technology, Harbin, China
[2]School of MPE, Nanyang Technological University, Singapore

ABSTRACT

Martensitic transformation process is a dynamical process. The well-known experimental fact is: once the nucleation barrier has been overcome, the excess free energy becomes so large that the martensite plate grows rapidly until it hits a barrier such as another plate, or a high angle grain boundary. In the present paper, the dynamic energy release rate was derived. To obtain the analytical expression of the energy release rate, two simple cases were considered: one is a semi-infinite one-dimensional bar under the action of an impact load, and the other example is a spherical martensite inclusion expanding along its radial direction. From the two examples, it can be found that the dynamic energy release rate decreases with increasing the propagation speed of transformation, and when the speed reaches some critical value, the deriving force becomes zero. That means that the propagation speed of martensitic transformation has an upper limit.
KEYWORDS: Energy Release Rate, Martensitic, Transformation

INTRODUCTION

Martensitic transformation process is a dynamical process even though most researchers have taken the quasistatic assumption in their analysis. The well-known experimental fact is this: once the nucleation barrier has been overcome, the excess free energy becomes so large that the martensite plate grows rapidly until it hits a barrier such as another plate, or a high angle grain boundary. A single plate of martensite in steel grows in 10^{-5} to 10^{-7} s to its full size, at velocities approaching the speed of sound. Using resistivity changes to monitor the growth of individual plates of martensite in, e.g. Fe-Ni alloys, researchers have measured speeds of 800-1100 m/s (Nishiyama, 1978). In fact, such a dynamic propagation process of the martensite is always observed in practice even if the applied load is quasistatic. Under impact, or other dynamic loading, the propagation of martensite becomes even more difficult to analyze. Because of the great speed of martensite growth, it is extremely difficult to study this transformation experimentally.
There have been extensive analytical, experimental and numerical investigations of static and quasistatic behaviors of martensitic transformations. On the other hand, the existing study of dynamic behaviors of martensitic transformation is rather limited. Escobar and

Q.P. Sun (ed.), IUTAM Symposium on Mechanics of Martensitic Phase Transformation in Solids, 111–120.

Clifton (1993) carried out pressure-shear plate impact experiments on Cu-14.44Al-4.19Ni single crystals to study the kinetics of the stress-induced phase transformation. By using the experimental data obtained in their work, Abeyaratne and Knowles (1997) determined the values of phase boundary velocity and driving force, as well as the kinetic law which relates these two quantities. Chen and Lagoudas (2000) carried out a dynamic analysis of the impact induced phase transformation in a shape-memory alloy rod, with a special focus on the propagation of stress waves and phase transformation fronts in the rod. Wang et al. (1999) considered the dynamic growing spherical inclusion, and they obtained the analytical solution for the elastodynamic problem based on the work of Willis (1965) and Mikata and Nemat-Nasser (1988). Ball and James (1987) and James (1981, 1986) considered the equilibrium shock problems.

The Gibbs free energy of the system will be reduced during the martensitic transformation process, and the released energy will serve as the driving force for the phase transformation. The Gibbs free energy of the system can be divided into the chemical free energy, the interfacial free energy and elastic energy. The elastic energy is the most complicated part to evaluate. In dynamic fracture mechanics, generally speaking, the energy release rate associated with crack propagation will decrease with the velocity of crack propagation. And when the velocity of crack propagation reaches some critical value, the energy release rate becomes zero. Since we know that energy dissipation is inevitable in the crack propagation process, crack propagation at velocities exceeding the critical velocity is impossible in the continuum model. Using basic ideas of dynamic fracture mechanics, we can reason that released elastic energy associated with the formation of martensitic phase should also depend on the propagation speed of martensitic transformation. By setting the energy release rate equal to zero, one can find a maximum limiting velocity for martensitic transformation in materials.

Therefore in this investigation, we will derive the elastic energy release rate of the martensitic transformation as a surface integral. Two examples are considered to calculate the energy release rate as a function of the transformation speed; one is a one-dimensional bar under impact loading, the other is a spherical transformation domain expanding along the radial direction. For such simple cases, we did find the energy release rate decreases with increasing propagation speed of the transformation domain.

DRIVING FORCE FOR THE PROPAGATION OF MARTENSITIC TRANSFORMATION

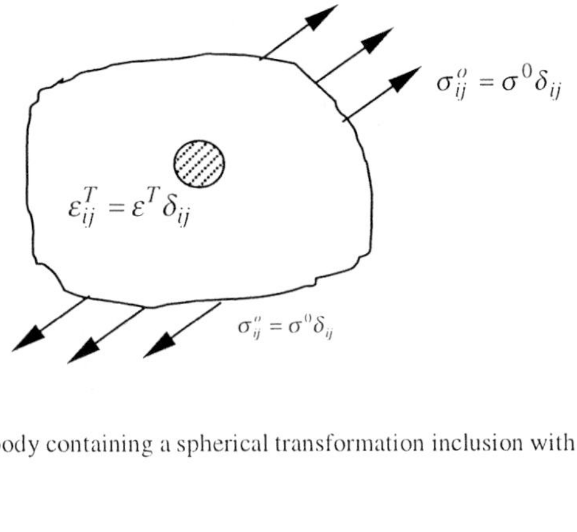

Figure 1. A finite elastic body containing a spherical transformation inclusion with a dilatational eigenstrain

Abeyaratne and Knowles (1990) presented a general formulation of the driving traction on a surface of strain discontinuity in a continuum undergoing a general thermomechanical process. In this part, we focus on the martensitic transformation problem, and use a more direct approach to derive the mechanical energy release rate accompanying the transformation process. We consider that in a parent material, which can be in the austenitic phase, there is a general shaped nucleus of martensite (Fig.1). And the martensitic phase has a transformation strain ε_{ij}^T. The changing rate of the mechanical energy in the martensitic phase can be written in the form as

$$\dot{W}_1 = \frac{d}{dt} \iiint_{\Omega(t)} (U + K) dv = \frac{d}{dt} [\frac{1}{2} \iiint_{\Omega(t)} (\sigma_{ij} e_{ij} + \rho \dot{u}_i^2) dv], \tag{1}$$

where U and K are specific elastic energy and kinetic energy, ρ, u_i are the mass density and displacement component, and e_{ij}, σ_{ij} are elastic strain and stress components, and given by

$$\begin{aligned} e_{ij} &= \frac{\partial u_i}{\partial x_j} - \varepsilon_{ij}^T \\ \sigma_{ij} &= C_{ijkl} e_{kl} = C_{ijkl} (\frac{\partial u_k}{\partial x_l} - \varepsilon_{kl}^T) \end{aligned}, \tag{2}$$

where C_{ijkl} is the elastic modulus tensor of the martensite. Since the integrand and the integral region in equation (1) are all the function of time t, one can use the Reynolds' transport theorem (Gurtin, 1981) to obtain its changing rate as

$$\dot{W}_1 = \iiint_{\Omega(t)} \frac{d}{dt} (U + K) dv + \iint_{S^+} (U + K) V_i n_i ds, \tag{3}$$

where S^+ is the inner surface of $\Omega(t)$, V_i is the speed of the surface S, and n_i is the outward normal vector of S.

Substituting equation (2) into the first term in equation (3), one can derive

$$\begin{aligned} \iiint_{\Omega(t)} (\dot{U} + \dot{K}) dv &= \iiint_{\Omega(t)} \sigma_{ij} (\frac{\partial \dot{u}_i}{\partial x_j} - \dot{\varepsilon}_{ij}^T) dv + \iiint_{\Omega(t)} \rho \dot{u}_i \ddot{u}_i dv \\ &= \iiint_{\Omega(t)} \frac{\partial}{\partial x_j} (\sigma_{ij} \dot{u}_i) dv - \iiint_{\Omega(t)} \sigma_{ij} \dot{\varepsilon}_{ij}^T dv - \iiint_{\Omega(t)} \rho \dot{u}_i \ddot{u}_i dv + \iiint_{\Omega(t)} \rho \dot{u}_i \ddot{u}_i dv \\ &= \iint_{S^+} \sigma_{ij} \dot{u}_i n_j ds - \iiint_{\Omega(t)} \sigma_{ij} \dot{\varepsilon}_{ij}^T dv \end{aligned}, \tag{4}$$

Generally speaking, the last term in equation (4) can be taken as zero since the transformation strain is expected to be unchanged with time. In the same way, one can derive the mechanical energy change rate of the outside region as

$$\dot{W}_2 = \frac{d}{dt}\iiint_{V-\Omega(t)} (U+K)dv = \iint_{\Gamma} \sigma_{ij}\dot{u}_i n_j ds - \iint_{S^-} \sigma_{ij}\dot{u}_i n_j ds - \iint_{S^-} (U+K)V_i n_i ds \qquad (5)$$

where S^- is the outer surface of $\Omega(t)$, and Γ is the boundary of the whole material. On the boundary Γ, the rate of the mechanical work is

$$\dot{F} = \iint_{\Gamma} \sigma_{ij}\dot{u}_i n_j ds \,, \qquad (6)$$

Therefore the energy release rate equals the rate of the mechanical work minus the rate of the mechanical energy of the system as follows,

$$\dot{G}_d = \iint_{S} \{[U+K]V_j + [\sigma_{ij}\dot{u}_i]\} n_j ds \,, \qquad (7)$$

where $[A]=A^0-A^I$, denotes the jump of A across the surface S. Equation (7) is very similar to the dynamic J-integral expression in dynamic fracture mechanics. The released energy given by equation (7) will serve as the driving force for the propagation of the martensitic transformation to compensate for the surface energy increase and other energy dissipation.

1. MARTENSITIC TRANSFORMATION ALONG A ONE-DIMENSIONAL BAR SUBJECTED TO AN IMPACT LOAD

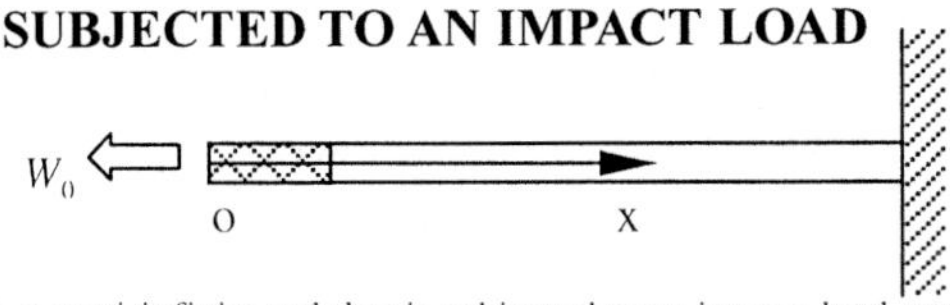

Figure 2. a semi-infinite rod that is subjected to an impact load at the end starting from an initial time

We consider a semi-infinite rod that is subjected to an impact load at the end, at time t=0. At t=0, the rod occupies the positive real axis, and is composed of austenite (Fig.2). For simplicity, we consider that the bar is in an isothermal environment. Suppose that this semi-infinite rod is subjected to an impact loading in which its boundary $x=0$ is given a constant velocity W_0 for time $t \geq 0$. As a result of this loading, a martensitic phase boundary is nucleated at the instant of impact t=0 at the impact face x=0, and it then propagates along the positive part of the x-axis. By properly controlling the single crystal specimen orientation, we can keep the martensite habit plane perpendicular to the axis of the bar (Abeyaratne and Knowles, 1997). If the displacement component of a material particle along the x-direction is denoted by u, and the transformation strain by ε^T, the normal stress along the x-axis is given by

$$\sigma_{xx} = E\{\frac{\partial u}{\partial x} - \varepsilon^T H[s(t)-x]\} \,, \qquad (8)$$

In deriving equation (8), we have taken the small deformation assumption; $H[s(t)-x]$ is the Heaviside step function,

$$H[s(t)-x]=\begin{cases}1 & x\le s(t)\\ 0 & x>s(t)\end{cases}, \tag{9}$$

where *s(t)* is the location of the phase boundary, If the martensite boundary propagates with uniform speed *V*, then *s(t)=Vt*. With the assumption that σ_{xx} is the only non-vanishing stress component, the equation of motion in the x direction is

$$\frac{\partial^2 u}{\partial x^2}-\frac{1}{c^2}\frac{\partial^2 u}{\partial t^2}=-\varepsilon^T\delta[s(t)-x], \tag{10}$$

where $c=(E/\rho)^{\frac{1}{2}}$ is the longitudinal propagation speed, and E,ρ are Young's modulus and the mass density, respectively. The initial and boundary conditions are given by

$$\begin{aligned} &t=0,\ x\ge 0,\quad u=\frac{\partial u}{\partial t}=0,\\ &x=0,\ t>0,\quad \frac{\partial u}{\partial t}=W_0 \ . \end{aligned} \tag{11}$$

The solution of the inhomogeneous equation (10) for the given initial and boundary condition (11) can be set as $u=\phi_1+\phi_2$, where ϕ_1,ϕ_2 are the solutions of the following two subproblems,

$$\frac{\partial^2 \phi_1}{\partial x^2}-\frac{1}{c^2}\frac{\partial^2 \phi_1}{\partial t^2}=0, \tag{12}$$

subject to the inhomogeneous initial and boundary condition

$$\begin{aligned} &t=0,\ x\ge 0,\quad \phi_1=\frac{\partial \phi_1}{\partial t}=0,\\ &x=0,\ t>0,\quad \frac{\partial \phi_1}{\partial t}=W_0 \ . \end{aligned} \tag{13}$$

and

$$\frac{\partial^2 \phi_2}{\partial x^2}-\frac{1}{c^2}\frac{\partial^2 \phi_2}{\partial t^2}=-\varepsilon^T\delta[Vt-x], \tag{14}$$

subject to the homogeneous initial and boundary conditions

$$\begin{aligned} &t=0,\ x\ge 0,\quad \phi_2=\frac{\partial \phi_2}{\partial t}=0,\\ &x=0,\ t>0,\quad \frac{\partial \phi_2}{\partial t}=0 \ . \end{aligned} \tag{15}$$

In what follows, we will obtain the solutions for the two subproblems.

(1) Solution of subproblem 1

Equations (12) and (13) is the ordinary one-dimensional impact problem, its solution can be written in the form (Johnson, 1972)

$$\phi_1 = \begin{cases} W_0(t - \frac{x}{c}) & x < ct \\ 0 & x > ct \end{cases} . \tag{16}$$

(2) Solution of subproblem 2

We can use the Laplace transform technique to solve the inhomogeneous equations (14) and (15). If the Laplace transform ϕ_2 is $\tilde{\phi}_2$, where

$$\tilde{\phi}_2 = \int_0^\infty e^{-pt}\phi_2(t)dt , \tag{17}$$

the Laplace transform of equation (14) with respect to time t is

$$p^2\tilde{\phi}_2 - c^2\frac{\partial^2\tilde{\phi}_2}{\partial x^2} = \frac{c^2\varepsilon^T}{V}e^{-\frac{px}{V}} , \tag{18}$$

where we have used the initial conditions, equation (15). The solution of equation (18) can be obtained as

$$\tilde{\phi}_2 = \Delta p^{-2}(e^{-\frac{px}{V}} - e^{-\frac{px}{c}}) \tag{19}$$

where $\Delta = c^2V\varepsilon^T/(c^2 - V^2)$.

By taking the inverse Laplace transformation of equation (19), one can obtain

$$\phi_2 = \phi_2^1 + \phi_2^2 , \tag{20}$$

where,

$$\phi_2^1 = \begin{cases} \Delta(t - \frac{x}{c}) & t > \frac{x}{c} \\ 0 & 0 < t < \frac{x}{c} \end{cases} , \tag{21}$$

$$\phi_2^2 = \begin{cases} -\Delta(t - \frac{x}{V}) & t > \frac{x}{V} \\ 0 & 0 < t < \frac{x}{V} \end{cases} . \tag{22}$$

Generally speaking, the velocity V of the phase boundary movement is smaller than the propagation speed c of the longitudinal wave. We can combine equations (16) and (20) to obtain the displacement of the bar

$$u(x,t) = \begin{cases} \Delta(\frac{x}{V} - \frac{x}{c}) + W_0(t - \frac{x}{c}) & x < Vt \\ W_0(t - \frac{x}{c}) + \Delta(t - \frac{x}{c}) & Vt < x < ct \\ 0 & x > ct \end{cases} . \tag{23}$$

The strain and stress fields are given by

$$\varepsilon = \frac{\partial u}{\partial x} = \begin{cases} \Delta(\frac{1}{V} - \frac{1}{c}) - \frac{W_0}{c} & x < Vt \\ -\frac{1}{c}(\Delta + W_0) & Vt < x < ct \\ 0 & x > ct \end{cases} . \tag{24}$$

$$\sigma = \begin{cases} E[\Delta(\frac{1}{V} - \frac{1}{c}) - \frac{W_0}{c} - \varepsilon^T] & x < Vt \\ -\frac{E}{c}(\Delta + W_0) & Vt < x < ct \\ 0 & x > ct \end{cases} . \tag{25}$$

Substituting equations (23), (24) and (25) into equation (7), one can derive the energy release rate:

$$\dot{G}_d = AV\{\frac{1}{2}\frac{E}{c^2}[\Delta^2(2r+1-r^2) + 2\Delta W_0(1+r)] - \frac{E}{rC^2}[(\Delta + W_0)^2 + r\Delta W_0 - W_0(\Delta + W_0)]\} \tag{26}$$

where $r = V/c$, and A is the area of the cross section of the bar. The energy release rate can be expressed in the following form,

$$\dot{G}_d = \frac{\partial G_d}{\partial L} \cdot \frac{\partial L}{\partial t} = Vg_d . \tag{27}$$

where L is the length occupied by the martensitic phase, and g is the energy release rate per unit increase of the length occupied by the martensitic phase, it is given by

$$g_d = A\{\frac{1}{2}\frac{E}{c^2}[\Delta^2(r-1)^2 + 2\Delta W_0(1-r)] - Er[(\Delta + W_0)^2 - x\Delta W_0 + W_0(\Delta + W_0)]\} . \tag{28}$$

By setting V=0, one can obtain the static energy release rate from equation (28) as

$$g_s = -AE\varepsilon^T \frac{W_0}{c} . \tag{29}$$

By giving $W_0/c = -0.2, \varepsilon^T = 0.04$, the energy release rate g_d/g_s versus the propagation velocity of the phase boundary r=V/c is given in Fig.3. It is very interesting to notice that the energy release rate decreases with increasing propagation speed of the transformation, just as in dynamic fracture mechanics; the energy release rate also decreases with increasing propagation speed of the crack. One can also find that when the propagation speed of the transformation reaches some critical value, which is smaller than the longitudinal wave speed c, the energy release rate is zero. Since the energy dissipation accompanying the transformation is inevitable, the propagation velocity of the transformation must have an upper limit. As indicated by the continuum model, the propagation velocity of the transformation cannot be over the limit when subjected to any boundary conditions. Note that the predicted upper limit of the velocity may be much higher than the observed transformation speed; in dynamic fracture mechanics, the theoretical critical propagation speed for a crack is much higher than its experimental value.

2. AN EXPANDING SPHERICAL TRANSFORMATION INCLUSION

It is a well-known fact that due to the crystallographic dependence of martensite formation, the martensite nuclei within a grain grow in a limited number of orientations. Thus the final morphology of martensites usually takes the form of thin plates. To derive the analytical result for the mechanical energy release rate, and discuss its physical meaning, , we assume now that a spherical martensite nucleus grows self-similarly along its radial direction. Although the assumption is far from engineering reality, the method can be used to analyze practical martensite growth problems.

Consider a finite elastic body containing a spherical transformation inclusion with a dilatational eigenstrain $\varepsilon_{ij}^T = \varepsilon^T \delta_{ij}$. The whole body is subjected to a uniform static load $\sigma_{ij}^o = \sigma^0 \delta_{ij}$ (Fig.3). Under the action of the applied load, the transformation inclusion expands along its radial direction self-similarly with a constant speed V. Based on the solution obtained by the authors (Wang, *et al*, 1999), the radial component of the displacement for interior points and exterior points of the transformation inclusion can be written in the following form:

$$u_r^I(t) = \frac{3\lambda + 2\mu}{3\rho a(a+V)^2}(a+2V)r\varepsilon^T + u_r^0 \tag{30}$$

$$u_r^E(t) = \frac{(3\lambda + 2\mu)\varepsilon^T}{3\rho a r^2 (a^2 - V^2)^2}[rV - a(r_0 + Vt)]^2[2rV + a(r_0 + Vt)] + u_r^0, \tag{31}$$

where $a^2 = (\lambda + 2\mu)/\rho$, λ, μ are Lame's constants of the material, ρ is the density of the material, r_0 is the initial radius of the transformation inclusion, and u_r^0 is the displacement field created by the external load in the body without the transformation inclusion. Note that the displacement components given by equations (30) and (31) are the only non-zero components for this spherically symmetric problem. By using equations (30) and (31), one can derive the strain and stress components both inside the inclusion and outside the inclusion, respectively.

After cumbersome mathematical derivation, the dynamic energy release rate can be given by equation (7) as follows,

$$\begin{aligned} g_d = -\frac{1}{2a(a-V)^2(a+V)^2\rho^2}\{\varepsilon^T(3\lambda+2\mu)[-4\varepsilon^T V^3(3\lambda+2\mu)\rho + 3a^5(\varepsilon^T - 2\varepsilon^0)\rho^2 - \\ 2a^3\rho(-6\varepsilon^0 V^2\rho + \varepsilon^T(3\lambda + 2\mu + 3V^2\rho)) + \\ a(-6\varepsilon^0 V^4\rho^2 + \varepsilon^T(3\lambda^2 + 8\lambda\mu + 4\mu^2 + 15V^2\lambda\rho + 10V^2\mu\rho + 3V^4\rho^2))]\} \end{aligned} \tag{32}$$

where ε^0 is the hydrostatic strain field in the homogeneous matrix created by the applied field $\sigma_{ij}^o = \sigma^0 \delta_{ij}$.

By setting V=0, one can recover the static energy release rate as follows,

$$g_s = \frac{3E}{(1-2\gamma)}\varepsilon^0\varepsilon^T - \frac{E}{(1-\gamma)}(\varepsilon^T)^2 \quad . \tag{33}$$

where E is the Young's modulus.

By giving the material constants as follows: $E = 165Gpa$, $\gamma = 0.25$, $\rho = 2500\ kg/m^3$, $\varepsilon^0 = 0.01$, , $\varepsilon^T = 0.04$, $r_0 = 5\times10^{-6}m$, $t = 10^{-7}s$ we find the energy release rate g_d/g_s versus the propagation velocity of the phase boundary V/a as given in Fig.4. The energy release rate decreases with increasing propagation speed of the transformation, just as in the one-dimensional case. One can also find that when the propagation speed of the transformation reaches some critical value, which is much smaller than the dilational wave speed a, the energy release rate is zero. Since there must be energy dissipation accompanying the transformation, the propagation velocity of the transformation must have an upper limit.

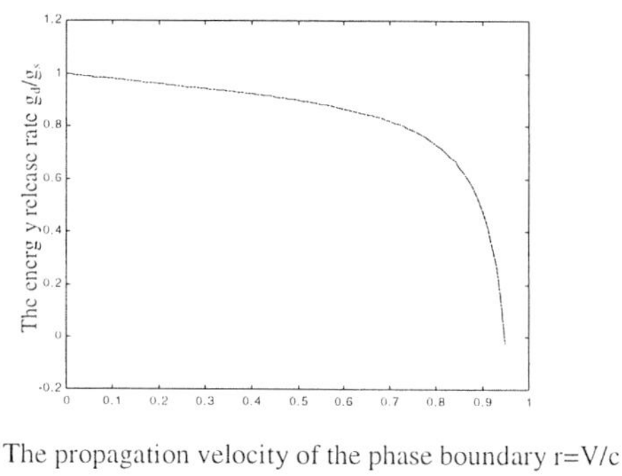

Fig.3. Dependence of the energy release rate on the velocity of propagation of the phase boundary

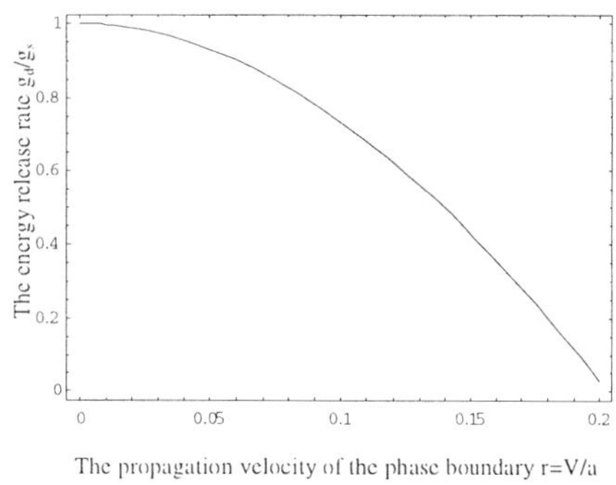

Figure 4. the energy release rate g_d/g_s versus the propagation velocity of the phase boundary V/a

3. CONCLUSION

The present investigation intends to derive the driving force, i.e., the dynamic energy release rate accompanying martensitic transformation, and discuss the effect of the propagation speed of transformation. Two examples were considered in detail: a semi-infinite rod subjected to an impact load at the end, an expanding spherical transformation inclusion in a matrix. In the two examples, we found the dynamic energy release rate decreases with increasing propagation speed of transformation, and when the speed reaches some critical value, the deriving force becomes zero: the propagation speed of martensitic transformation has an upper limit.

Acknowledgements: The authors express their gratitude to Professor Q. P. Sun of Hong Kong University of Science and Technology for his stimulating discussion of this paper. The paper was supported by the National Science Foundation of China with the Excellent Young Investigator Award.

References:

1.Nishiyama,Z., (1978), Martensitic Transformation, New York, Academic Press.
2.Escobar,J.C. and Clifton,R.J.,(1993), On Pressure-Shear Plate Impact for Studying the Kinetics of Stress-Induced Phase Transformations. *J.Mater.Sci.Engng,A170, pp.125-142*

3.Abeyaratne,R. and Knowles,J.K., (1997), On the Kinetics of an Austenite->Martensite Phase Transformation Induced by Impact in a Cu-Al-Ni Shape-Memory Alloy. *Acta Mater. Vol 45, No 4, pp.1671-1683.*

4.Chen,Y.C. and Lagoudas, D.C. (2000), Impact Induced Phase Transformation in Shape Memory Alloys. *J.Mech.Phys.Solids.Vol.48 , Issue. 2, pp.275-300*

5.Wang,B.,Sun,Q.,Xiao,Z., (1999), On the Dynamic Growth of a Spherical Inclusion With Dilatational Transformation Strain in Infinite Elastic Domain, ASME *J. Appl. Mech. Vol 66,pp.879-884.*

6.Willis,J.R.,(1965), Dislocations and Inclusions, *J.Mech.Phys.Solids.Vol.13,pp.377-395*

7.Mikata,Y., and Nemat-Nasser,S., (1988), First-Order Dynamic Eshelby Tensor. *Proceedings of 25th Annual Technical Meeting Society of Engingeering Science.* CA

8.Ball,J.M. and James, R.D., (1987), Fine Phase Mixtures as Minimizers of Energy, *Archs ration. Mech. Analysis, Vol 100,13*

9.James,R.D., (1981), Finite Deformation by Mechanical Twinning, *Archs ration. Mech. Analysis, Vol 77,143*

10.James, R.D., (1986), Displacive Phase Transformations In Solids, *J. Mech. Phys. Solids. Vol.34, pp.359*

11.Abeyaratne,R, and Knowles, J.K., (1990), On the Driving Traction on a Surface of Strain Discontinuity in a Continuum, *J. Mech.Phys.Solids.Vol.38, No.3, pp.345-360.*

12.Gurtin,M.E., (1981), *An Introduction to Continuum Mechanics*, New York, Academic Press.

13.Johnson,W., (1972), *Impact Strength of Materials*, London, Edward Arnold

MARTENSITIC TRANSFORMATIONS IN CONSTRAINED FILMS

A.L. ROYTBURD AND J. SLUTSKER*
Department of Materials and Nuclear Engineering
University of Maryland, College Park, MD 20742, USA

**Materials Science and Engineering Laboratory, National Institute of Standards and Technology, Gaithersburg, MD 20899, USA*

1. Introduction

Polydomain micro- and nano-structures are natural products of phase transformations in solids. The trend to minimize the energy of internal long-range fields (magnetic, electric or mechanical) leads to the formation of the arrangement of domains of different phases or differently oriented domains of the same phase [1-3]. Transformations of bulk materials usually result in formation of complex irregular polydomain structures, which are difficult to control. However, the advance of thin film technology makes it possible to obtain well- controlled polydomain structures with desirable properties. The objective of this paper is to show how controlled structures of elastic domain in constrained single crystalline layers are developed and how these structures can be designed by engineering of a constraint.

As an example, we consider an austenite-martensite transformation during the cooling of a film coupled with a relatively thick substrate. The phase transformation in the film leads to a characteristic bending of a film/substrate couple, which can be used as an actuator or a sensor sensitive to temperature changes [4, 5].

Thermodynamic analysis presented in [5] describes microstructure and stress evolution in the film during an actuator performance, particularly, the hysteresis between direct and reverse transformations. In this paper we use special phase diagrams constructed in coordinates misfit strain/ temperature to analyze the austenite-martensite transformation under cooling. To illustrate basic principles of the phenomenon a two-dimensional model of the transformation is considered first. The results obtained for a 2-D model are expanded to a 3-D martensitic transformation in a biaxialy constrained film. The possibility of obtaining the stress-free film at a changing temperature is discussed.

Q.P. Sun (ed.), IUTAM Symposium on Mechanics of Martensitic Phase Transformation in Solids, 121–130.

2. Formulation of the Problem

We consider a constrained crystalline film which can transform from one phase to another. The relative stability of the phases is determined by the difference of their free energies:

$$\Delta f = f_2^0 - f_1^0 \quad (1)$$

where f_1^0 and f_2^0 are the free energies of undistorted initial and product phases, respectively. Here Δf is determined by the temperature T and changes its sign at T_0, where T_0 is an equilibrium temperature (Fig.1). It is convenient to consider Δf as an "effective temperature".

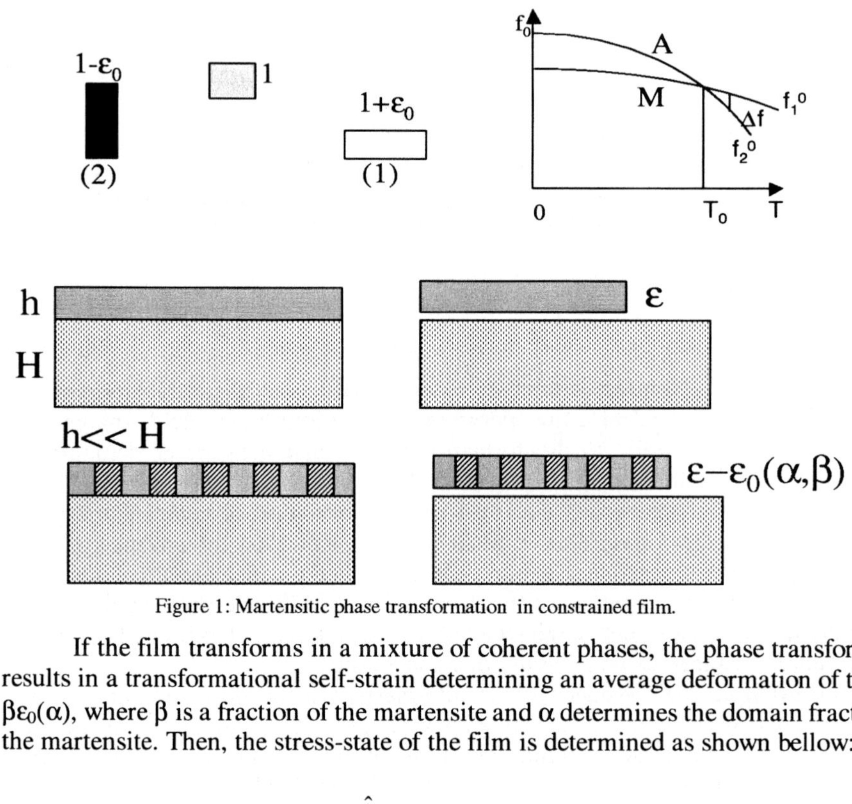

Figure 1: Martensitic phase transformation in constrained film.

If the film transforms in a mixture of coherent phases, the phase transformation results in a transformational self-strain determining an average deformation of the film $\beta\varepsilon_0(\alpha)$, where β is a fraction of the martensite and α determines the domain fractions in the martensite. Then, the stress-state of the film is determined as shown bellow:

$$\hat{\sigma} = \hat{\hat{G}}(\hat{\varepsilon} - \beta\hat{\varepsilon}_0(\alpha)) \quad (2)$$

where $\hat{\varepsilon}$ is a misfit between the substrate and the film in an austenite state and $\hat{\hat{G}}$ is an in-plane elastic modulus [1, 6]. The equilibrium microstructure is determined by the minimum of the free energy of the two-phase film:

$$f(\beta,\alpha)=\beta\Delta f+\frac{1}{2}\hat{\hat{G}}(\varepsilon-\beta\varepsilon_0(\alpha))^2+e_{\text{int}}+O(\sqrt{l/h}) \tag{3}$$

where the second term is a constraint energy due to the film/substrate elastic interaction, the third term is an energy of internal stresses due to incompatibility between the phases in the film and the last term includes the energy of interfaces which can be neglected if the film is not very thin. The thermodynamic analysis leads to the conclusion that under mechanical constraint a typical equilibrium microstructure of the film is a polydomain, i.e. an alternation of the plane-parallel layers, or domains, of the parent and product phases with a special crystallographic orientation of interfaces between domains [6]. The relative fractions of the domains of the different phases as well as twin-domains in the martensite are determined by the equilibrium equation :

$$\partial f/\partial\beta=\partial f/\partial\alpha=0 \tag{4}$$

and depend on external conditions (an average strain and temperature).

The task is to find the evolution of an equilibrium microstructure under cooling and heating of the relatively thin film on the thick substrate. It corresponds to the usual experimental studies of the performance of SMA actuators.

3. Two- Dimensional Model of Martensitic Transformation in a Constrained Film

To emphasize the principal characteristics of martensitic transformation in a constrained film consider a two-dimensional model of the transformation of a phase with a square lattice ("austenite") to a phase with a rectangular lattice ("martensite") (Fig.1). A symmetry relation between the phases dictates a formation of two equivalent martensite variants (twin domains) with the self-strains:

$$\hat{\varepsilon}_0^{\,1}=\varepsilon_0\begin{pmatrix}1 & 0\\ 0 & -1\end{pmatrix};\quad \hat{\varepsilon}^2=\varepsilon_0\begin{pmatrix}-1 & 0\\ 0 & 1\end{pmatrix} \tag{5}$$

The constraint elastic energy can be diminished if the mixture of the two domains is formed with an average self-strain of the martensite

$$\hat{\varepsilon}_0(\alpha)=(1-\alpha)\hat{\varepsilon}_0^{\,1}+\alpha\hat{\varepsilon}_0^{\,2}=\varepsilon_0\begin{pmatrix}1-2\alpha & 0\\ 0 & -(1-2\alpha)\end{pmatrix} \tag{6}$$

where α is a fraction of the domain 2. The free energy of the constrained film in an austenite-martensite state is

$$f(\alpha,\beta)=f_1^0+\beta\Delta f+\frac{1}{2}E[\varepsilon-\beta(1-2\alpha)\varepsilon_0]^2 \tag{7}$$

where E is the Young's modulus. The internal energy in Eq.(7) is absent because the phases and the domains in this example are compatible. Minimization of the free energy ($\partial f/\partial\alpha=\partial f/\partial\beta=0$) leads to a different equilibrium microstructure at $\Delta f<0$ ($T<T_0$) and $\Delta f>0$, $T>T_0$.

If $\Delta f<0$, the minimum of the free energy corresponds to the maximum of β, i.e. $\beta=1$. Then $f(\beta=1,\alpha)=f_2^0+\frac{1}{2}E(\varepsilon-(1-2\alpha)\varepsilon_0)^2$ and at equilibrium $\alpha_0=\frac{1}{2}(1-\frac{\varepsilon}{\varepsilon_0})$, $\sigma=E(\varepsilon-(1-2\alpha_0)\varepsilon_0 = 0$, $f=f_2^0$ if $-\varepsilon_0<\varepsilon<\varepsilon_0$, while $\alpha=0$ or 1 and $f=f_2^0+\frac{1}{2}E(\varepsilon-\varepsilon_0)^2$ if $|\varepsilon|>\varepsilon_0$.

If $\Delta f>0$, the minimum of the free energy (Eq.(7)) is obtained for minimum β and maximum $(1-2\alpha)$, i.e. $\alpha=0$ for $\varepsilon>0$ or $\alpha=1$ for $\varepsilon<0$. The free energy (7) is reduced to

$$f(\beta)=f_1^0+\beta\Delta f+\frac{1}{2}E(\varepsilon-\beta\varepsilon_0)^2 \tag{8}$$

and it has the minimum at

$$\beta_0=-\frac{\Delta f}{E\varepsilon_0^2}+\frac{\varepsilon}{\varepsilon_0} \tag{9}$$

(it is considered $\varepsilon>0$, hereafter)

The equations, $\beta_0=0$ and $\beta_0=1$, determine the boundaries of the two-phase areas on Δf-ε diagram (Fig.2). The line $\varepsilon/\varepsilon_0=\Delta f/E\varepsilon_0^2$ corresponds to the onset of the austenite-martensite transformation; the line $\varepsilon/\varepsilon_0=\Delta f/E\varepsilon_0^2+1$ shows the end of the transformation. Inside of the two-phase region of the phase diagram. the stress is $\sigma=\Delta f/\varepsilon_0$, while outside of this region $\sigma=E\varepsilon_0$ in the austenite region, and $\sigma=E(\varepsilon-\varepsilon_0)$ in the martensite region.

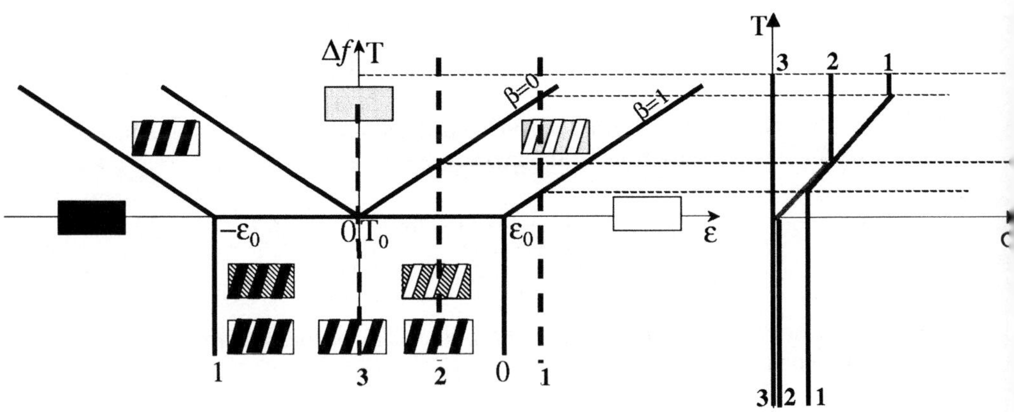

Figure 2: Phase diagram and temperature dependence of stress at constant misfit.

Thus, under cooling the intermediate two-phase polydomain state containing the initial austenite and one of the variants of the martensite, is formed at the beginning of transformation. Then, the transformation completes with the formation of a single domain martensite if $\varepsilon > \varepsilon_0$. At the beginning of the transformation $df/dT=0$, i.e., the point of the first order unconstrained transformation splits into the two second order transformation points. For constraint $\varepsilon<\varepsilon_0$ the residual austenite transforms discontinuously into the polytwin martensite at $\Delta f=0$. An "eutectic" microstructure arises which then is reconstructed in a stable uniform polytwin structure.

The thermodynamics analysis used for construction of the phase diagram above, as well as in the following section, contains some assumptions which need to be discussed:

1. The elastic moduli of the phases are assumed to be identical. This assumption simplifies calculations but does not affect the principal results obtained.
2. Polydomain phases are considered as homogeneous, i.e. the domain period, D, is much smaller than a film thickness, h. Then, the energy of interdomain interfaces and the energy of non-uniform microstresses near the film/substrate interface can be neglected. These two energies form together an effective interface energy, which has minimum $\Gamma \cong e_0 D_0$ at an equilibrium domain period, $D_0 = \sqrt{lh}$ (where $l \cong \gamma / E\varepsilon_0^{\,2}$ is a microscopic characteristic length and γ is a specific surface energy of the interdomain interfaces). Therefore, the contributions of the interface energy and the volume elastic energy to the energy density are in the ration or $\sqrt{l}/\sqrt{h}$ [6] that means the interface energy can be neglected when $D_0/h \ll 1$. This assumption corresponds to the common approximation of homogenization theories.
3. The self-strains are small, so geometric non-linear effects can be neglected. The average strain of polydomain phase can be considered as an additive sum of the domain self-strains (Eq.(6)) only if the non-linear terms which is result of the mutual rotations of the domains are neglected.

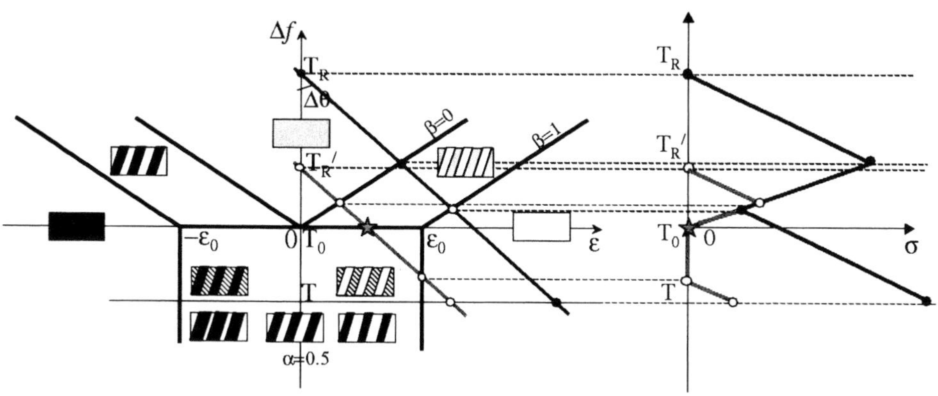

Figure 3: Evolution of stress in the film with changing temperature at cooling.

The phase diagram allows one to follow the evolution of stress and microstructure in the film during cooling (or heating). The change of misfit strain with temperature is presented on the diagram by a straight line

$$\varepsilon=\Delta\Theta(T-T_R) \tag{10}$$

where $\Delta\Theta$ is a difference of coefficient of thermal expansions of a film and a substrate. The misfit ε is absent at the temperature T_R (T_R is close to the temperature of the deposition of a film). Along this line the stress is $\sigma=E\varepsilon$ in the austenite region of the diagram, $\sigma=\Delta f/\varepsilon_0$ in the two-phase region and $\sigma=0$ in the polytwin region (Fig.3).

Two different scenarios of evolution of microstructure are possible. If $\varepsilon(T_0)=\Delta\Theta(T_0-T_R)>\varepsilon_0$, the stress-induced martensite transformation proceeds through the formation of the two-phase state which transfers to a single domain martensite at $T>T_0$. The stress is always positive (Fig.4a).

If $\varepsilon(T_0)<\varepsilon_0$, the intersection of the line $\varepsilon(T)$ with the axes $\Delta f=0$ corresponds to the abrupt transformation of the remaining austenite into the "eutectic" mixture of the martensite domains. The stress reaches zero and zero level is maintained until the two-domain state transforms to a single domain one. This "invar" stress-free state is a result of a compensation of the thermal induced misfit between the film and the substrate by the changing self-strain of the polytwin martensite phase (Fig.4b).

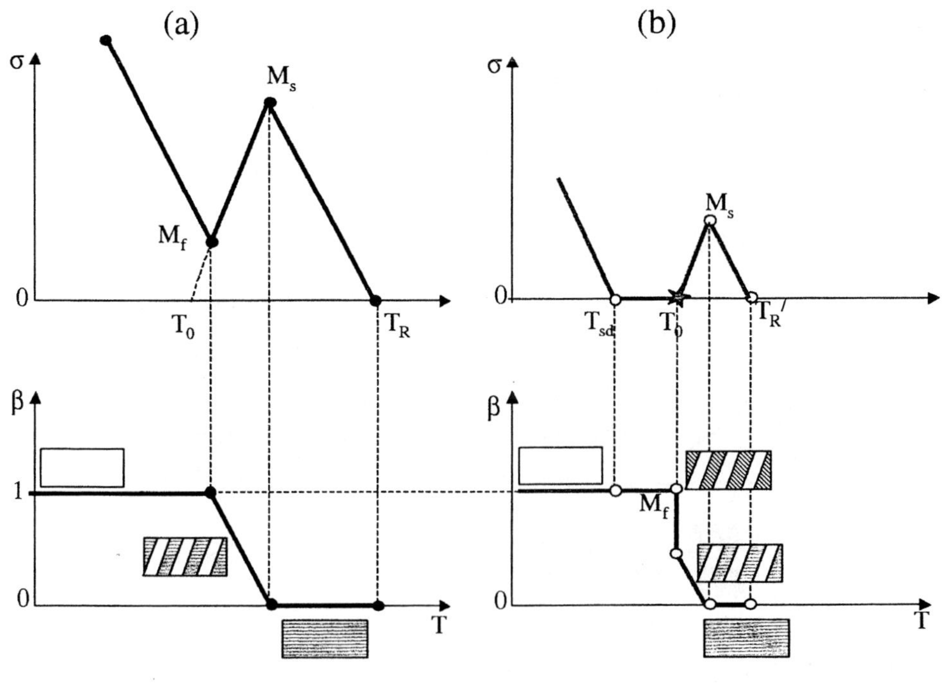

Figure 4: Microstructure and stress evolution in film at cooling.

4. Cubic-Tetragonal Martensitic Transformation in a Constrained Film

As known, many martensitic transformations between close-packed structures can be modeled as a cubic-tetragonal transformation. The three domains of a tetragonal martensite correspond to the following self-strains (Fig.5):

$$\hat{\varepsilon}_0^1 = \varepsilon_0 \begin{pmatrix} 2 & 0 & 0 \\ 0 & -1 & 0 \\ 0 & 0 & -1 \end{pmatrix};\ \hat{\varepsilon}_0^2 = \varepsilon_0 \begin{pmatrix} -1 & 0 & 0 \\ 0 & 2 & 0 \\ 0 & 0 & -1 \end{pmatrix};\ \hat{\varepsilon}_0^3 = \varepsilon_0 \begin{pmatrix} -1 & 0 & 0 \\ 0 & -1 & 0 \\ 0 & 0 & 2 \end{pmatrix} \tag{11}$$

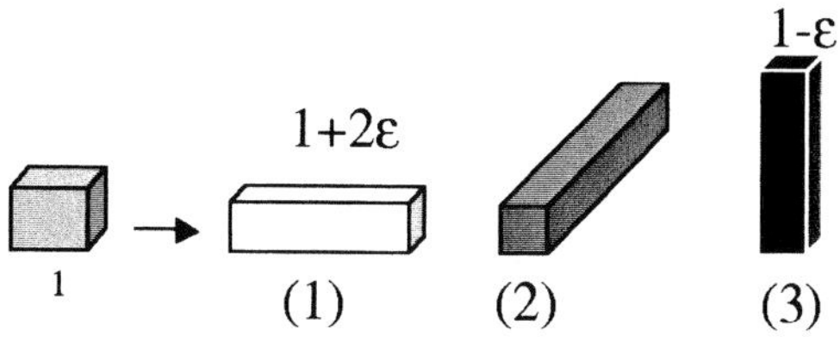

Figure 5: Domains of tetragonal phase.

For simplicity, a transformation without a volume change is considered. We will show that this transformation in an isotropically extended film can be described by using a phase diagram, which is very similar to the one considered above for a 2-D transformation.

The free energy of the two-phase microstructure is as following

$$f(\beta,\alpha) = f_1^0 + \beta\Delta f + \frac{E}{1-\nu}(\varepsilon - \beta\varepsilon_0\ (\alpha))^2 \tag{12}$$

where the constraint energy is expressed through the isotropic biaxial stress, $\sigma = \frac{E}{1-\nu}(\varepsilon - \beta\varepsilon_0(\alpha))$, with in-plane components of misfit, ε, and an average self-strain $\beta\varepsilon_0(\alpha)$, ν is the Poisson's ratio. Due to isotropy of the misfit the self-strain corresponding to an equilibrium has to be isotropic also and the microstructure should contain equal fractions of domain (1) and (2) as follows

$$\hat{\varepsilon}_0(\alpha) = (1-\alpha)\frac{1}{2}(\hat{\varepsilon}_0^1 + \hat{\varepsilon}_0^2) + \alpha\hat{\varepsilon}_0^3 \tag{13}$$

with in-plane components $\varepsilon_{11}(\alpha) = \varepsilon_{22}(\alpha) = \varepsilon_0(\alpha) = \frac{1}{2}(1-3\alpha)\varepsilon_0$ where α is a fraction of domain 3. As in the 2-D case, the internal elastic energy is absent, because the martensite phase forms polytwin plates compatible to the austenite.

If $\Delta f<0$, the minimum of the free energy corresponds to the maximum $\beta=1$, and at the equilibrium $\alpha_0 = \frac{1}{3} - \frac{2}{3}\frac{\varepsilon}{\varepsilon_0}$ and the stress equals to zero if $-\varepsilon_0 < \varepsilon < \frac{1}{2}\varepsilon_0$ ($0<\alpha<1$).

If $\Delta f>0$, the minimum of the free energy is obtained for maximum of $(1-3\alpha)$, i.e. when $\alpha=0$ ($\varepsilon>0$). At the equilibrium fraction, $\beta_0 = -2(1-\nu)\frac{\Delta f}{E\varepsilon_0^2} + 2\frac{\varepsilon}{\varepsilon_0}$, the stress is $\sigma = \Delta f / \varepsilon_0$. The equations, $\beta_0=0$ and $\beta_0=1$, determine the borders of two-phase area on Δf-ε diagram (Fig. 6a).

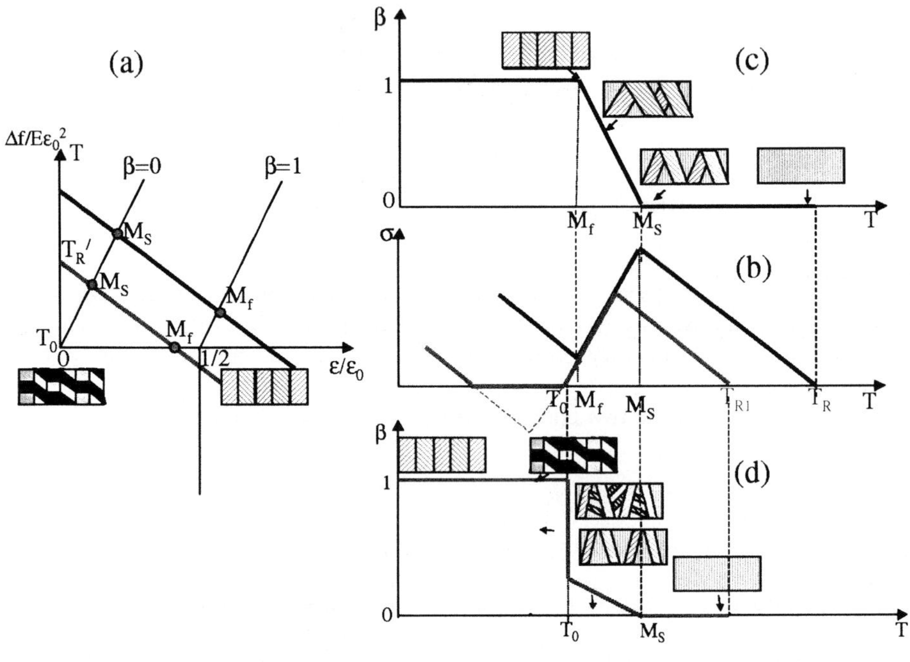

Figure 6: Stress and microstructure evolution in the constrained film under cooling. Phase diagram (a), temperature dependence of stress (b) and microstructure evolution for $\varepsilon(T_0)=\Delta\Theta(T_0-T_R)>\varepsilon_0$ (c) and $\varepsilon(T_0)=\Delta\Theta(T_0-T_R)<\varepsilon_0$ (d).

Two scenarios are possible at cooling the film/substrate couple as in the 2-D case (Fig.6b). If $\varepsilon(T_0)=\Delta\Theta(T_0-T_R)>\varepsilon_0$, the stress-induced martensite transformation proceeds through the formation of the two-phase state containing the polytwin martensite plates formed by domains 1 and 2. The final product of transformation is a martensite consisting of equal fractions of domains 1 and 2. The stress is always positive (Fig.6c).

If $\varepsilon(T_0)=\Delta\Theta(T_0-T_R)<\varepsilon_0$, the stress in the film is vanished before the transformation is complete. The remaining austenite transforms to the "eutectic" mixture of martensite plates containing all three domains with zero in-plane components of the average self-strain. The fraction of domain 3 decreases with temperature maintaining zero level of the stress until the domain 3 disappears (Fig. 6d).

5. Discussion and Conclusions

The existence of two different paths of microstructure evolution during martensitic transformations in constrained film is the main conclusion that we would like to note in this paper. The realization of one of them depends on the relative position of the temperature of phase equilibrium, T_0, and the temperature of a reference stress-free state, T_R, which usually close to the temperature of deposition or annealing.

If the temperatures T_0 and T_R are far from each other, the transformation under cooling is developed being induced by the thermal stresses due to the difference of thermal contractions of the film and the substrate. The equilibrium microstructure contains only restricted number of martensite domain variants dictated by the film stress state (e.g. for the cubic tetragonal transformation in the (001) film considered in the section 4, these are domain (1) and (2)). The stress in the film decreases with progressing transformation but does not reach a zero level.

If T_R and T_0 approach each other the complete relaxation of thermal stresses becomes possible, and the stress-induced regime of transformation changes to the stress-assistant one that should proceed with the minimum of an average self-strain in the film plane. At this state of the transformation all possible variants of the martensite domain can be involved in the microstructure development. As a result, a very flexible martensitic film is formed that enables to demonstrate the “invar” behavior.

These theoretical predictions are supported by experimental studies including recent observations of invar effect in Ni-Ti-Pd films [7]. The analysis of available experimental data allows us to conclude that the thermodynamical theory can explain the main peculiarities of martensitic transformations in constrained films.

We have considered here a simple case of constrained martensite transformation when the phases remain compatible during cooling. As has been shown in [5] the phases become incompatible during the reverse transformation at heating as well as at the transformation in the film under contraction. The interactions of incompatible plates determine the internal elastic energy of the film (term e_{int} in Eq.(3)). Including this energy and taking into account the crystallographic characteristic of specific martensitic transformations, the theory will be able to describe quantitatively the transformations in single crystalline and highly textured constrained films provided the films are sufficiently perfect to form near-equilibrium martensitic microstructures.

Acknowledgments

We are gratefully acknowledging the support of AFOSR (grant # F496-200-110-110) and NSF (grant # DMR 990-32-79). Dr. Slutsker is grateful for the support of National Research Council Postdoctoral Assistantship.

References

1. Roitburd, A.L. : Martensitic transformation as a typical phase transformation in solids. *Solid State Physics* **vol.33**., eds. H.Ehrenreich, F.Seitz and D.Turnbull , Academic Press, New York, 1978, pp. 317-390.

2. Khachaturyan, A.G: *Theory of Structural Transformation in Solids*, John Wiley & Sons, New York, 1983.

3. Roytburd, A.L: Elastic domains and polydomain phases in solids, *Phase Transitions*, **45**, .(1993) , 1-33.

4. George, E.P., Gotthardt R., Otsuka K, Troiler-McKinstry, S. and Wun-Fogle, M. (eds) ., Materials for Smart System II, MRS Proc , vol.459, MRS, Pittsburgh, 1997.

5. Roytburd, A.L., Kim, T.S., Quanmin Su, Slutsker, J. and Wuttig, M.: Martensitic Transformation in Constrained Films, *Acta Materilia*, **46**, (1998), 5095-5107.

6.Roytburd, A.L., Thermodynamics of Polydomain Heterostrucutres. Parts I and II., *J.Appl. Phys*. **83**, (1998), 228-250.

7. Mathews, S.A., Processing and in-situ Characterization of Thin Films (Ti-Ni, Ti-Pd) Shape Memory Allows., *Ph.D. Thesis*, University of Maryland, 2000.

INFLUENCE OF MICROSTRUCTURAL PARAMETERS ON SHAPE MEMORY ALLOYS BEHAVIOR

E. PATOOR, C. NICLAEYS, S. ARBAB CHIRANI, T. BEN ZINEB
Laboratoire de Physique et Mécanique des Matériaux (UMR CNRS 7554)
ENSAM Metz, 4, rue Augustin Fresnel, 57070 Metz, France.
e-mail : etienne.patoor@metz.ensam.fr

1. Introduction

Modeling the functional behavior of shape memory alloys (SMA) is a challenge for the development of industrial applications using these materials. Modern concepts of scale transition and homogenization methods have been developed these last years in mechanics of materials. Several works had established that these concepts are very suitable to model the thermomechanical behavior of SMAs (Ono & Sato, 1988, Sun & Hwang, 1993, Patoor *et al.*, 1988, Gall & Sehitoglu, 1999). Strong predictive capacities are related to these approaches based upon a description of physical mechanisms at the origin of the observed macroscopic behavior and where kinematical description and thermodynamical analysis are performed at a so-called microscopic level which is the pertinent scale to describe strain mechanisms.
One advantage in these frameworks is their ability to give an estimation of the internal stress field building by strain incompatibilities. Two major sources of incompatibilities are found in SMAs. The first one deals with the nucleation and growth of several variants of martensite inside a unique parent phase crystal. The second is related to the granular structure of polycrystalline materials. Strain incompatibilities from grains to grains build an intergranular strain field added to the intragranular one originated from the first source. Due to its displacive character, the martensitic transformation is very sensitive to the stress state and internal stress field have to be taken into account for behavior modeling. Two successive homogenizations are then required considering the two representative scales involved.
In this paper, the influence of the microstructural parameters associated to these two representative scales is underlined. In the first part we examine, at the parent phase single crystal scale, the relation between the crystallographical modeling of the transformation strain and the determination of the interaction matrix. This point is of importance because the interaction matrix plays an important role into the description of the SMA behavior. The particular case of NiTi alloys is addressed in this section. In the second part we deal with the influence of the initial crystallographical texture on the overall behavior of polycrystalline materials. In this part, results obtained on CuZnAl alloy for different crystallographic textures are discussed.

Q.P. Sun (ed.), IUTAM Symposium on Mechanics of Martensitic Phase Transformation in Solids, 131–138.

2. Interaction energy in single crystal

In scale transition scheme, the assumption of the existence of a Representative Volume Element (RVE) is a key point. To compute the overall response of polycrystalline materials, grains are very often chosen as RVE and kinematical and thermodynamical analysis are then performed at this scale, assuming that grains in polycrystal behave as single crystals. Thermodynamical potential over the RVE is composed by chemical, interfacial and elastic energies (Ortin & Planes, 1988, Patoor *et al.*, 1988). Chemical energy depends to the temperature and to the volume fraction of martensite. A linear approximation around the equilibrium temperature is commonly used. Interfacial energy is often considered as negligible with respect to the other contributions, due to the oblate shape of martensite plates. The determination of the elastic energy is much more complex, because we have to consider the non uniformity in the transformation strain field associated to the nucleation and growth of several variants of martensite inside a same parent phase single crystal of volume V. The existence of this microstructure built an internal stress field denotes by $\tau(r)$ and split the elastic energy into two contributions: one related to the applied stress σ and one, called the interaction energy, associated to the internal stress field (Mura, 1987, Patoor *et al.,* 1997).

$$W_{elas} = \frac{1}{2}\sigma_{ij}\, M_{ijkl}\, \sigma_{kl} - \frac{1}{2V}\int_V \tau_{ij}(r)\,\varepsilon_{ij}^T(r)\, dV \qquad (1)$$

In equation 1, the four-order tensor M, denotes the elastic stiffness, which is assumed in this work to be uniform in the two phases, and $\varepsilon^T(r)$ is the transformation strain field. The evaluation of the interaction energy is the keystone of the thermomechanical analysis. Neglecting this contribution leads to a large underestimation of the applied stress required to transform the material and an overestimation of the macroscopic transformation strain (Patoor *et al.,* 1997). According to previous micromechanical analysis (Fassi Fehri *et al.*, 1987, Agouram *et al.*, 1998, Siredey *et al.*, 1999), this energy, W_{int} can be approximated by a quadratic function of the volume fractions f^n of the different variant of martensite and requires the definition of an interaction matrix H^{nm} which gives the nature of the interaction (compatible or incompatible) between the different variants involved.

In these analyses the transformation strain associated to a given variant is considered as a stress free strain and according to the definition of habit plane variant (HPV) given in the WLR theory (Wescher *et al.*, 1953) it is expressed as:

$$\varepsilon_{ij}^{Tn} = \frac{1}{2} g \left(n_i\, m_j + n_j\, m_i\right) \qquad (2)$$

where vector **n** denotes the normal to the habit plane, **m** gives the transformation direction and g is the transformation displacement (see table 1 for NiTi alloy data). From this definition, the transformation strain field is considered as a piecewise uniform function having a constant value ε^{Tn} inside each variant. The interaction energy by unit volume, may be expressed as:

$$W_{int} = -\frac{1}{2V}\int_V \tau_{ij}(r)\,\varepsilon_{ij}^{T}(r)\,dV \;=\; -\;\frac{1}{2V}\sum_{n} \bar{\varepsilon}_{ij}^{n}\,\bar{\tau}_{ij}^{n}\,V^{n} \qquad (3)$$

where $\bar{\tau}^{n}$ and $\bar{\varepsilon}^{n}$ are average values of internal stress and transformation strain over the volume V^{n} occupied by a mixture of austenite and variant n. Siredey *et al.*, (1999) proposed to evaluate this quantity using the interfacial operator technique. According to Walpole (1967) and Hill (1983), across an interface between two domains n and m, having the same elastic constants, jumps in the stress and strain tensors are related by an interfacial operator called Q^{nm} which is a function to the normal of the interface and to the elastic constants (see Siredey *et al.*, 1999 for details):

$$\bar{\tau}_{ij}^{n} \;=\; \bar{\tau}_{ij}^{m} \;-\; Q_{ijkl}^{nm}\left(\bar{\varepsilon}_{kl}^{n} - \bar{\varepsilon}_{kl}^{m}\right) \qquad (4)$$

From equations (3) and (4), interaction energy can be expressed as (with $F^{n} = V^{n}/V$):

$$W_{int} \;=\; -\frac{1}{2}\sum_{n,m} F^{n}\,F^{m}\,\left(\bar{\varepsilon}_{ij}^{n} - \bar{\varepsilon}_{ij}^{m}\right)Q_{ijkl}^{nm}\left(\bar{\varepsilon}_{kl}^{n} - \bar{\varepsilon}_{kl}^{m}\right) \qquad (5)$$

From the observation that in polycrystalline material stress-induced transformation develops mainly two variants inside each grain, the minimization of the interaction energy is performed considering only interactions between couple of variants:

$$W_{int} \;=\; -\frac{1}{2}F^{n}\,F^{m}\,\left(\bar{\varepsilon}_{ij}^{n} - \bar{\varepsilon}_{ij}^{m}\right)Q_{ijkl}^{nm}\left(\bar{\varepsilon}_{kl}^{n} - \bar{\varepsilon}_{kl}^{m}\right) \qquad (6)$$

Minimization of expression (6) for every couple of two variants allows defining a 24x24 interaction matrix taking into account all the possible interactions between two variants. Two kinds of interaction are obtained: compatible interaction when the minimum value takes by W_{int} is close to zero and incompatible one, when this value is large. This determination is very suitable for Cu-based shape memory alloys and gives good agreement when it is introduced into a self-consistent scheme to compute the overall behavior of these alloys. At the crystal scale, the four-variants self-accommodating groups experimentally observed are also determined in that way for CuZnAl and CuAlBe alloys. This interaction matrix was introduced in several crystallographical approaches (Gall & Sehitoglu, 1999, Gao *et al.*, 2000). But unfortunately experimental observations disagree with this interaction matrix in the very important case of NiTi alloys. In these alloys, 48 self-accommodated groups are observed, each one grouping three variants around a (0 0 1) pole (Miyazaki *et al.*, 1989) and most of these groups are found to involve incompatible interaction in the sense gives by the minimization of equation 6 (see figure 2). So it is no longer possible to use this expression to model the interaction energy in this class of alloy.

1	2	3	4	5	6	7	8	9	10	11	12
1(+)	1(-)	1’(+)	1’(-)	2(+)	2(-)	2’(+)	2’(-)	3(+)	3(-)	3’(+)	3’(-)
1-2	1-2’	1’-2	1’-2’	2-1	2-1’	2’-1	2’-1’	3-4	3-2’	3’-4	3’-4’

13	14	15	16	17	18	19	20	21	22	23	24
4(+)	4(-)	4’(+)	4’(-)	5(+)	5(-)	5’(+)	5’(-)	6(+)	6(-)	6’(+)	3’(-)
4-3	4-3’	4’-3	4’-3’	5-6	5-6’	5’-6	5’-6’	6-5	6-5’	6’-5	6’-5’

$$\mathbf{n}: \begin{pmatrix} -0.8888 \\ 0.4044 \\ 0.2150 \end{pmatrix} \qquad \mathbf{m}: \begin{pmatrix} 0.4315 \\ 0.4874 \\ 0.7574 \end{pmatrix} \qquad g = 0.13$$

Table 1. Crystallographical data for the martensitic transformation in NiTi alloy (Matsumoto *et al.*, 1987), and the HPVs/CVs correspondence (Madangopal *et al.*, 2000)

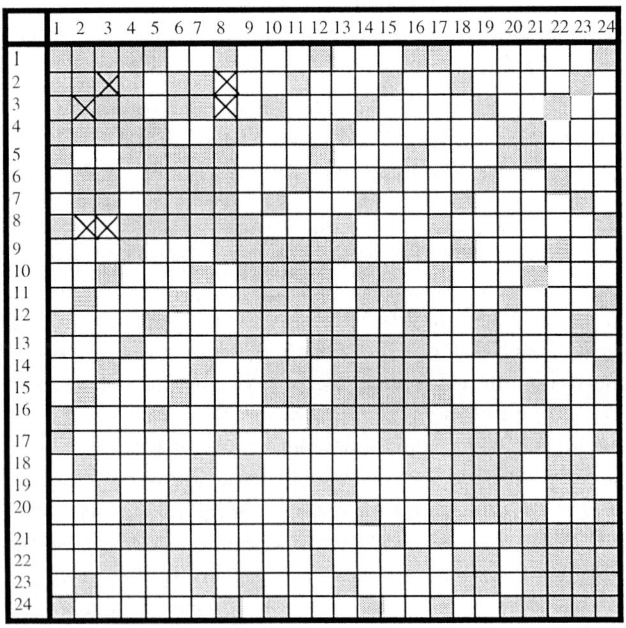

Figure 2. Interaction matrix for Cu-based SMAs, gray boxes means compatibility between two variants. A similar result was obtained for NiTi alloy when only HPVs are considered. In this case, many self-accommodating groups are not accounted. Boxes marked with crosses gives the example of the self-accommodated group 1(-) 1’(+) 2’(-) experimentally observed but not described in that way.

In order to find a correct approximation for the interaction energy in NiTi, we propose to reexamine this problem starting from crystallographic considerations. The WLR theory was originally proposed to model the martensitic transformation in ferrous alloys (Fe-Ni, Fe-C) where the strain accommodation between the parent austenitic phase and the Bain strain in the martensite is assumed to be made by a lattice invariant strain (LIS), which is a plastic shear. According to this phenomenological crystallographical theory, the shape strain matrix associated to a martensitic variant is given by:

$$S = \Phi \; P \; B \tag{7}$$

B is the lattice distortion matrix, the so-called Bain strain; P is the LIS matrix and Φ is a lattice rotation matrix. B is related to the lattice parameters of parent and product phases and to the lattice correspondence. P is defined in such a way the strain accommodation occurs along the habit plane. In the phenomenological theory this condition is achieved when:

$$\det\left(F\,F^{t} - I\right) = 0 \tag{8}$$

where F is defined by $F = P\,B$. Equation (8) allows to determine the habit plane normal **n**, the shape strain direction **m** and the magnitude g of the shear used in relation (2) to define the transformation strain in so-called habit plane variant (HPV). This theory was extended to Cu-based SMA where the microscopic accommodation was obtained through stacking faults and to NiTi where it was produced by twinning. In this last case, resolution of equation (8) gives the twin width and NiTi HPVs result from the combination of two domains having different Bain strain which are twin related and are called correspondence variant (CV). The notion of HPV derived from this scheme is in many works considered as the pertinent microscopic scale to describe the martensitic transformation in scale transition approaches. In fact, good results were obtained in that way to describe the superthermoelastic behavior in copper-based alloys and also TRIP effect in steels. In these alloys, the LIS occurs in a homogeneous way inside HPVs. This is no longer the case for twinning, which gives a microstructural pattern and produces a heterogeneous inelastic strain field inside the HPV. In Ni-Ti alloys accommodation by microdomains in the sense given by Roytburd (1993) has to be considered. To verify the microscopic heterogeneity is responsible to the discrepancy observed with the interaction matrix in NiTi alloys, we propose to determine the interaction energy between two NiTi HPVs considering each HPV is composed by its two CVs.

In equation (6), this new description of the microstructure modify the physical meaning given to strain tensors $\overline{\varepsilon}^{n}$ and $\overline{\varepsilon}^{m}$. These are now considered as the mean value of the transformation strain over the mixture formed by the combination of the two CVs that compose the HPV considered:

$$\overline{\varepsilon}^{n}_{ij} = \theta\, \varepsilon^{n1}_{ij} + (1-\theta)\, \varepsilon^{n2}_{ij} \quad \text{and} \quad \overline{\varepsilon}^{m}_{ij} = \theta\, \varepsilon^{m1}_{ij} + (1-\theta)\, \varepsilon^{m2}_{ij} \qquad (9)$$

In equations (9), ε^{n1} and ε^{n2} are the strain tensors of the CVs that compose the HPV n (resp. ε^{m1} and ε^{m2} for HPV m) and θ is the twin ratio (table 2). Interfacial operator Q^{nm} involved in relation (6) keeps the same meaning. A new interaction matrix is defined in that way in the NiTi case from the minimization of expression (6) with respect to equation (9). In this matrix new compatible solutions are obtained and all the 48 three-variants self-accommodated groups are now described (figure 3).

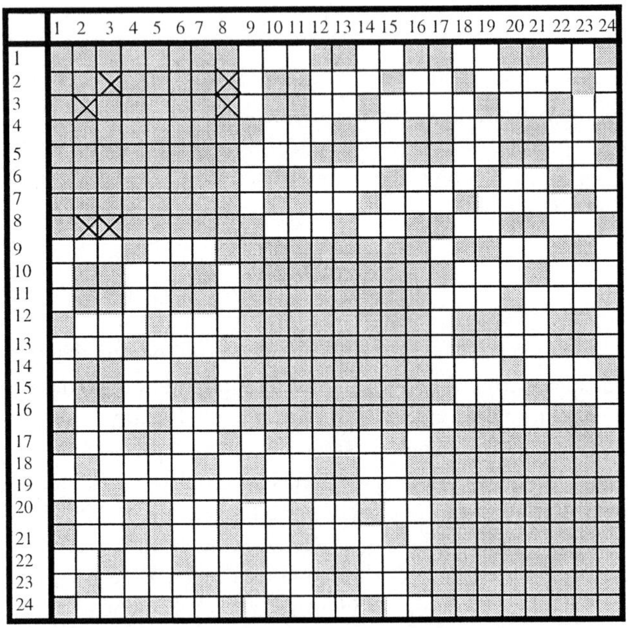

Figure 3. Interaction matrix determined in the NiTi case considering the existence of two CVs inside each HPV. All the three-variants self-accommodated groups are now accounted.

$$\overline{\varepsilon}^{n1} = \begin{pmatrix} -4.09 & 4.35 & -4.35 \\ 4.35 & 2.72 & -6.03 \\ -4.35 & -6.03 & 2.71 \end{pmatrix} 10^{-2}$$

Table 2. Green-Lagrange strain tensor for CVs in Ni-Ti alloy (Madangopal *et al.*, 2000)

3. Texture effect in superelasticity

In polycrystalline material second major source of incompatibility in the strain field comes from the granular structure itself. Bhattacharya & Kohn (1996) have underlined

the effect of texture on the occurrence of the SMA effect. In this paper we show it is also important on the mechanical behavior using a self-consistent approximation to compute the overall response. In this scheme, the transformation mechanism is defined at the RVE scale (see previous section) and the granular structure must be described. Each grain is represented by its shape, its volume fraction and its crystallographic orientation, given by the three Euler's angles (see Patoor *et al.*, 1997 for details). In this section we address the shape of transformation surface in the stress space as a function of initial texture and we compare the effect of three different one on a CuZnAl alloys.
The shape of the transformation surface is a very important feature for the mechanical behavior associated to it. Behaviors used in structure calculation code are often based upon transformation criterion defined from this surface (Gillet et al. 2000).
Experimental determination of this surface is very complex and numerical simulation is very useful. In order to determine the influence of the initial crystallographical texture we compute it for an isotropic texture having a random distribution of the grains orientation, a rolling one and a drawing one. In these last two cases, initial orientation of grains is rotated by a 50% plastic strain.
For these three textures the transformation surface is defined at the same strain level, varying the loading direction (figure 4), 140 different loadings are compute for each surface. For all textures there is no symmetry according to the tension or compression tests. Important shape changes were observed according to the initial crystallographic texture. This clearly indicates that the orientation of austenitic grains strongly influence the way the different grains transform and by consequence the kinetics of the martensitic transformation. This kind of analysis allows the optimization of the texture according to the loading conditions imposed.

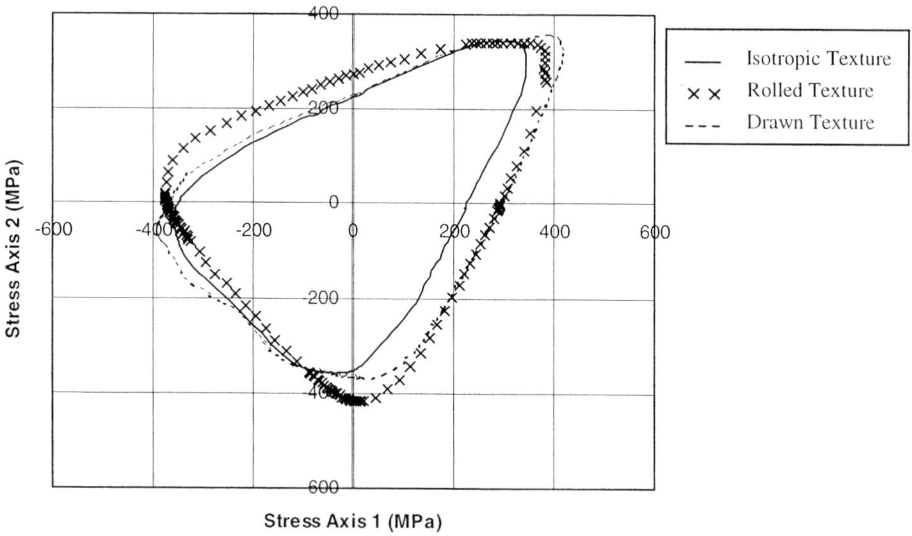

Figure 4. Influence of the initial crystallographic texture on the transformation surface defined at 0.4 % transformation strain for a CuZnAl superelatic alloys (M_s = -100 C).

4. Conclusion

The two examples presented in this paper illustrate the great sensibility of SMA behavior to microstructural parameters as the transformation strain of the martensitic domains (HPVs or CVs) and the orientation of the crystal lattice of the grains with respect to the mechanical loading axis.
It is remarkable to see how a little difference, at the microscopic scale, between the way the martensitic transformation proceed in Cu-based and Ni-Ti SMAs strongly modify the modeling approach at more macroscopic scales.

References

Agouram, S., Abdou, L. and BenSalah, M.O. (1998) Comportement thermomecanique en plasticite de transformation, *Eur. Phys. J. AP*, 341-346.

Bhattacharya, K and Kohn, R.V (1996), Symmetry, Texture and the Recoverable Strain of Shape Memory Polycrystals, *Acta mater.*, **44**, 529-542.

Fassi Fehri, O., Hihi, A. and Berveiller, M. (1987) Elastic interactions between variants in pseudoelastic single crystals. *Scripta Met.* ***21**, 771.*

Gall, K and Sehitoglu, H. (1999) The role of texture in tension-compression asymmetry in polycrystalline NiTi, *International Journal of Plasticity*, **15**, 69-92.

Gao, X., Huang, M. and Brinson, C. (2000) A multivariant micromechanical model for SMAs. Part I. Crystallographic issues for single crystal model. *International Journal of. Plasticity*, **16**, 1345-1369.

Gillet, Y., Patoor, E. and Berveiller, M. (1998) Calculation of pseudoelastic elements using a non-symetrical thermomechanical transformation criterion and associated rule, *Journal of intelligent material systems and structures*, **9**, 366-378.

Hill, R. (1983) Interfacial operators in the mechanics of composite media, *J. Mech. Phys. Solids*, **31**, 347-357.

Madangopal, K. and Singh, J.B. (2000) A novel B19' martensite in nickel titanium shape memory alloys, *Acta mater.*, **48**, 1325-1344.

Matsumoto, O., Miyazaki, S., Otsuka, K. and Tamura, H. (1987) Crystallography of martensitic transformation in Ti-Ni single crystals, *Acta metall.*, ***35**, 2137-2144*

Mura, T.(1987) *Micromechanics of defects in solids*. Martinus Nijhoff, The Hague.

Miyazaki, S., Otsuka, K. and Wayman, M. (1989) The shape memory mechanism associated with the martensitic transformation in Ni-Ti alloys. I :Self accommodation, *Acta metall.*, **37**, 1873-1884.

Ono, N. and Sato, A. (1988) Plastic deformation governed by the stress induced martensitic transformation in polycristals. *Trans. Japan Inst. of Metal*, ***29**, 267-273.*

Ortin, J. and Planes, T. (1988) Thermodynamics analysis of thermal measurements in thermoelastic martensitic transformations. *Acta metall.*, ***36**, 1873-1889.*

Patoor, E., Eberhardt, A. and Berveiller, M. (1988) Thermomechanical behaviour of shape memory alloys. *Arch. of Mech.*, **40**, 775-794.

Patoor, E. and Berveiller, M. (1997) Micromechanichal modelling of thermomechanical behavior of shape memory. *In CISM Lecture Notes*, ***368**, pp. 121-188*, eds. F.D. Fisher & M. Berveiller, Springer Wien New York.

Roytburd, A.L. (1993) Elastic domains and polydomain phases in solids. *Phase Transitions*, **45**, 1-33.

Siredey, N., Patoor, E., Berveiller and Eberhardt, A. (1999) Constitutive equations for polycrystalline thermoelastic shape memory alloys. Part I. Intragranulaire interactions and behavior of the grain, *International Journal of Solids and Structures*, **36**, 4289-4315.

Sun, Q.P. and Hwang, K. C. (1993) Micromechanical modelling for the constitutive behavior of polycrystalline shape memory alloys, I and II. *J. Mech. Phys. Solids*, ***41**, 1-17 and 19-33.*

Walpole, L.J. (1967) The elastic field of an inclusion in an anisotropic medium. *P. Roy. Soc. Lond. A Mat.*, ***300***, 270-289.

Weschler, M.S., Lieberman, D.S. and Read, T.A. (1953) On the theory of the formation of martensite, *Transaction of AIME*, **197**, 1503-1515.

ELASTIC DEFORMATIONS IN SHAPE MEMORY ALLOY FIBER REINFORCED COMPOSITES

H. TSAI AND X. FAN
Materials Science and Mechanics, Michigan State University
East Lansing, MI 48824 U.S.A.

1. Introduction

The unique property of martensitic transformations in shape memory alloys (SMA) has prompted many recent attempts to construct adaptive structures and systems with these materials [1]. Most of the applications focus on embedding active elements in conventional structures to achieve certain control, sensing or actuating abilities. A typical configuration involves composites reinforced with shape memory alloy fibers. This setting utilizes the unidirectional transformation properties of the active fiber in corporation with an elastic matrix to harvest the shape memory effects in the axial direction. In this setting, the interaction between the phase transformation and the deformations in the matrix determines the overall mechanical properties of the structure and hence plays a crucial role in the design and analysis of such composites.

Analysis of the deformations associated with an inclusion undergoing phase transformation in an elastic host can be traced back to the works by Eshelby [2, 3], in which the entire inclusion transforms. Whereas in the case of shape memory alloy fiber reinforced composites, the fiber has to go through a stage when only part of it undergoes phase transformation. During this transition stage, regions of high strain phase and low strain phase coexist. They are separated by sharp phase boundaries across which the strains and stresses suffer finite jumps while the displacements and the traction remain continuous. To accommodate this transformation strain, the matrix is forced to deform accordingly in order to maintain the bonding at the interface. The shear load distribution at the interface and, in particular, the possible stress concentrations near the phase boundary are major concerns in the design of SMA fiber reinforced composites.

Q.P. Sun (ed.), IUTAM Symposium on Mechanics of Martensitic Phase Transformation in Solids, 139–146.

In this report, we summarize the results of our recent study on the elastic deformations associated with SMA fiber reinforced composites. In order to better understand the interaction between the SMA fiber and the matrix, we concentrate on a simple problem involving a single infinite fiber embedded in an infinite elastic matrix. A portion of the fiber is assumed to undergo phase transformation along the axial direction so that its length is changed by the corresponding transformation strain (typically a few percentages), while the matrix remains linearly elastic and isotropic.

The interaction between the SMA fiber and the matrix is determined by the bonding conditions at the interface. We consider two models: The "perfect bonding" model requires continuities of displacements and traction across the interface. The "spring bonding" model treats the interface as coupled by elastic "spring" which maintains the continuities of normal displacement and normal stress across the interface but relates the shear stress to the relative tangential displacement by a linear relation. Both models are formulated as axisymmetric deformations following the framework of Muki and Sternberg [4]. The integral form of the exact elasticity solutions to these problems are found by using Love's stress function and Fourier transform. It is found that, for the "perfect bonding" model, there exist stress singularities along the interface in the sense that the normal stresses suffer finite jumps while the shear stress blows up at the phase boundary with a logarithmic singularity. The results also indicate that the jumps of the normal stresses and the intensity of singularity of the shear stress are entirely determined by the Young's moduli, the Poisson's ratios and the transformation strain. They are independent of other geometric parameters in the problem.

For the "spring bonding" model, there exists no singularity in stresses. The shear stress concentrates near the phase boundary. The maximum stress occurs near the intersection of the phase boundary and the fiber-matrix interface. The dependence of the stress concentration factor on the elastic properties of the constituents (the matrix, the SMA fiber and the spring) as well as the geometry of the composites is obtained by numerical evaluation.

2. General Formulation and Exact Solutions

Let $R^{(1)}$ and $R^{(2)}$ be the three-dimensional regions occupied by the matrix and the fiber of a continuous fiber reinforced composite in its undeformed configuration, respectively (Fig. 1). Throughout this paper, the superscript "(1)" indicates quantities associated with matrix and the superscript "(2)" with fiber, respectively. Assume the fiber have a circular cross-section of radius a. Choose cylindrical coordinates (r, θ, z) with z-direction along the

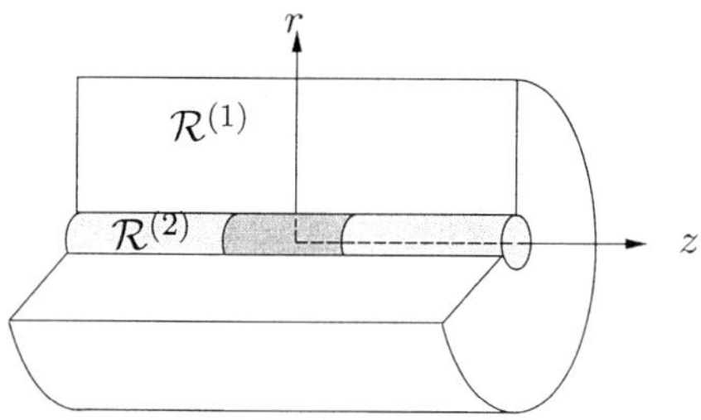

Figure 1. A cylindrical fiber embedded in an infinite matrix

longitudinal axis of the fiber. The region $R^{(i)}$ are then given by the revolution about z-axis of the following two-dimensional domains $\Omega^{(i)}$:

$$\begin{aligned} \Omega^{(1)} &= \{(r,z)|a < r < \infty, -\infty < z < \infty\}, \\ \Omega^{(2)} &= \{(r,z)|0 \le r < a, -\infty < z < \infty\}. \end{aligned} \tag{1}$$

For axisymmetric deformations with respect to the z-axis, the non-trivial displacement components are $u_r^{(i)}$ and $u_z^{(i)}$, with the strain components $\gamma_{rr}^{(i)}$, $\gamma_{\theta\theta}^{(i)}$, $\gamma_{zz}^{(i)}$, and $\gamma_{rz}^{(i)}$. All these components are functions of r and z only.

The matrix is made of a homogeneous, isotropic, and linearly elastic material with Young's Modulus $E^{(1)}$ and Poisson's ratio $\nu^{(1)}$. The fiber is capable of undergoing a displacive phase transformation along the axial direction. The transformation strain is denoted by γ^T. Within each phase, we assume the fiber is linearly elastic and isotropic with Young's Modulus $E^{(2)}$ and Poisson's ratio $\nu^{(2)}$. We model the deformations in a SMA composite with a partially transformed fiber as follows.

Assume that a segment of length $2L$ of the fiber transforms with uniform transformation strain γ^T along z-direction. After the phase transformation, the length of this segment becomes $2L(1+\gamma^T)$. The stress-strain relations are given by

$$\begin{aligned} \gamma_{rr}^{(i)} &= \frac{1}{E^{(i)}}[\sigma_{rr}^{(i)} - \nu^{(i)}(\sigma_{\theta\theta}^{(i)} + \sigma_{zz}^{(i)})], \\ \gamma_{\theta\theta}^{(i)} &= \frac{1}{E^{(i)}}[\sigma_{\theta\theta}^{(i)} - \nu^{(i)}(\sigma_{rr}^{(i)} + \sigma_{zz}^{(i)})], \\ \gamma_{zz}^{(i)} &= \frac{1}{E^{(i)}}[\sigma_{zz}^{(i)} - \nu^{(i)}(\sigma_{rr}^{(i)} + \sigma_{\theta\theta}^{(i)})] + \delta_{i2}\gamma^*, \\ \gamma_{rz}^{(i)} &= \frac{1+\nu^{(i)}}{E^{(i)}}\sigma_{rz}^{(i)} \quad \text{on } \Omega^{(i)}, \end{aligned} \tag{2}$$

where

$$\gamma^* = \gamma^*(z) \equiv \begin{cases} \gamma^T & |z| \leq L\,, \\ 0 & |z| > L. \end{cases} \tag{3}$$

Thus, the stress components $\sigma_{rr}^{(i)}$, $\sigma_{\theta\theta}^{(i)}$, $\sigma_{zz}^{(i)}$, and $\sigma_{rz}^{(i)}$ are functions of r and z. In this setting, the equilibrium equations (in the absence of body forces) reduce to

$$\begin{aligned} &\frac{\partial \sigma_{rr}^{(i)}}{\partial r} + \frac{\partial \sigma_{rz}^{(i)}}{\partial z} + \frac{\sigma_{rr}^{(i)} - \sigma_{\theta\theta}^{(i)}}{r} = 0, \\ &\frac{\partial \sigma_{rz}^{(i)}}{\partial r} + \frac{\partial \sigma_{zz}^{(i)}}{\partial z} + \frac{\sigma_{rz}^{(i)}}{r} = 0 \quad \text{on } \Omega^{(i)}. \end{aligned} \tag{4}$$

A standard way of solving the above elastostatic problem is to introduce the Love's stress functions $\Phi^{(i)} = \Phi^{(i)}(r, z)$ [5, 6], such that the stresses are given by

$$\begin{aligned} \sigma_{rr}^{(i)} &= \frac{\partial}{\partial z}\left[\nu^{(i)}\nabla^2\Phi^{(i)} - \frac{\partial^2\Phi^{(i)}}{\partial r^2}\right], \\ \sigma_{\theta\theta}^{(i)} &= \frac{\partial}{\partial z}\left[\nu^{(i)}\nabla^2\Phi^{(i)} - \frac{1}{r}\frac{\partial\Phi^{(i)}}{\partial r}\right], \\ \sigma_{zz}^{(i)} &= \frac{\partial}{\partial z}\left[(2-\nu^{(i)})\nabla^2\Phi^{(i)} - \frac{\partial^2\Phi^{(i)}}{\partial z^2}\right], \\ \sigma_{rz}^{(i)} &= \frac{\partial}{\partial r}\left[(1-\nu^{(i)})\nabla^2\Phi^{(i)} - \frac{\partial^2\Phi^{(i)}}{\partial z^2}\right] \quad \text{on } \Omega^{(i)}, \end{aligned} \tag{5}$$

The equilibrium equations (4) are satisfied by (5) provided that the stress functions $\Phi^{(i)}$ are biharmonic:

$$\nabla^2\nabla^2\Phi^{(i)} = 0 \quad \text{on } \Omega^{(i)}. \tag{6}$$

The strains and displacements can all be expressed in terms of the functions $\Phi^{(i)}$. The solution to the biharmonic function is obtained via the Fourier transform:

$$\tilde{\Phi}(\eta) \equiv \mathcal{F}[\Phi] = \int_{-\infty}^{\infty} \Phi(z)e^{i\eta z}dz, \quad -\infty < \eta < \infty. \tag{7}$$

After Fourier transform, the general solution of (6) can be found to be of the following form

$$\tilde{\Phi}^{(i)}(r;\eta) = A^{(i)}(\eta)R_0^{(i)}(|\eta|r) + B^{(i)}(\eta)|\eta|rR_1^{(i)}(|\eta|r) \tag{8}$$

on $\tilde{\Omega}^{(i)}$, where $R_0^{(1)} = K_0$ and $R_1^{(1)} = K_1$ are the modified Bessel's functions of the second kind of order zero and one, respectively; while $R_0^{(2)} = I_0$, and $R_1^{(2)} = I_1$ are the modified Bessel's functions of the first kind of order zero and one, respectively. This form is consistent with the standard requirements that all fields are bounded at $r = 0$ and as $r \to \infty$. The unknown functions $A^{(i)}(\eta)$ and $B^{(i)}(\eta)$ are determined by the bonding conditions.

Two bonding conditions are considered. The "perfect bonding" model requires continuity of traction and displacements across the interface so that

$$[\![u_r]\!] = [\![u_z]\!] = [\![\sigma_{rr}]\!] = [\![\sigma_{rz}]\!] = 0 \quad \text{on } \Pi \tag{9}$$

where $\Pi = \{(a, z)| - \infty < z < \infty\}$ is the interface between the fiber and the matrix. The "spring bonding" model relaxes the continuity of u_z and replaces it by

$$[\![u_r]\!] = [\![\sigma_{rr}]\!] = [\![\sigma_{rz}]\!] = 0, \quad \sigma_{rz}^{(i)} = k[\![u_z]\!] \quad \text{on } \Pi \tag{10}$$

where k is the stiffness of the "spring". The exact solutions for the stress functions $\Phi^{(i)}$ can then be easily obtained by inverse Fourier transform. In particular, the stresses are given by the following expressions:

$$\begin{aligned}
\sigma_{rr}^{(n)}(r, z) &= i\frac{E^{(1)}}{2\pi(1+\nu^{(1)})}\int_{-\infty}^{\infty}\frac{1}{\Delta}\bigg\{\Big[R_0^{(n)}(|\eta|r) - \frac{(-1)^n}{|\eta|r}R_1^{(n)}(|\eta|r)\Big]\Delta_A^{(n)}\\
&-\Big[(-1)^n(1-2\nu^{(n)})R_0^{(n)}(|\eta|r) + |\eta|rR_1^{(n)}(|\eta|r)\Big]\Delta_B^{(n)}\bigg\}\tilde{\gamma}^*(\eta)e^{-iz\eta}d\eta,\\
\sigma_{\theta\theta}^{(n)}(r, z) &= \frac{E^{(1)}}{2\pi(1+\nu^{(1)})}\int_{-\infty}^{\infty}\frac{(-1)^n}{\Delta}\bigg\{\frac{1}{|\eta|r}R_1^{(n)}(|\eta|r)\Delta_A^{(n)}\\
&-(1-2\nu^{(n)})R_0^{(n)}(|\eta|r)\Delta_B^{(n)}\bigg\}\tilde{\gamma}^*(\eta)e^{-iz\eta}d\eta,\\
\sigma_{zz}^{(n)}(r, z) &= \frac{E^{(1)}}{2\pi(1+\nu^{(1)})}\int_{-\infty}^{\infty}\frac{1}{\Delta}\bigg\{-R_0^{(n)}(|\eta|r)\Delta_A^{(n)}\\
&+\Big[(-1)^n 2(2-\nu^{(1)})R_0^{(n)}(|\eta|r) + |\eta|rR_1^{(n)}(|\eta|r)\Big]\Delta_B^{(n)}\bigg\}\tilde{\gamma}^*(\eta)e^{-iz\eta}d\eta,\\
\sigma_{rz}^{(n)}(r, z) &= i\frac{E^{(1)}}{2\pi(1+\nu^{(1)})}\int_{-\infty}^{\infty}\frac{\text{sign}(\eta)}{\Delta}\bigg\{(-1)^{n+1}R_1^{(n)}(|\eta|r)\Delta_A^{(n)}\\
&+\Big[(-1)^n|\eta|rR_0^{(n)}(|\eta|r) + 2(1-\nu^{(1)})R_1^{(n)}(|\eta|r)\Big]\Delta_B^{(n)}\bigg\}\tilde{\gamma}^*(\eta)e^{-iz\eta}d\eta\\
&\qquad \text{on } \Omega^{(n)}, n = 1 \text{ or } 2
\end{aligned} \tag{11}$$

where Δ, $\Delta_A^{(n)}$ and $\Delta_B^{(n)}$ are functions of η and are dependent of the material properties and the geometry of the composites. Explicit expressions of these

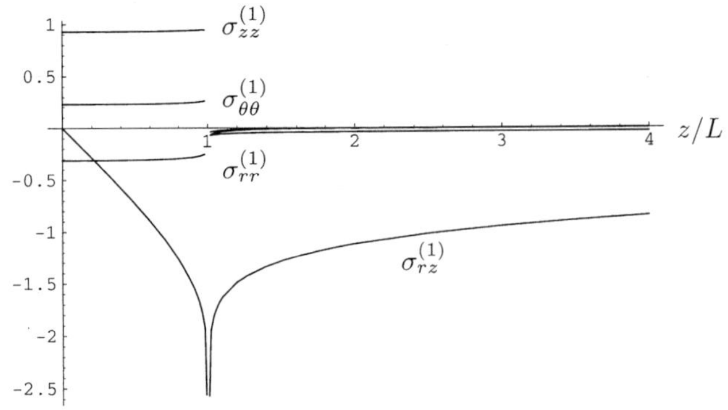

Figure 2. The stress distributions in the matrix on the fiber-matrix interface for the perfect bonding model. $G^{(1)}/G^{(2)} = 0.000096$, $\nu^{(1)} = 0.35$, $\nu^{(2)} = 0.29$, and $\alpha = 10$.

functions for the "perfect bonding" model and the "spring bonding" model are obtained [7], but too lengthy to be presented here in full. Note that $\tilde{\gamma}^*$ is the Fourier transform of the transformation strain field,

$$\tilde{\gamma}^* = \tilde{\gamma}^*(\eta) = \frac{2\gamma^T}{\eta} \sin(L\eta). \tag{12}$$

3. Numerical Examples

Numerical integration is performed to evaluate the displacement, strain and stress fields for both models. Of particular interest are the stress distributions along the fiber-matrix interface. Fig. 2 shows the interfacial stress distributions in the matrix for the perfect bonding model. The stresses are normalized by $E^{(1)}\gamma^T$. Notice that the normal stresses all suffer finite jumps, while the shear stress blows up at the location of the phase boundary ($\bar{z} = z/L = 1$). Fig. 3 shows the interfacial stress distributions for the spring bonding model. The stresses are now all continuous and bounded. The material constants used in these calculations are as follows. For perfect bonding model, the shear modulus ratio $G^{(1)}/G^{(2)} = .000096$, the Poisson's ratios $\nu^{(1)} = 0.35$, $\nu^{(2)} = 0.29$, and the aspect ratio $\alpha = L/a = 10$. For the spring bonding model, $E^{(1)} = 1\text{GPa}$, $E^{(2)} = 100\text{GPa}$, $\nu^{(1)} = \nu^{(2)} = 0.3$, $\bar{k} = ak/E^{(2)} = 10^{-5}$, and $\alpha = L/a = 10$.

4. Asymptotic Analysis of the Interfacial Stresses

We developed a procedure to analyze the interfacial stress distributions by asymptotic analysis. The approximate forms of the Bessel functions are

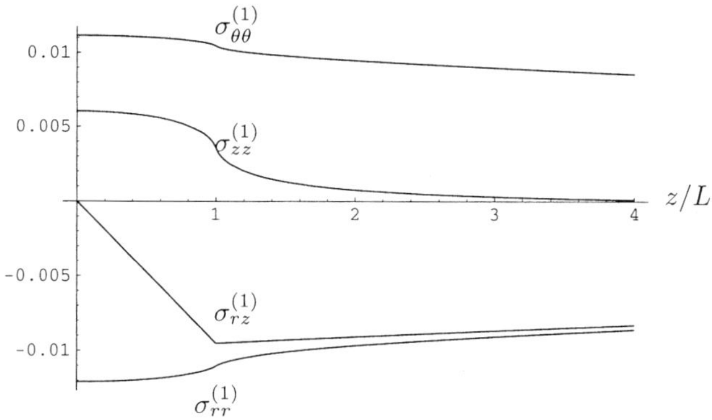

Figure 3. The stress distributions in the matrix on the fiber-matrix interface for the spring bonding model. $E^{(1)} = 1\text{GPa}$, $E^{(2)} = 100\text{GPa}$, $\nu^{(1)} = \nu^{(2)} = 0.3$, $\bar{k} = 10^{-5}$, and $\alpha = 10$.

used to isolate the terms that are responsible for the singularities in the perfect bonding model. It is found that for the prefect bonding model, the interfacial stresses can be separated into two parts

$$\begin{aligned}
\bar{\sigma}_{rr}^{(n)}(1,\bar{z}) &= \frac{2E^{(1)}\gamma^T}{(1+\nu^{(1)})\pi} C_{rr}^{(n)} \int_0^\infty \frac{\sin(\alpha\bar{\eta})\cos(\alpha\bar{z}\bar{\eta})}{\bar{\eta}} d\bar{\eta} + f_{rr}^{(n)}, \\
\bar{\sigma}_{\theta\theta}^{(n)}(1,\bar{z}) &= \frac{2E^{(1)}\gamma^T}{(1+\nu^{(1)})\pi} C_{\theta\theta}^{(n)} \int_0^\infty \frac{\sin(\alpha\bar{\eta})\cos(\alpha\bar{z}\bar{\eta})}{\bar{\eta}} d\bar{\eta} + f_{\theta\theta}^{(n)}, \\
\bar{\sigma}_{zz}^{(n)}(1,\bar{z}) &= \frac{2E^{(1)}\gamma^T}{(1+\nu^{(1)})\pi} C_{zz}^{(n)} \int_0^\infty \frac{\sin(\alpha\bar{\eta})\cos(\alpha\bar{z}\bar{\eta})}{\bar{\eta}} d\bar{\eta} + f_{zz}^{(n)} \\
\bar{\sigma}_{rz}^{(n)}(1,\bar{z}) &= \frac{2E^{(1)}\gamma^T}{(1+\nu^{(1)})\pi} C_{rz}^{(n)} \int_0^\infty \frac{\sin(\alpha\bar{\eta})\sin(\alpha\bar{z}\bar{\eta})}{\bar{\eta}} d\bar{\eta} + f_{rz}^{(n)} \qquad (13)
\end{aligned}$$

where the second terms ($f^{(n)}$'s) are continuous functions so that the integrals are responsible for the singularities in stresses. The coefficients C's are dependent of the Poisson's ratios and the shear modulus ratio $G^{(1)}/G^{(2)}$. The jumps in the normal stresses across the location of the phase boundaries ($z = \pm L$) are then given by

$$\begin{aligned}
[\![\bar{\sigma}_{rr}^{(n)}]\!] &= \frac{E^{(1)}\gamma^T}{1+\nu^{(1)}} C_{rr}^{(n)}, \\
[\![\bar{\sigma}_{\theta\theta}^{(n)}]\!] &= \frac{E^{(1)}\gamma^T}{1+\nu^{(1)}} C_{\theta\theta}^{(n)},
\end{aligned}$$

$$[\![\bar{\sigma}_{zz}^{(n)}]\!] = \frac{E^{(1)}\gamma^T}{1+\nu^{(1)}} C_{zz}^{(n)}, \tag{14}$$

whereas the shear stress approaches infinity with the intensity

$$\bar{\sigma}_{rz}^{(n)}(1,\bar{z}) \sim \frac{E^{(1)}\gamma^T}{(1+\nu^{(1)})\pi} C_{rz}^{(n)} \ln\left|\frac{1+\bar{z}}{1-\bar{z}}\right|, \qquad (|\bar{z}| \to 1). \tag{15}$$

On the other hand, the same procedure applied to the spring bonding model confirms that all stresses are bounded and continuous.

5. Concluding Remarks

The procedure established in this work provides an important tool to analyze the local deformations and stress concentration in a SMA composite. In the two models described above, it is indicated that the accommodation of the transformation strain in a coherent bonding condition results in infinite shear stress. The spring bonding model seems to depict a more reasonable stress state where all stresses are bounded. In this case, the stiffness of the spring k plays a crucial role. In reality, the material often exhibits fine mixtures of phases near external boundaries in order to accommodate the transformation strain. Further study is needed to investigate the relation between the interfacial microstructure and the stiffness of the bonding.

Parameter studies performed on the spring bonding model indicates that the strength of the matrix must be much lower than that of the fiber in order to maintain the small strain assumption used in the above analysis. Strong correlation is observed between the maximum shear stress and the aspect ratio of the transformed region. High stress concentration is generally associated with large aspect ratio.

References

1. Redmond, J. and Main, J. ed. (2000) *Adaptive Structures and Material Systems*, American Society of Mechanical Engineers, AD-Vol. 60, New York.
2. Eshelby, J. D. (1957) The determination of the elastic field of an ellipsoidal inclusion and related problems, Proceedings of the Royal Society, *Series A Mathematical and Physical Sciences* **241**, 376–396.
3. Eshelby, J.D. (1959) The elastic field outside an ellipsoidal inclusion, Proceedings of the Royal Society, *Series A Mathematical and Physical Sciences* **252**, 561–569.
4. Muki, R. and Sternberg, E. (1969) On the diffusion of an axial load from an infinite cylindrical bar embedded in an elastic medium, *International Journal of Solids and Structures* **5**, 587–605.
5. Love, A.E.H. (1944) *A Treatise on the Mathematical Theory of Elasticity*, Dover Publications, Inc., New York.
6. Timoshenko, S. and Goodier, J.N. (1951) *Theory of Elasticity*, McGraw-Hill Book Company, Inc., New York.
7. Fan, X. (2001) *Elastic Deformations in Shape Memory Alloy Fiber Reinforced Composites*, Ph.D. Thesis, Michigan State University, East Lansing, Michigan.

MODELING OF SUPERELASTIC ADAPTIVE COMPOSITES

J. SLUTSKER* AND A.L. ROYTBURD
Department of Materials and Nuclear Engineering
University of Maryland, College Park, MD 20742, USA

**Materials Science and Engineering Laboratory, National Institute of Standards and Technology, Gaithersburg, MD 20899, USA*

1.Introduction

An adaptive composite containing a shape memory alloy as an active component is the subject of the paper. This composite utilizes the unique properties of shape memory alloys (SMA), which change their crystalline structure in response to the change temperature and stress. Previous research shows that the combination of a shape memory alloy as an active and the elastic material as a passive components allows one to improve the properties of the SMA itself (create two-way shape memory) and to engineer a composite material with controlled superelastic properties [1,2].

In previous works [3,4] we have shown that the main source of the instability during stress- or strain- induced transformation is the incompatibility of the austenite and martensite phases. It appeared as a negative slope on stress-strain curve. The instability of martensitic transformation leads to the hysteretic behavior of the shape-memory material. In [1,2], the adaptive elastically homogeneous composite has been studied. In these papers we have shown that by varying volume fraction of the active component it is possible to control the instability, i.e., to control the hysteresis of the composite. However, materials with elastic modulus close to that of the SMA, (metals) are not the best choices for passive component in the designing of superelastic composites, because metals could not elongate elastically beyond 0.01%. In this case, the superelastic effect will be very small. To design the composite with large superelastic effect it is necessary that the passive component should be able to deform elastically to a sufficient strain. Therefore, polymer materials can serve as passive elements in the adaptive composite. However, because the elastic moduli of polymers and shape memory alloys are sufficiently different, it requires investigation of the heterogeneous adaptive composites.

2. General theory of layer heterogeneous adaptive composites

Consider a composite consisting of active and passive layers (Fig.1). The elastic energy of internal stress due to interaction between layers is

Q.P. Sun (ed.), IUTAM Symposium on Mechanics of Martensitic Phase Transformation in Solids, 147–154.

$$e = \gamma(1-\gamma)e_I(\hat{\hat{G}}_1, \hat{\hat{G}}_2, \gamma, \hat{\varepsilon}^M)$$
$$\hat{\hat{G}}_{1,2} = \hat{\hat{C}}_{1,2} - \hat{\hat{C}}_{1,2}\vec{n}(\vec{n}\hat{\hat{C}}_{1,2}\vec{n})^{-1}\vec{n}\hat{\hat{C}}_{1,2} \tag{1}$$

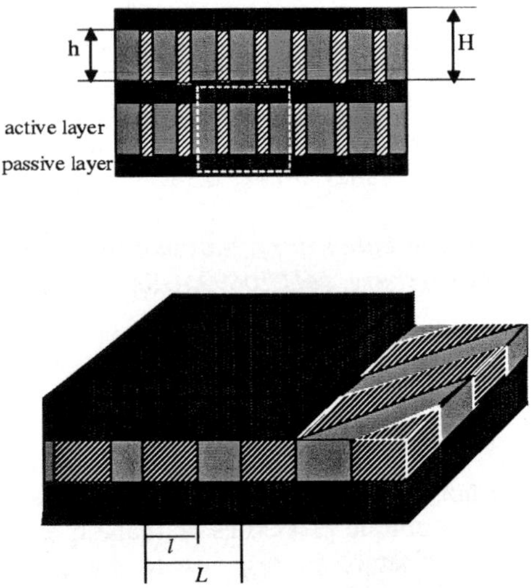

Figure 1. Multilayer composite containing transformable, active, and non-transformable, passive, layers. γ=h/H is the fraction of the active layer, β=*l*/L is the fraction of the martensite phase in the active layer, α=d/D is the fraction of the twin domain in the martensite phase, D is the interdomain period, d is a intertwin period.

γ is a fraction of an active layer , $\hat{\hat{C}}_1$ and $\hat{\hat{C}}_2$ are tensors of elastic moduli for passive (1) and active (2) layers, respectively, $\vec{n}$ is orientation of the normal of the interface, $\hat{\varepsilon}^M$ is the misfit strain characterizing incompatibility between layers, e_I is an energy of interaction between layers:

$$e_I = \frac{1}{2}\hat{\varepsilon}^M\ \hat{\hat{G}}_1(\overline{G})^{-1}\hat{\hat{G}}_2\hat{\varepsilon}^M \tag{2}$$
$$\overline{G} = (1-\gamma)\hat{\hat{G}}_1 + \gamma\hat{\hat{G}}_2$$

For isotropic materials of layers and if the difference between Poisson's ratios of active and passive layers is negligible, the energy of interaction between layers is :

$$e_I = \frac{\gamma(1-\gamma)}{2(1-\nu^2)}\frac{E_1E_2}{(1-\gamma)E_1+E_2}\left[(\varepsilon_1^M)^2 + (\varepsilon_2^M)^2 + 2\nu\varepsilon_1^M\varepsilon_2^M\right] \tag{3}$$

Where ε_1^M and ε_2^M are the principal misfit strains in the interface plane, E_1,and E_2 are Young's moduli of passive and active layers, respectively and ν is Poisson ratio.

After transformation, the austenite-martensite microstructure is formed in the active layers. Here the cubic into three variants (domains) of tetragonal transformation is considered (Fig.2a).

$$\hat{\varepsilon}_{0,1} = \varepsilon_0\begin{pmatrix}1 & 0 & 0\\ 0 & -\chi & 0\\ 0 & 0 & -\chi\end{pmatrix};\ \hat{\varepsilon}_{0,2} = \varepsilon_0\begin{pmatrix}-\chi & 0 & 0\\ 0 & 1 & 0\\ 0 & 0 & -\chi\end{pmatrix};\ \hat{\varepsilon}_{0,3} = \varepsilon_0\begin{pmatrix}-\chi & 0 & 0\\ 0 & -\chi & 0\\ 0 & 0 & 1\end{pmatrix} \tag{4}$$

Assuming that the structure is the simplest equilibrium one (containing mixture of two domains) presented in Fig.2b, the free energy of the composite is

$$F = e_I(\varepsilon^M + \beta\varepsilon_0(\alpha)) + \gamma\left[\beta\Delta f + \beta(1-\beta)\hat{\hat{G}}_2(\varepsilon_0(\alpha))^2\right] + o(\frac{D}{H}, \frac{d}{D}) \quad (5)$$

Where β is the fraction of martensite, $\varepsilon_0(\alpha)$ is the self-strain accompanying formation of martensite phase with domain fraction α, $\hat{\hat{G}}_2(\varepsilon_0(\alpha))^2$ is the energy of incompatibility between martensite and austenite, Δf is the difference of the free energies of the undistorted austenite and martensite phases, D is the period of the martensite phase, d is the twin spacing, H is the thickness of the active layer. The following assumptions have been made:

(1) The difference of the elastic moduli between austenite and martensite is much less than that between austenite and passive layers and can be neglected.

(2) The ratios D/h and d/D are small, thus the martensite and austenite/martensite mixture can be considered as homogeneous mixtures. It means that surface energy of interfaces and the energy of oscillating microstresses can be neglected.

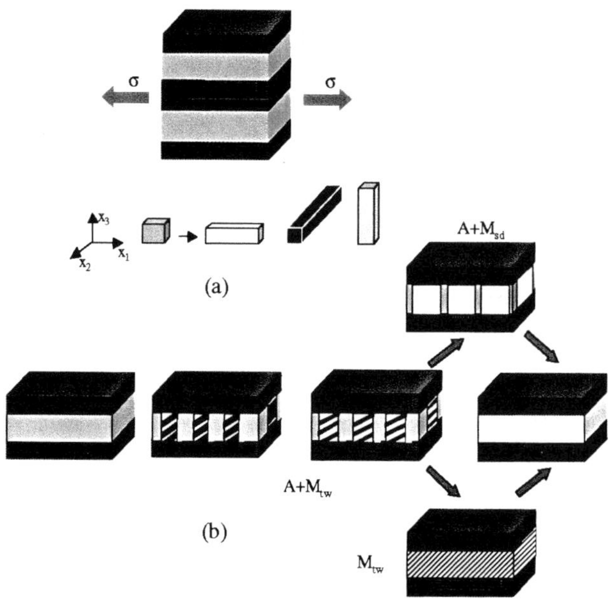

Figure 2. Evolution of an equilibrium microstructure in the active layer at uniaxial deformation. (a) Cubic-tetragonal transformation in the active layer. (b) Equilibrium microstructure containing a mixture of austenite and polydomain martensite. The possible paths of evolution of the microstructure are shown: mixture of austenite and single-domain martensite (A+M_{sd}) and twinned martensite (M_{tw})

3. Uniaxial deformation of the composite

In this section, we analyze the multilayer composite consisting of the active and passive layers under uniaxial external stress applied along x_1 (Fig.2). Our task is to obtain the stress-strain relations and the conditions of stability of superelastic deformation of the composite. Elastic properties of the active and passive layers are assumed to be isotropic with equal Poisson's ratios.

Since the external stress is uniaxial, the martensitic phase contains a major fraction of domain 1 which contributes to the maximum elongation along the extension axis. Thus, the polydomain martensitic phase should be a mixture of domain 1 and domain 2 which provide the maximal work due to transformation strain and minimum misfit with passive layer. Its self-strain is:

$$\varepsilon = (1-\alpha)\hat{\varepsilon}_{0,1} + \alpha\hat{\varepsilon}_{0,2} = \varepsilon_0 \begin{pmatrix} 1-\alpha(1+\chi) & 0 & 0 \\ 0 & -\chi+\alpha(1+\chi) & 0 \\ 0 & 0 & -\chi \end{pmatrix} \qquad (6)$$

Under external stress, misfit between the layers is described as follow

$$\hat{\varepsilon}^M(\sigma) = \begin{pmatrix} \beta\varepsilon_0^1(\alpha) + (\frac{1}{E_2} - \frac{1}{E_1})\sigma & 0 & 0 \\ 0 & \beta\varepsilon_0^2(\alpha) - \nu(\frac{1}{E_2} - \frac{1}{E_1})\sigma & 0 \\ 0 & 0 & \beta\varepsilon_0^3(\alpha) - \nu(\frac{1}{E_2} - \frac{1}{E_1})\sigma \end{pmatrix} \qquad (7)$$

where σ is an external stress ($\sigma \equiv \sigma_{11}$). Then, the free enthalpy (Gibbs free energy) (Eq.(5)) is equal to

$$\Phi = -(1-\gamma)\frac{1}{2}\frac{1}{E_1}\sigma^2 - \gamma\frac{1}{2}\frac{1}{E_2}\sigma^2 +$$

$$\gamma(1-\gamma)\frac{1}{2}\frac{E_1E_2}{(1-\gamma)E_1+\gamma E_2}\left[\frac{(\beta\varepsilon_0(1-\chi)+(1-\nu)\Delta S\sigma)^2}{2(1-\nu)} + \frac{(\varepsilon_0(1+\chi)(1-2\alpha)+(1+\nu)\Delta S\sigma)^2}{2(1+\nu)}\right] - \gamma\beta\sigma\varepsilon_0(1-\alpha(1+\chi)) + \qquad (8)$$

$$\gamma\beta\Delta f + \gamma\beta(1-\beta)\frac{1}{2}E_2\varepsilon_0^2(-\chi+\alpha(1+\chi))^2$$

The first two terms are the enthalpy of a composite before transformation; the third term is the misfit elastic energy, the forth term is the work of external stress on the self-strain of transformation. The last two terms determine the internal free energy of the active two-phase layer.

After introducing the dimensionless parameters: $\varphi = \frac{\Phi}{E_2\varepsilon_0}; \phi = \frac{\Delta f}{E_2\varepsilon_0}; \overline{\sigma} = \frac{\sigma}{E_2\varepsilon_0}; \xi = \frac{E_1}{E_2}; g = \frac{1}{(1-\gamma)\xi+\gamma}$, the Gibbs free energy can be written as follow:

$$\varphi = -\frac{1}{2}g\overline{\sigma}^2 + \gamma(1-\gamma)\frac{1}{2}g\xi\left[\frac{(1-\chi)^2}{2(1-\nu)} + \frac{(1+\chi)^2}{2(1+\nu)}(1-2\alpha)^2\right]\beta^2 -$$

$$\gamma g(1-\alpha(1+\chi))\overline{\sigma} + \gamma\beta\phi + \gamma\beta(1-\beta)\frac{1}{2}[-\chi+\alpha(1+\chi)] \qquad (9)$$

The parameter, $\phi = q\frac{T-T_0}{T_0 E_2 \varepsilon_0}$ (q is latent heat, T_0 is the temperature of phase equilibrium), can be considered as an "effective temperature".

The equilibrium microstructure (α_0,β_0) is determined by the equation:

$$\frac{\partial \varphi}{\partial \alpha} = \frac{\partial \varphi}{\partial \beta} = 0 \tag{10}$$

The stress-strain relations and effective compliance of the composite can be obtained from follows

$$\varepsilon = -\frac{d\varphi}{d\overline{\sigma}}\Big|_{\alpha_0,\beta_0}, S^* = -\frac{d^2\varphi}{d\overline{\sigma}^2}\Big|_{\alpha_0,\beta_0} \tag{11}$$

The equilibrium solutions, β_0, α_0, corresponds to the stable or unstable (saddle points) states depending on the sign of D:

$$D = \frac{\partial^2\varphi}{\partial\alpha^2}\frac{\partial^2\varphi}{\partial\beta^2} - \left(\frac{\partial^2\varphi}{\partial\alpha\partial\beta}\right)^2 \tag{12}$$

The average strain and the effective compliance of the composite at the transformation in the active layer is determined by Eqs.(11) and are as follow:

$$\varepsilon = g\overline{\sigma} + \gamma g\beta_0[1-\alpha_0(1+\chi)] \tag{13}$$

$$S_{eff} = g + \gamma g\frac{\partial\beta_0}{\partial\overline{\sigma}}[1-\alpha_0(1+\chi)] - \gamma g\beta_0(1+\chi)\frac{\partial\alpha_0}{\partial\overline{\sigma}} \tag{14}$$

where α_0, β_0 are determined by Eqs(15a,b).

Solution of the set of equations (Eqs.(10)) shows that there are three different paths of the microstructure evolution depending on the temperature, ϕ, and constraint parameter, Γ ($\Gamma=(1-\gamma)g\xi$)

(1) The evolution of austenite-martensite microstructure proceeds through the formation and growth of single-domain martensite, no polydomain martensitic phase formed.

(2) Transformation starts with formation and growth of polydomain martensitic phase. However, the fraction of twin domain, α, disappears before austenite transformation completes. Further superelastic deformation proceeds through the increasing of fraction of single-domain martensite.

(3) The transformation starts with formation and growth of polydomain martensitic phase. As deformation proceeds, the fraction of twin domain, α, decreases. The austenite transformation completes before twin structure in the martensitic phase disappears and further deformation proceeds through the twinning of the martensite (superplastic deformation).

In Fig. 3 the three regimes of microstructure evolution at superelastic deformation is shown in the diagram (ϕ, Γ). The temperatures, ϕ^*, ϕ^{**} which separate different regimes of microstructural evolution are

$$\phi^* = \chi - \frac{\chi^2}{2} \quad \phi^{**} = \frac{\chi^2}{2} + \Gamma\frac{(1-\chi)(\chi-\nu)}{1-\nu^2} \tag{15}$$

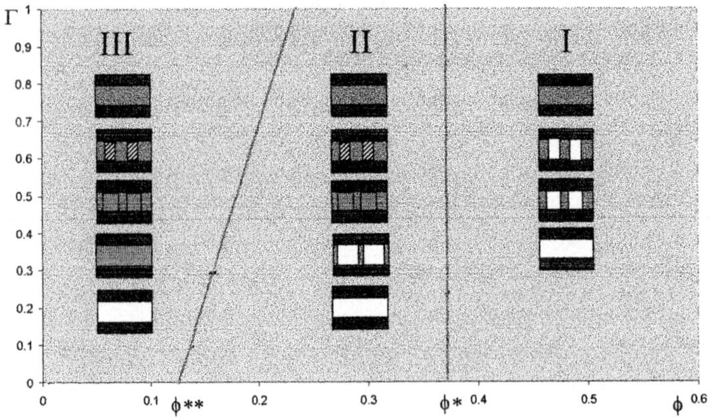

Figure 3. Three regimes of evolution of an equilibrium microstructure during superelastic deformation of a composite (φ is an "effective temperature", Γ is a constraint parameter, Γ= (1-γ) gξ).

The typical stress-strain curves in each regime for different fraction of the active layer, γ, are shown in Figs.4-6. The stress-strain curve for unconstraint active material (γ=1) is shown for comparison. The negative slope on the stress-strain curves (the negative effective Young's modulus) corresponds to the unstable solutions of Eq.(12) (D<0) and to the hysteresis at stress controlled deformation. To obtain a stable composite it is necessary to avoid the negative effective elastic modulus of

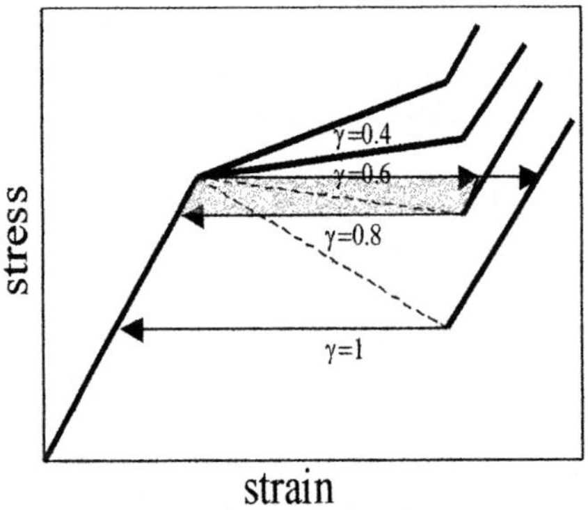

Figure 4. Stress-strain curves for φ>φ*, at different fractions of the active layer, γ, ξ=1.

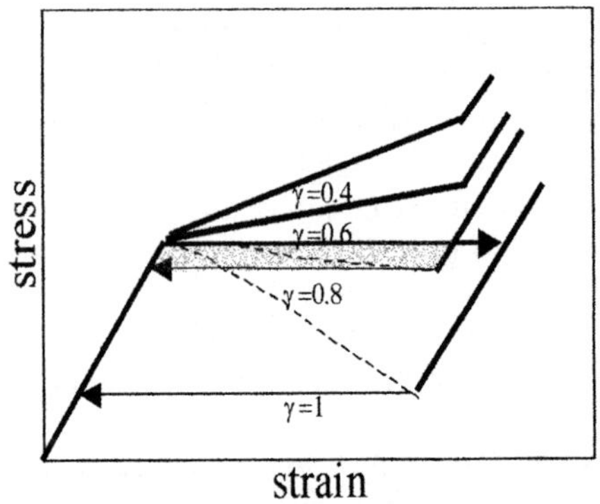

Figure 5. Stress-strain curves for φ**<φ <φ*, at different fractions of the active layer, γ, ξ=1.

superelastic deformation at all stages of transformation. The conditions of stable superelastic deformation should be found for each regime of microstructural evolution (φ<φ**, φ**<φ<φ*, φ>φ*). The regions of stability of the composite at the beginning of transformation (yield stress) and at all stages of transformation (general stability) are shown in Fig.7 (φ, Γ). These conditions can be found either directly by calculating the effective elastic compliance of the composite (Eq.(15)) or by using the condition of stability (Eq.(13)). The areas where the superelastic deformation of the composite is stable at all regimes of transformation, stable at yield stress, or unstable are shown by different colors. As can be seen from this diagram it is possible to have composite which is stable at the beginning of transformation, but unstable at the changing regimes of transformation.

4. Discussion and conclusions

The problem of the composite containing shape memory materials as an active component has recently attracted great interest for development of smart materials and structures [5-8]. One of the main obstacles for use of shape memory materials for wider engineering applications is the stress-strain hysteresis. In this paper the principles of design of adaptive composites with a controlled stress-strain hysteresis are proposed. However, to obtain the stable superelastic deformation of the composite it is necessary not only to eliminate negative effective modulus, but also to avoid any changes of microstructure regimes of transformation (since the nucleation of the new microstructure can create additional instabilities). Therefore, the temperature, ϕ^{**}, at which austenite transformation completes and twin structure disappears simultaneously, is the best suitable parameters for designing the composite with minimum hysteresis. Thus, all fractions of an active layer, γ, and ratios between Young's moduli of the layers, ξ, which lie on the line ϕ^{**} (Fig.7), are suitable for designing of the composite without thermodynamic hysteresis. However, in order to obtain the maximal superelastic deformation it is desirable to have a larger fraction of an active layer as possible. The point in the diagram at which the superelastic deformation of the composite is stable at

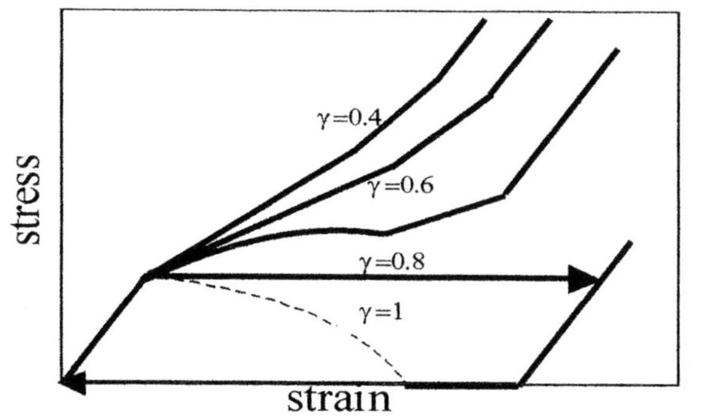

Figure 6. Stress-strain curves for $\phi<\phi^{**}$ at different fractions of the active layer, γ, $\xi=1$.

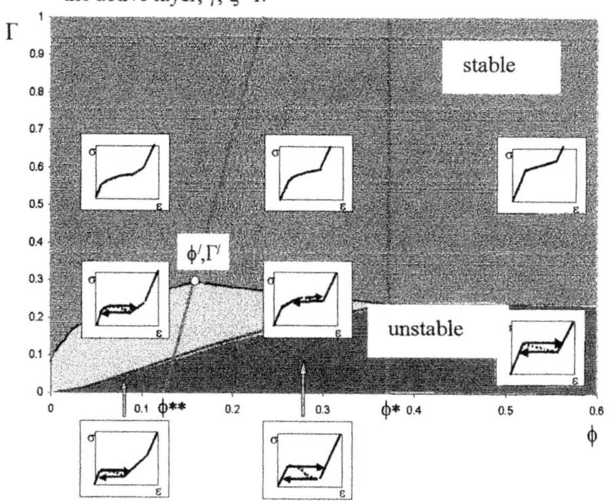

Figure 7. Diagram of stability of superelastic deformation of a composite (ϕ is an "effective temperature", Γ is a constraint parameter), light gray area represents the superelastic deformation stable at yield stress).

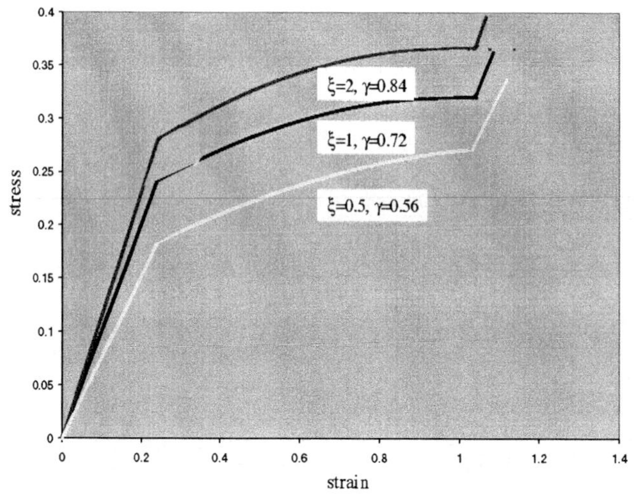

Figure 8. The stable superelastic deformation of a composite with maximal elongation due to transformation in the active layer at different ratios of Young's moduli of layers, ξ (ϕ=0.15, Γ=0.28).

maximum elongation due to transformation corresponds to (ϕ'=0.15, Γ'=0.28). The fraction of the active layer corresponding to this point is

$$\gamma = \frac{\xi(1-\Gamma')}{\xi+\Gamma'(1-\xi)} \qquad (16)$$

The stress-strain curves for this point at different ratios of elastic moduli, ξ, are shown in Fig.8. These curves correspond to the stable superelastic deformation, the austenite and twins disappear simultaneously and there are no changing microstructure regimes of transformation. Thus, using the diagram in Fig.7 and Eq.(16) it is possible to obtain the fraction of the active component ,γ, for the composite with elastic properties of the layers, ξ, which gives stable superelastic deformation with maximum elongation due to transformation.

Acknowledgments

The support of National Research Council Postdoctoral Assistantship is greatly appreciated by J. Slutsker.
A. Roytburd is grateful to the support of AFOSR, Grant # F496-200-110-110.

References

1. Roytburd, A.L. and Slutsker, J. (1999) Deformation of adaptive materials. Part II. Adaptive composite. *Journ.Mech. & Phys. Solids*, **47**, 2331-2349.
2. Slutsker, J. and Roytburd, A.L. (2001) Control of intrinsic instability of superelastic deformation, *International Journal of Plasticity*, submitted.
3. Roytburd, A.L. and Slutsker, J., (2001) Deformation of adaptive materials. Part III. Deformation of crystals with polytwin product phases, *J. Mech. & Phys. Solids*, **49**, 1795-1822.
4. Roytburd, A.L. and Slutsker, J. (1999) Deformation of adaptive materials. Part I. Constrained deformation of polydomain crystals. *Journ. Mech. & Phys. Solids* **47**, 2299-2329.
5. Roytburd, A.L., Slutsker, J. and Wuttig, M. (2000) Smart composites with shape memory alloys, *in Comprehensive Composite Materials*, eds. A.Kelly and C. Zweben, Pergamon, Oxford, 2000, pp.507-522.
6. Cherkaoi, M., Sun, Q.P. and Song, G.Q., (2000) Micromechanics modeling of composite with ductile matrix and shape memory alloy reinforcement, *Inter. Journal of Solids and Structures* **37**, 1577-1594.
7.Sittner, P. and Stalmans, R. ,(2000) Developing hybrid polymer composites with embedded shape-memory alloy wires. *JOM-Journ. Minerals Metals. & Mater. Soc.*, **52**, 15-20.
8.Stalmans, R., Delaey, L. and Van Humbeeck, J., (1997) Modeling of adaptive composite materials with embedded shape memory alloy wires. *Mat. Res. Soc. Symp. Proc.* vol. 459, pp. 119-130.

INFLUENCE OF PLASTICITY ON TRANSFORMATION BEHAVIOUR OF MARTENSITE IN NiTi

YINONG LIU
Department of Mechanical and Materials Engineering, The University of Western Australia, Crawley, WA 6009, Australia

1. Introduction

Thermoelastic martensitic transformation is a diffusionless process of structural transformation involving a lattice distortion. Owing to the diffusionless nature, the lattice distortion of individual atoms is accumulated and manifested as lattice strains. In this context, a thermoelastic martensitic transformation is in effect a combined process of a thermal phase transformation and a mechanical distortion. During a thermally induced transformation in a polycrystalline material, the mechanical aspect is relatively ineffective due to the self-accommodation of martensite variants. However, when a macroscopic deformation is involved, such as in the cases of stress-induced martensitic transformation, martensite reorientation or two-way memory effect, transformation behaviour becomes altered due to the participation of plasticity in the process. Typical observations of the influence of deformation on transformation behaviour include the increase of the critical temperature for reverse transformation of deformed martensite [1-3], dependence of mechanical hysteresis on the magnitude of deformation in stress-induced transformations [4-5] and the phenomenon of two-way memory effect itself.

In the past, understanding of thermoelastic martensitic transformations has largely been established on the basis of theories and experimental evidences of thermally induced transformations. One of the fundamental understandings of thermoelastic martensitic transformation is expressed in the free energy balance equation [6]:

$$\Delta G = \Delta H - T\Delta S + \Delta E_{el} + \Delta E_{ir} \tag{1}$$

This equation expresses the thermoelastic behaviour (transformation temperature interval) of the transformation in terms of a reversible non-chemical elastic energy ΔE_{el} and the hysteretic behaviour in terms of an irreversible energy ΔE_{ir} [6,7] as:

$$\Delta T_t = (T_f - T_s) = \frac{\Delta E_{el}}{\Delta S} \tag{2}$$

$$\Delta T_h = T_{m-a} - T_{a-m} = \frac{\Delta E_{ir}}{\Delta S} \tag{3}$$

where ΔT_t and ΔT_h are the transformation temperature interval and hysteresis, T_f and T_s are the finishing and starting temperatures of the transformation and $T_{a\text{-}m}$ and $T_{m\text{-}a}$ are the characteristic temperatures of the forward A-M transformation and the reverse M-A

Q.P. Sun (ed.), IUTAM Symposium on Mechanics of Martensitic Phase Transformation in Solids, 155–162.

transformation, respectively. The critical temperature for a thermoelastic martensitic transformation can also be determined using this equation at $\Delta G=0$, as:

$$T(f_T) = \frac{\Delta H + \Delta E_{el}(f_T) + \Delta E_{ir}(f_T)}{\Delta S} \tag{4}$$

where f_T denotes the fraction of transformation and $T(f_T)$ is the temperature of the transformation at the moment corresponding to f_T. In these equations, ΔE_{el} accounts for the mechanical strain energy associated with transformation lattice distortion and is regarded variant structure dependent. ΔE_{ir} represents the summation of all irreversible energies, such as the creation of structural defects, dissipation of heat and emission of sound waves, and is dependent on the path of a transformation process.

A thermoelastic martensitic transformation can also be induced by the application of a stress. Experimental observations of the mechanical behaviour of thermoelastic martensite have largely been interpreted using a Clausius-Clapeyron type relation [8]:

$$\frac{d\sigma}{dT} = -\frac{\rho \Delta S}{\varepsilon_t} = const. \tag{5}$$

where ρ is the density of the material and ε_t is the crystallographic strain of the transformation in the direction of the stress. This equation relates the effect of temperature on the transformation to stress. Based on this understanding, transformation parameters determined from a thermally induced transformation can be related to parameters measured from a mechanically induced transformation, such as:

$$\Delta\sigma = (\frac{d\sigma}{dT})\Delta T \tag{6}$$

for either transformation intervals or transformation hysteresis and

$$\sigma(f_T) = \frac{d\sigma}{dT}[T - T(f_T)] \tag{7}$$

for critical stresses and temperatures of the transformation. A specific situation of equation (7) is at $f_T=0$ when $\sigma=0$ for the forward A-M transformation:

$$T = T(f_T = 0) = M_s \tag{8}$$

Such co-relations are attractive for the determination of some essential thermodynamic parameters, modelling of the thermomechanical behaviour of the transformations and application and design of shape memory apparatuses. However, discrepancies often occur between theoretical predictions and experimental observations and in comparisons of stress and temperature parameters between mechanically induced transformations and thermally induced transformations. These include:

- the disagreement between temperature hysteresis of a thermally induced transformation and the stress hysteresis of a stress-induced transformation when related by the Clausius-Clapeyron relation [9]; and
- the inaccuracy in the prediction of the critical temperatures for pseudoelasticity and shape memory effect [9,10].*

---

* It is generally regarded that pseudoelastic recovery starts at $T=A_s$ and completes at $T=A_f$. Experimental evidences indicate that the reverse transformation of deformed martensite will not occur until a higher temperature, $T=A_s+\Delta T$, is reached.

These discrepancies stem from the fact that the Clausius-Clapeyron relation was developed on the basis of thermodynamic analysis of a perfect thermoelastic martensitic transformation in a perfect single crystal, without any consideration of the influence of plasticity in a real crystal [11]. Thus, investigation and clarification of the influences of deformation, plasticity and polycrystallinity on transformation behaviour of shape memory alloys are of fundamental significance for both the development of theories of thermoelastic martensitic transformations and practical application of materials exhibiting the transformations.

2. Experimental Techniques

Two near-equiatomic commercial polycrystalline NiTi alloys of Ti-50.15at%Ni and Ti-50.2at%Ni were used. These alloys exhibit relatively simple heat treatment responses and transformation behaviour. As-received materials were cold rolled and annealed at various temperatures for different properties. Deformation was performed in uniaxial tension or simple shear in austenitic state via stress-induced martensitic transformation. Transformation behaviour was studied using differential scanning calorimetry (DSC).

3. Results and Discussion

3.1. EFFECT OF DEFORMATION ON TRANSFORMATION TEMPERATURES

Figure 1 shows the effect of deformation via stress-induced martensitic transformation on the transformation behaviour of Ti-50.2at%Ni. The alloy had been cold rolled and annealed at 1073 K, and then deformed in tension to 6% of strain, as indicated by the stress-strain curve shown in the inset. It is seen that there had been no pseudoelastic recovery upon unloading, implying that the stress-induced martensite structure had been preserved after the deformation. Transformation behaviour after the deformation was measured on DSC using samples cut form the middle section of the tensile specimen. Prior to DSC measurement the sample had been cooled in liquid nitrogen and the measurement started on heating. The deformed sample exhibited a distinct difference in transformation behaviour between the first thermal transformation cycle and

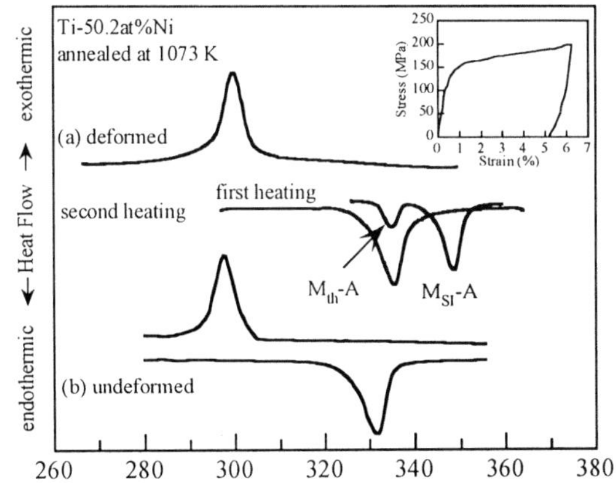

Figure 1. Effect of deformation via stress-induced martensitic transformation on transformation behaviour of NiTi. For the undeformed sample, M_s=305 K, M_f=288 K, A_s=320 K and A_f=337K.

the following thermal transformation cycles, as shown in curve (a). On the first heating the sample exhibited two separate endothermic transformations. The first endothermic peak corresponded to the reverse transformation of thermally induced martensite ($M_{th} \rightarrow A$). This martensite was formed from the residual austenite resulting from incomplete stress-induced martensitic transformation during deformation. The second peak corresponded to the reverse transformation of stress-induced martensite ($M_{SI} \rightarrow A$). In the second cycle, however, only one endothermic transformation was observed, which occurred at the same position as the $M_{th} \rightarrow A$ transformation on the first heating. No drastic difference was observed for the forward transformation on cooling between the first and the following cycles. For comparison, DSC measurement of the transformation behaviour of an undeformed sample is also shown in the figure (curve (b)). It is seen that the transformation behaviour of the deformed sample in subsequent cycles after the first heating conforms to that of the undeformed sample.

The increase of the critical temperature for reverse transformation after deformation indicates that oriented martensite produced by stress-induced martensitic transformation is stabilised relative to self-accommodating martensite formed thermally on cooling. Similar stabilisation effect is also observed after deformation via martensite reorientation [3]. These observations suggest that the stabilisation effect be due to the mechanical effect of deformation. It has been suggested that formation of oriented martensite in a polycrystalline matrix be inevitably accompanied by internal plastic deformation [3,12]. This is due to the need to co-ordinate among different microscopic strains of preferentially oriented martensite variants in neighbouring grains at grain boundaries. The production of this internal plastic deformation imposes a net mechanical resistance to a stress-induced martensitic transformation, in addition to any other possible resistive forces to the transformation. The reverse transformation of the oriented martensite in a re-shaped (deformed) matrix is a process in analogue to the forward stress-induced martensitic transformation, but driven by intrinsic thermal energy. Therefore, similar mechanical resistance arises during the reverse transformation when the microstrains of preferentially oriented martensites are reversed and negated. This leads to the increase of the critical temperature for the reverse transformation of the deformed martensite as relative to the self-accommodating martensite.

An interesting question is the critical temperature for the formation of oriented martensite in a polycrystalline matrix, as relative to that for the transformation of self-accommodating martensite. The above discussion suggests that it occurs at a lower temperature, due to the need of extra thermal driving force to overcome the extra mechanical resistance associated with the transformation that produces a macroscopic deformation, as illustrated in Figure 2. In the figure □ , ⊠ and ▱ represent austenite, self-accommodating martensite and oriented martensite, respectively. Curves (a) and (b) are DSC spectra of heating and cooling transformations, respectively, of a Ti-50.2at%Ni alloy annealed at 1025 K. Curve (c) is the DSC spectrum of the reverse transformation of the same alloy after tensile deformation via stress-induced martensitic transformation. Curve (d) in dashed line is Curve (b) shifted to a lower temperature as a prediction of the transformation of oriented martensite of the polycrystalline alloy on cooling. The magnitude of the shift is arbitrary. Such a transformation is impossible in an isotropic polycrystalline matrix, and thus no direct experimental evidence is available for verification. However, indirect information may be derived from mechanical testing.

Figure 3 shows measurements of the critical stress for stress-induced martensitic transformation as a function of testing temperature for Ti-50.2at%Ni. The samples were annealed at different temperatures, as indicated in the figure. The lines have essentially the same slope: $d\sigma/dT \cong 7.9$ MPa/K. $M_{s,th}$ marks the M_s temperature determined by DSC of the samples, which exhibited essentially the same transformation behaviour [7]. Extrapolations of the linear relationships to zero stress, which is commonly taken as the M_s, are different for the samples. These temperatures are denoted $M_{s,mech}$ to indicate that they are in fact critical temperatures for the formation of mechanically induced martensite at zero stress.

A comparison between $M_{s,mech}$ determined at σ=0 from the above measurement and $M_{s,th}$ determined from DSC measurement is shown in Figure 4. It is seen that $M_{s,mech}$ and $M_{s,th}$ were similar for specimens annealed at T<1000 K. For specimens annealed at T>1000 K, $M_{s,mech}$ deviated to lower values compared to $M_{s,th}$. This suggests that in specimens annealed at T<1000 K, the mechanical resistance to the formation of

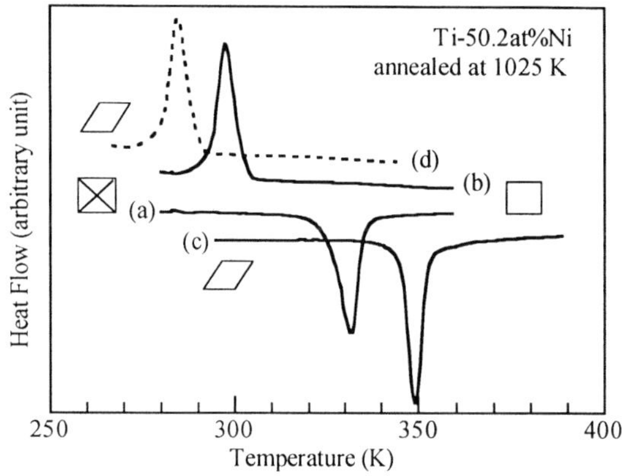

Figure 2. Prediction of critical temperature for the formation of oriented martensite.

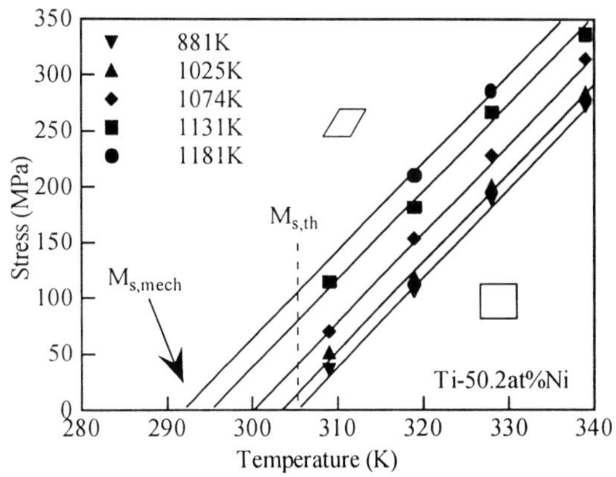

Figure 3. Critical stress for stress-induced martensitic transformation as function of temperature.

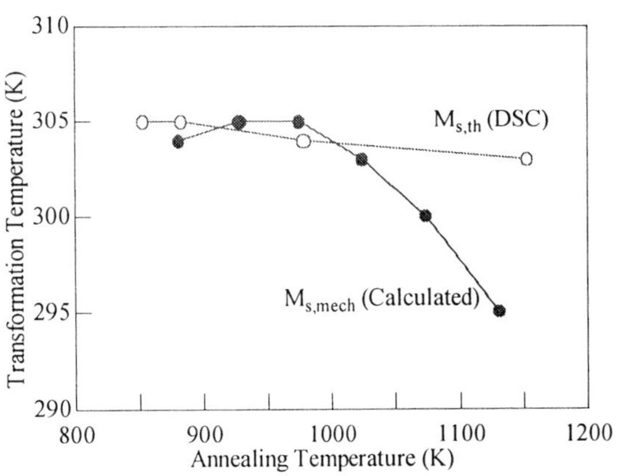

Figure 4. Comparison of critical temperatures for martensitic transformation determined by DSC and mechanical measurement.

oriented martensite is compatible to the frictional resistance to the formation of self-accommodating martensite. In specimens annealed at T>1000 K, the mechanical resistance is greater. Whereas it is beyond the scope of this article to discuss the physical causes of the differences between $M_{s,mech}$ and $M_{s,th}$, it needs to be pointed out that they are of two different mechanical processes and that their values are not comparable on the same basis.

3.2. TRANSFORMATION HYSTERESIS

Figure 5 shows the thermal transformation behaviour of Ti-50.2at%Ni in full and partial transformation cycles. Curve (a) shows the partial forward transformation and the reverse transformation of the partially formed martensite. Curve (b) shows the partial reverse transformation and the formation of martensite from the partially reverted austenite. It is seen that the reverse processes of the partial transformations occurred at the same temperatures with their respective full transformations in both cases, demonstrating that the transformation hysteresis was independent of the fraction of transformation. More partial transformations were performed to facilitate the measurement of transformation hysteresis for different fractions of transformation.

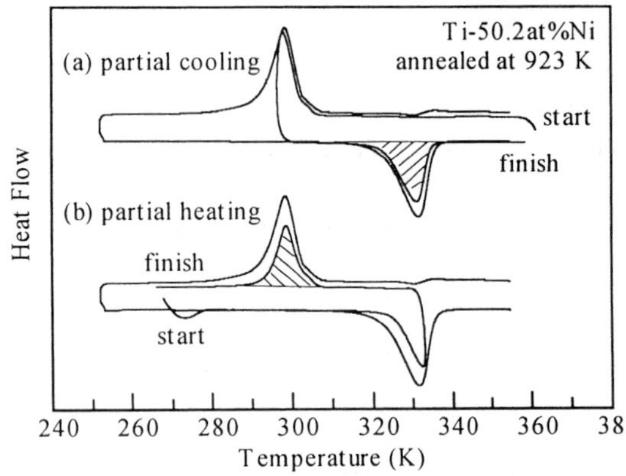

Figure 5. Thermal transformation behaviour in full and partial transformation cycles.

Figure 6 shows pseudoelastic stress-strain behaviour of a Ti-50.15at%Ni plate specimen in shear. The specimen had been annealed at 705 K and deformed in pseudoelasticity to 9% of apparent strain for 70 cycles prior to the measurements to stabilise the stress-strain behaviour. After the initial 70 cycles, a permanent strain of ~3% had been accumulated. To facilitate the determination of stress hysteresis at the middle strain of each pseudoelastic loop and the determination of the transformation strain, elastic deformation has been removed using a shear modulus of 25 GPa for all the curves shown in the figure.

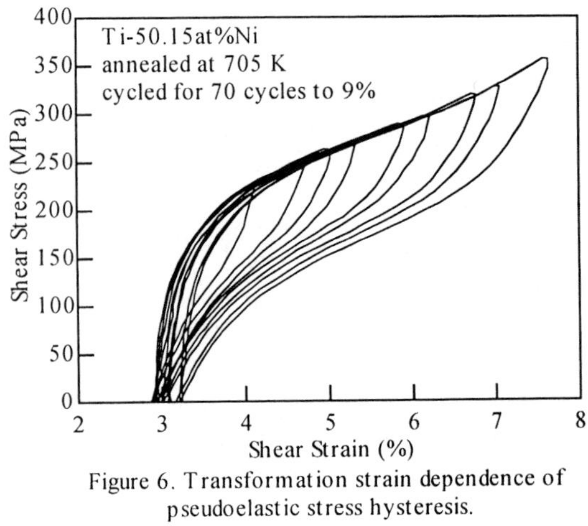

Figure 6. Transformation strain dependence of pseudoelastic stress hysteresis.

Figure 7 shows the transformation fraction

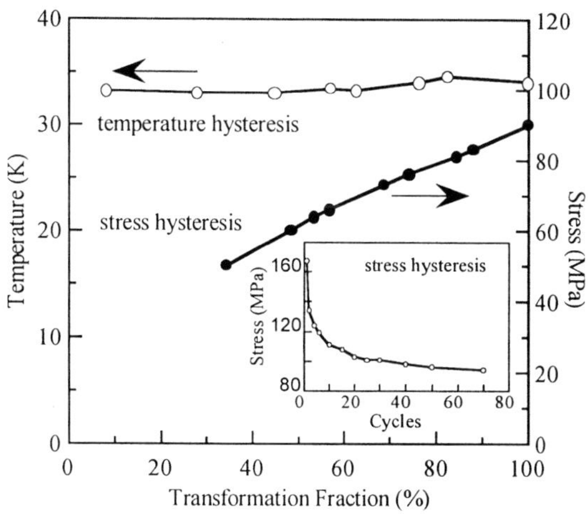

Figure 7. Transformation fraction-dependence of temperature and stess hystereses of martensitic transformation.
Inset: evolution of pseudoelastic hysteresis during initial cycling.

dependence of transformation hystereses for the thermally induced transformations shown in Figure 5 and the stress-induced transformations shown in Figure 6. For the thermal transformations, transformation fraction is determined as the ratio of the heat of a partial transformation, as indicated as the shaded area in Figure 5, to the heat of the corresponding full transformation. For the stress-induced transformations, transformation fraction is determined as the ratio of the strain span of a minor stress-strain loop to the strain span of the largest stress-strain loop. It is seen that the temperature hysteresis of the thermal transformations was independent of the fraction of transformation whereas the stress hysteresis of the stress-induced martensitic transformation was dependent on transformation fraction. It needs to be pointed out that there is no direct comparison between the magnitudes of the two hystereses shown in the figure, even with the aides of the Clausius-Clapeyron equation. This is because of the following reasons. Firstly, the two materials had been annealed at different temperatures, thus exhibit different transformation behaviour. Secondly, the specimen for the measurement of stress hysteresis had been cycled in pseudoelasticity for 70 cycles, during which the stress hysteresis had decreased dramatically from an original magnitude of 167 MPa to 90 MPa, as shown in the inset. Thirdly, the two transformations (the thermal transformation and the stress-induced transformation) were two completely different mechanical processes.

3.3. THERMODYNAMIC CONSIDERATIONS

It is seen from the above discussion that a stress-induced martensitic transformation is different from a thermally induced martensitic transformation both in terms of the critical temperature or stress for the transformation and the hysteretic behaviour. This is due to the differences in ΔE_{el}, which is variant accommodation structure-dependent (state-dependent), and ΔE_{ir}, which is transformation path-dependent, involved in the transformations. Therefore, whereas equation (4) still holds true in principle, the critical temperature determined for a thermal transformation,

$$M_{s,th} = \frac{\Delta H + \Delta E_{el,th} + \Delta E_{ir,th}}{\Delta S} \tag{9}$$

is unequal to the critical temperature determined from equation (8), which is effectively:

$$M_{s,mech} = \frac{\Delta H + \Delta E_{el,mech} + \Delta E_{ir,mech}}{\Delta S} \tag{10}$$

Recognising the difference between $\Delta E_{ir,th}$ and $\Delta E_{ir,mech}$, it is easy to acknowledge that the temperature hysteresis of a thermal transformation, as determined from euqation (3),

$$\Delta T_{h,th} = \frac{\Delta E_{ir,th}}{\Delta S} \tag{11}$$

is unequal to the equivalent temperature hysteresis of the stress hysteresis as determined using the Clausius-Clapeyron equation (equation (6)) :

$$\Delta T_{h,mech} = (\frac{dT}{d\sigma})\Delta\sigma = \frac{\Delta E_{ir,mech}}{\Delta S} \tag{12}$$

The transformation fraction-dependence of $\Delta\sigma$ implies that

$$\Delta E_{ir,mech} = \Delta E_{ir,mech}(\varepsilon) \tag{13}$$

4. Conclusions

The experimental evidences and discussions presented above may be summarised as following. The critical temperatures and derived parameters of thermally induced thermoelastic martensitic transformations are non-equivalent when related via Clausius-Clapeyron equation to the critical stresses and derived parameters of stress-induced martensitic transformations. This is due to the difference in the non-chemical free energy contributions involved in the two processes, which are, from a mechanical viewpoint, two different processes.

REFERENCES

1. Lin, H.C., Wu, S.K., Chou, T.S. and Kao, H.P.: The effects of cold rolling on the martensitic transformation of an equiatomic TiNi alloy, *Acta metall. mater.* **39** (1991), 2069-2080.
2. Piao, M., Otsuka, K., Miyazaki, S. and Horikawa, H.: Mechanism of the As temperature increase by pre-deformation in thermoelastic alloys, *Mater. Trans. JIM* **34** (1993), 919-929.
3. Liu, Y. and Favier, D.: Stabilisation of martensite due to shear deformation via variant reorientation in polycrystalline NiTi, *Acta Materialia* **48** (2000), 3489-3499.
4. Orgéas, L., Liu, Y. and Favier, D.: Experimental study of mechanical hysteresis of NiTi during ferroelastic and superelastic deformation, *J. de Physique IV* **7** (1997), C5-477-482.
5. Liu, Y., Houver, I., Xiang, H., Bataillard, L. and Miyazaki, S.: Strain-dependence of pseudoelastic hysteresis of NiTi, *Metall. Mater. Trans. A* **30A** (1999), 1275-1282.
6. Ortin, J. and Planes, A.: Thermodynamic analysis of thermal measurements in thermoelastic martensitic transformations, *Acta metall.* **36** (1988), 1873-1889.
7. Liu, Y. and McCormick, P.G.: Thermodynamic analysis of the martensitic transformation in TiNi - I. Effect of heat treatment on transformation behaviour, *Acta metall. mater.* **42** (1994), 2401-2406.
8. Wollants, P. Roos, J.R. and Delaye, L.: Thermally- and stress-induced thermoelastic martensitic transformations in the reference frame of equilibrium thermodynamics, *Progress in Mat. Sci.* **37** (1993), 227-288.
9. Liu, Y. and Galvin, S.P.: Criteria for pseudoelasticity in near-equiatomic NiTi shape memory alloys, *Acta Materialia* **45** (1997), 4431-4439.
10. Liu, Y., Liu, Y. and Van Humbeeck, J.: Two-way shape memory effect developed by martensite deformation in NiTi, *Acta Materialia* **47** (1999), 199-209.
11. Wollants, P., De Bonte, M. and Roos, J.R.: A thermodynamic analysis of the stress-induced martensitic transformation in a single crystal, *Z. Metallkd.* **70** (1979), 113-117.
12. Jourdan, C., Guénin, G., Belkahla, S., Gastaldi, J. and Grange, G.: X-ray diffraction effects given by the austenitic phase during the training process of shape memory Cu-Zn-Al polycrystals, J. de Phys. IV 7 (1997), C5-489-494.

TO DESCRIBE THE DEFORMATION FROM DO_3 AUSTENITE TO *18R* MARTENSITE IN SHAPE MEMORY ALLOYS USING GROUP THEORY

J. J. ZHU, W. M. HUANG AND K. M. LIEW
CANES, School of MPE
Nanyang Technological University, Singapore 639798

1. INTRODUCTION

Martensitic transformation is a first-order diffusionless phase transformation from high-temperature austenite phase to low-temperature martensite phase. In Shape Memory Alloys (SMAs), the lattice structure of austenite has higher order of symmetry than that of martensite. As such, more than one martensite variants may be induced from one austenite. Martensite variants are identical crystal lattice but orient along different directions with regard to austenite. The relationship among these lattice structures is essential for understanding better the mechanism behind the surprising shape memory phenomenon, and furthermore, for modeling the sophisticated thermo-mechanical behavior under various loading conditions. So far significant progress has been made both theoretically (Wechsler *et al.*, 1953, Saburi and Nenno, 1981, Ball and James, 1987, Bhattacharya, 1992, Pitteri and Zanzotto, 1998, Hane, 1999) and experimentally (Saburi *et al.*, 1976, 1980, Saburi and Wayman, 1979, Zhang *et al.*, 1999a, 1999b). It was found that self-accommodation plays an important role in shape recovery. However, a systematic description of lattice relationship and the corresponding deformation are still missing. The purpose of this paper is to present an explicit formula for DO_3 to $18R$ martensitc transformation by group theory. This method has been successfully used in the estimation of the deformation energy in martensitic transformation (Zhu *et al.*, 2001). Sections 2 and 3 present the mathematical expression of habit plane variants and the orientation relationship among martensite variants, respectively. Section 4 describes the self-accommodation group. Section 5 presents applications in CuZnAl. Section 6 is conclusions.

2. HABIT PLANE VARIANTS

As shown in Figure 1, for $18R$ martensite, let the austenite lattice coordinate system be $[\mathbf{i}_1, \mathbf{i}_2, \mathbf{i}_3]$ (called $c-$basis), and an orthogonal coordinate system be $[\mathbf{i}'_1, \mathbf{i}'_2, \mathbf{i}'_3]$ (called $o-$basis). The relation between these two systems is given by

Q.P. Sun (ed.), IUTAM Symposium on Mechanics of Martensitic Phase Transformation in Solids, 163–170.

$$\begin{pmatrix} \mathbf{i}'_1 \\ \mathbf{i}'_2 \\ \mathbf{i}'_3 \end{pmatrix} = \begin{pmatrix} \sqrt{2}/2 & 0 & \sqrt{2}/2 \\ 0 & 1 & 0 \\ -\sqrt{2}/2 & 0 & \sqrt{2}/2 \end{pmatrix} \begin{pmatrix} \mathbf{i}_1 \\ \mathbf{i}_2 \\ \mathbf{i}_3 \end{pmatrix}. \tag{1}$$

It is convenient to formulate the change of inter-atomic relation in phase transformation using $o-$basis (Saburi *et al.*, 1976). The transformation from austenite to long-period stacking ordered structure is assumed to take place by contraction along $\mathbf{i}'_2$ and expansions along $\mathbf{i}'_1$ and $\mathbf{i}'_3$. This results in a close-packed plane with normal direction along $\mathbf{i}'_3$. According to crystallographic theory, deformation gradient $\mathbf{F}$ may be expressed by

$$\mathbf{F} = \mathbf{RPB}. \tag{2}$$

Here, $\mathbf{R}$ is arbitrary rigid rotation, $\mathbf{P}$ lattice invariant shear and $\mathbf{B}$ Bain strain. In $o-$basis, they are given by

$$\mathbf{B} = \begin{bmatrix} \eta_1 & 0 & 0 \\ 0 & \eta_2 & 0 \\ 0 & 0 & \eta_3 \end{bmatrix}, \quad \mathbf{P} = \begin{bmatrix} 1 & 0 & g \\ 0 & 1 & 0 \\ 0 & 0 & 1 \end{bmatrix} \tag{3}$$

where

$$\eta_1 = \sqrt{2}a/a_0, \ \eta_2 = b/a_0, \ \eta_3 = 2\sqrt{2}c\sin\beta/(18a_0) \tag{4}$$

Here, a_0 is the lattice parameter of DO_3 austenite, a, b, c, and β are the lattice parameters of $18R$ martensite (Figure 2). According to Hadamard jump condition, at austenite and martensite interphase there are vectors $\mathbf{m}$ and $\mathbf{b}$ that satisfy

$$\mathbf{F} - \mathbf{I} = \mathbf{b} \otimes \mathbf{m} \tag{5}$$

where $\mathbf{I}$ is unit matrix, $\mathbf{m}$ normal vector of habit plane, and $\mathbf{b}$ shear vector. Let

$$\mathbf{C} = \mathbf{F}^T\mathbf{F} = (\mathbf{I} + \mathbf{m} \otimes \mathbf{b})(\mathbf{I} + \mathbf{b} \otimes \mathbf{m}). \tag{6}$$

Assume three eigenvalues of $\mathbf{C}$ are in the sequence as $\lambda_1 \le \lambda_2 \le \lambda_3$. According to Ball and James (1987), $\mathbf{C}$ may be expressed by Eqn.(6) in terms of two non-zero vectors $\mathbf{m}$ and $\mathbf{b}$, if and only if $\lambda_1 \ge 0$ and $\lambda_2 = 1$. In order to satisfy $\lambda_2 = 1$, one must have

$$\det(\mathbf{C} - \mathbf{I}) = 0. \tag{7}$$

Substituting Eqn. (2) into Eqn. (6) and then substituting Eqn. (6) into Eqn. (7) yields

$$g = \pm\sqrt{(1-\eta_1^2)(\eta_3^{-2}-1)}. \tag{8}$$

Thus, three eigenvalues of $\det(\mathbf{C} - \lambda\mathbf{I}) = 0$ are

$$\lambda_1 = \eta_2^2, \ \lambda_2 = 1, \ \lambda_3 = \eta_1^2\eta_3^2. \tag{9}$$

The eigenvector corresponding to $g = -\sqrt{(1-\eta_1^2)(\eta_3^{-2}-1)}$ may be expressed in $o-$basis as

$$\left.\begin{aligned} \mathbf{e}_1 &= (0, \ 1, \ 0) \\ \mathbf{e}_2 &= \left(\eta_1\sqrt{(\eta_3^2-1)/(\eta_1^2\eta_3^2-1)}, \ 0, \ \sqrt{(\eta_1^2-1)/(\eta_1^2\eta_1^3-1)}\right) \\ \mathbf{e}_3 &= \left(\sqrt{(\eta_1^2-1)/(\eta_1^2\eta_3^2-1)}, \ 0, \ -\eta_1\sqrt{(\eta_3^2-1)/(\eta_1^2\eta_3^2-1)}\right) \end{aligned}\right\}. \tag{10}$$

Similarly, the eigenvector corresponding to $g=\sqrt{\left(1-\eta_1^2\right)\left(\eta_3^{-2}-1\right)}$ may be expressed in $o-$basis as

$$\left.\begin{aligned}\mathbf{e}_1' &= \left(0, \quad -1, \quad 0\right)\\ \mathbf{e}_2' &= \left(\eta_1\sqrt{\left(\eta_3^2-1\right)/\left(\eta_1^2\eta_3^2-1\right)}, \quad 0, \quad -\sqrt{\left(\eta_1^2-1\right)/\left(\eta_1^2\eta_1^3-1\right)}\right)\\ \mathbf{e}_3' &= \left(\sqrt{\left(\eta_1^2-1\right)/\left(\eta_1^2\eta_3^2-1\right)}, \quad 0, \quad \eta_1\sqrt{\left(\eta_3^2-1\right)/\left(\eta_1^2\eta_3^2-1\right)}\right)\end{aligned}\right\}. \tag{11}$$

Let $\mathbf{R}_\mathbf{n}^\theta$ be rotation of angle θ about the axis $\mathbf{n}$, then $\mathbf{R}_\mathbf{n}^\theta$ can be expressed in component form as

$$R_{kl} = \delta_{kl}\cos\theta + \varepsilon_{klm}\sin\theta\, n_m + \left(1-\cos\theta\right)n_k n_l \tag{12}$$

where ε_{klm} is the component of permutation tensor, i.e.

$$\varepsilon_{klm} = \begin{cases} +1 & \text{if } (k,l,m)=(1,2,3) \text{ or } (2,3,1) \text{ or } (3,1,2)\\ -1 & \text{if } (k,l,m)=(1,3,2) \text{ or } (3,2,1) \text{ or } (2,1,3)\\ 0 & \text{otherwise}\end{cases}. \tag{13}$$

Let $\mathbf{r}\xrightarrow{\mathbf{R}_\mathbf{n}^\theta}\mathbf{r}^*$, or

$$\mathbf{r}^* = \mathbf{R}_\mathbf{n}^\theta\mathbf{r}\,. \tag{14}$$

It is obvious that $\left[\mathbf{e}_1',\mathbf{e}_2',\mathbf{e}_3'\right]$ can rotate to $\left[\mathbf{e}_1,\mathbf{e}_2,\mathbf{e}_3\right]$ (called $d-$basis) by $\mathbf{R}_{\mathbf{i}_1'}^\pi$. Similarly, $\left[\mathbf{e}_1,\mathbf{e}_2,\mathbf{e}_3\right]$ can rotate to $\left[\mathbf{e}_1',\mathbf{e}_2',\mathbf{e}_1'\right]$ by $\mathbf{R}_{\mathbf{i}_1'}^\pi$. According to Ball and James (1987), $\mathbf{m}$ and $\mathbf{b}$ may be expressed as

$$\left.\begin{aligned}\mathbf{m} &= \left[-\sqrt{\left(1-\lambda_1\right)/\left(\lambda_3-\lambda_1\right)}\mathbf{e}_1 + \kappa\sqrt{\left(\lambda_3-1\right)/\left(\lambda_3-\lambda_1\right)}\mathbf{e}_3\right]\\ \mathbf{b} &= \left(\sqrt{\lambda_3}-\sqrt{\lambda_1}\right)\left[\sqrt{\lambda_3\left(1-\lambda_1\right)/\left(\lambda_3-\lambda_1\right)}\mathbf{e}_1 + \kappa\sqrt{\lambda_1\left(\lambda_3-1\right)/\left(\lambda_3-\lambda_1\right)}\mathbf{e}_3\right]\end{aligned}\right\}. \tag{15}$$

Substituting Eqn. (10) into Eqn. (15) and using Eqn. (1), $\mathbf{R}_{\mathbf{i}_2'}^\pi$ and $\mathbf{C}$ may be expressed in $c-$basis as

$$\mathbf{m} = \sqrt{\frac{1}{2\left(\eta_1^2\eta_3^2-\eta_2^2\right)}}\begin{pmatrix}\kappa\left(\sqrt{\eta_1^2-1}+\eta_1\sqrt{\eta_3^2-1}\right)\\ -\sqrt{2\left(1-\eta_2^2\right)}\\ \kappa\left(\sqrt{\eta_1^2-1}-\eta_1\sqrt{\eta_3^2-1}\right)\end{pmatrix}, \tag{16}$$

$$\mathbf{b} = \sqrt{\frac{\eta_2^2}{2\left(\eta_1^2\eta_3^2-\eta_2^2\right)}}\left(\sqrt{\eta_1^2\eta_3^2}-\sqrt{\eta_2^2}\right)\begin{pmatrix}\kappa\left(\sqrt{\eta_1^2-1}+\eta_1\sqrt{\eta_3^2-1}\right)\\ \sqrt{2\eta_1^2\eta_3^2\left(1-\eta_2^2\right)/\eta_2^2}\\ \kappa\left(\sqrt{\eta_1^2-1}-\eta_1\sqrt{\eta_3^2-1}\right)\end{pmatrix} \tag{17}$$

where $\kappa=\pm1$, which correspond to two pairs of $\left(\mathbf{m},\mathbf{b}\right)$. One pair can be produced from the other by rotation $\mathbf{R}_{\mathbf{e}_1}^\pi$ (or $\mathbf{R}_{\mathbf{i}_2'}^\pi$). They all correspond to the same phase transformation eigenstrain $\mathbf{C}$.

3. ORIENTATION RELATIONSHIP BETWEEN VARIANTS

Let $\tilde{\mathbf{I}}$ be central inverse, i.e. $\tilde{\mathbf{I}} = -\mathbf{I}$. It can be proved that applying $\tilde{\mathbf{I}}$ on $(\mathbf{m}, \mathbf{b})$ produces another pair, which satisfies Eqn. (6) as well. It is well known that any eigenvector multiplied by -1 is also an eigenvector, which corresponds to the same eigenvalue. Therefore, if eigenvectors in Eqns. (10) and (11) are multiplied by -1, the conclusions presented in Section 2 are also valid. As shown in Figure 3, two pairs of $(\mathbf{m}, \mathbf{b})$, which are inverse to each other in nature, correspond to the same martensite variant.

Note that traditionally these two pairs of $(\mathbf{m}, \mathbf{b})$ are considered as one pair, because the habit planes of these two pairs are parallel. Furthermore, we may define point group

$$\mathcal{P}_4 = \left\{\mathbf{I}, \mathbf{R}^{\pi}_{\mathrm{i}'_1}, \mathbf{R}^{\pi}_{\mathrm{i}'_2}, \mathbf{R}^{\pi}_{\mathrm{i}'_3}\right\} \tag{18}$$

and

$$\tilde{\mathcal{P}}_8 = \left\{\mathbf{I}, \tilde{\mathbf{I}}\right\} \otimes \mathcal{P}_4 . \tag{19}$$

To summarize the discussion above, we may conclude that eight pairs of $(\mathbf{m}, \mathbf{b})$ can be obtained in $o-$ basis. If any one pair of $(\mathbf{m}, \mathbf{b})$ is given, operation $\tilde{\mathcal{P}}_8$ on this pair results in other seven pairs. If two pairs, which are inverse to each other, are considered as one, there are only four pairs of $(\mathbf{m}, \mathbf{b})$. These four pairs of $(\mathbf{m}, \mathbf{b})$ correspond to four different habit planes. If one pair of $(\mathbf{m}, \mathbf{b})$ is known, operation $\mathcal{P}_4$ gives other three pairs. These four pairs of $(\mathbf{m}, \mathbf{b})$ only correspond to two different phase transformation eigenstrains. Two pairs of $(\mathbf{m}, \mathbf{b})$, which can be produced from one to the other by rotation transformation $\mathbf{R}^{\pi}_{\mathrm{i}'_2}$, correspond to one $\mathbf{C}$. In fact, group $\mathcal{P}_4$ may be decomposed into

$$\mathcal{P}_4 = \left\{\mathbf{I}, \mathbf{R}^{\pi}_{\mathrm{i}'_1}, \mathbf{R}^{\pi}_{\mathrm{i}'_2}, \mathbf{R}^{\pi}_{\mathrm{i}'_3}\right\} = \left\{\mathbf{I}, \mathbf{R}^{\pi}_{\mathrm{i}'_1}\right\} \otimes \left\{\mathbf{I}, \mathbf{R}^{\pi}_{\mathrm{i}'_2}\right\}. \tag{20}$$

Group $\mathcal{P}_2 = \left\{\mathbf{I}, \mathbf{R}^{\pi}_{\mathrm{i}'_2}\right\}$ is a regular subgroup of $\mathcal{P}_4$. Phase transformation eigenstrain $\mathbf{C}$ is invariant in operation $\mathcal{P}_2$. Two pairs of $(\mathbf{m}, \mathbf{b})$ generated by operation $\mathcal{P}_2$ have the same base plane and lattice invariant shear direction. Four pairs of $(\mathbf{m}, \mathbf{b})$ generated by operation $\mathcal{P}_4$ have the same base plane. Thus $\mathcal{P}_4$ may be called Base Plane Group. Equation (20) may be expressed in $c-$ basis as

$$\mathcal{P}_4 = \left\{\mathbf{I}, \mathbf{R}^{\pi}_{(\mathrm{i}_1-\mathrm{i}_3)/\sqrt{2}}, \mathbf{R}^{\pi}_{\mathrm{i}_2}, \mathbf{R}^{\pi}_{(\mathrm{i}_1+\mathrm{i}_3)/\sqrt{2}}\right\} = \left\{\mathbf{I}, \mathbf{R}^{\pi}_{(\mathrm{i}_1-\mathrm{i}_3)/\sqrt{2}}\right\} \otimes \left\{\mathbf{I}, \mathbf{R}^{\pi}_{\mathrm{i}_2}\right\}. \tag{21}$$

It is known that point group of DO_3 (or $B2$) austenite $\mathcal{P}_{24}$ may be expressed as (Pitteri and Zanzotto, 1998)

$$\mathcal{P}_{24} = \left\{\begin{array}{l} \mathbf{I}, \mathbf{R}^{\pi}_{\mathrm{i}_1}, \mathbf{R}^{\pi}_{\mathrm{i}_2}, \mathbf{R}^{\pi}_{\mathrm{i}_3}, \mathbf{R}^{\pi/2}_{\mathrm{i}_1}, \mathbf{R}^{3\pi/2}_{\mathrm{i}_1}, \mathbf{R}^{\pi/2}_{\mathrm{i}_2}, \mathbf{R}^{3\pi/2}_{\mathrm{i}_2}, \mathbf{R}^{\pi/2}_{\mathrm{i}_3}, \mathbf{R}^{3\pi/2}_{\mathrm{i}_3}, \mathbf{R}^{\pi}_{(\mathrm{i}_1-\mathrm{i}_2)/\sqrt{2}}, \mathbf{R}^{\pi}_{(\mathrm{i}_1+\mathrm{i}_2)/\sqrt{2}} \\ \mathbf{R}^{\pi}_{(\mathrm{i}_1-\mathrm{i}_3)/\sqrt{2}}, \mathbf{R}^{\pi}_{(\mathrm{i}_1+\mathrm{i}_3)/\sqrt{2}}, \mathbf{R}^{\pi}_{(\mathrm{i}_2-\mathrm{i}_3)/\sqrt{2}}, \mathbf{R}^{\pi}_{(\mathrm{i}_2+\mathrm{i}_3)/\sqrt{2}}, \mathbf{R}^{2\pi/3}_{(\mathrm{i}_1+\mathrm{i}_2+\mathrm{i}_3)/\sqrt{3}}, \mathbf{R}^{4\pi/3}_{(\mathrm{i}_1+\mathrm{i}_2+\mathrm{i}_3)/\sqrt{3}} \\ \mathbf{R}^{2\pi/3}_{(\mathrm{i}_1+\mathrm{i}_2-\mathrm{i}_3)/\sqrt{3}}, \mathbf{R}^{4\pi/3}_{(\mathrm{i}_1+\mathrm{i}_2-\mathrm{i}_3)/\sqrt{3}}, \mathbf{R}^{2\pi/3}_{(\mathrm{i}_1-\mathrm{i}_2+\mathrm{i}_3)/\sqrt{3}}, \mathbf{R}^{4\pi/3}_{(\mathrm{i}_1-\mathrm{i}_2+\mathrm{i}_3)/\sqrt{3}}, \mathbf{R}^{2\pi/3}_{(\mathrm{i}_1-\mathrm{i}_2-\mathrm{i}_3)/\sqrt{3}}, \mathbf{R}^{4\pi/3}_{(\mathrm{i}_1-\mathrm{i}_2-\mathrm{i}_3)/\sqrt{3}} \end{array}\right\}. \tag{22}$$

It is noted that $\mathcal{P}_4$ is a subgroup of $\mathcal{P}_{24}$, i.e.,

$$\mathcal{P}_{24} = \left\{\mathbf{I}, \mathbf{R}^{\pi/2}_{\mathrm{i}_1}, \mathbf{R}^{\pi/2}_{\mathrm{i}_2}, \mathbf{R}^{\pi/2}_{\mathrm{i}_3}, \mathbf{R}^{2\pi/3}_{(\mathrm{i}_1+\mathrm{i}_2+\mathrm{i}_3)/\sqrt{3}}, \mathbf{R}^{4\pi/3}_{(\mathrm{i}_1-\mathrm{i}_2-\mathrm{i}_3)/\sqrt{3}}\right\} \otimes \mathcal{P}_4 . \tag{23}$$

In another words,

$$\tilde{\mathcal{P}}_{48} = \left\{\mathbf{I},\ \mathbf{R}_{\mathbf{i}_1}^{\pi/2},\ \mathbf{R}_{\mathbf{i}_2}^{\pi/2},\ \mathbf{R}_{\mathbf{i}_3}^{\pi/2},\ \mathbf{R}_{(\mathbf{i}_1+\mathbf{i}_2+\mathbf{i}_3)/\sqrt{3}}^{2\pi/3},\ \mathbf{R}_{(\mathbf{i}_1-\mathbf{i}_2-\mathbf{i}_3)/\sqrt{3}}^{4\pi/3}\right\} \otimes \tilde{\mathcal{P}}_8 \tag{24}$$

where $\tilde{\mathcal{P}}_{48} = \left\{\mathbf{I},\ \tilde{\mathbf{I}}\right\} \otimes \mathcal{P}_{24}$. Thus, only six $o-$bases are required in order to get all pairs of $(\mathbf{m},\ \mathbf{b})$. The corresponding $\mathbf{R}_i$ may be taken from the left coset of $\mathcal{P}_4$ in $\mathcal{P}_{24}$. In each $o-$basis, there are eight pairs of $(\mathbf{m},\ \mathbf{b})$. Hence, there are 48 pairs of $(\mathbf{m},\ \mathbf{b})$ in total. Another approach to produce all pairs of $(\mathbf{m},\ \mathbf{b})$ is to take only one pair of $(\mathbf{m},\ \mathbf{b})$ in $o-$basis, which corresponds to $\kappa = +1$ in Eqn. (15). Other 47 pairs of $(\mathbf{m},\ \mathbf{b})$ can be generated by operation $\tilde{\mathcal{P}}_{48}$ on this pair, i.e., $(\mathbf{R}_i \cdot \mathbf{m},\ \mathbf{R}_i \cdot \mathbf{b})$, $\mathbf{R}_i \in \tilde{\mathcal{P}}_{48}$. The resulting pairs may be denoted by the corresponding element in $\tilde{\mathcal{P}}_{48}$. It is noted that $\mathcal{P}_{24}$ can be decomposed into

$$\mathcal{P}_{24} = \begin{Bmatrix} \mathbf{I}, & \mathbf{R}_{\mathbf{i}_1}^{\pi/2}, & \mathbf{R}_{\mathbf{i}_2}^{\pi/2}, & \mathbf{R}_{\mathbf{i}_3}^{\pi/2}, & \mathbf{R}_{(\mathbf{i}_1+\mathbf{i}_2+\mathbf{i}_3)/\sqrt{3}}^{2\pi/3}, & \mathbf{R}_{(\mathbf{i}_1-\mathbf{i}_2-\mathbf{i}_3)/\sqrt{3}}^{4\pi/3}, \\ \mathbf{R}_{(\mathbf{i}_1-\mathbf{i}_3)/\sqrt{2}}^{\pi}, & \mathbf{R}_{(\mathbf{i}_1+\mathbf{i}_2-\mathbf{i}_3)/\sqrt{3}}^{4\pi/3}, & \mathbf{R}_{\mathbf{i}_3}^{\pi}, & \mathbf{R}_{(\mathbf{i}_1+\mathbf{i}_2-\mathbf{i}_3)/\sqrt{3}}^{2\pi/3}, & \mathbf{R}_{(\mathbf{i}_2-\mathbf{i}_3)/\sqrt{2}}^{\pi}, & \mathbf{R}_{\mathbf{i}_3}^{3\pi/2} \end{Bmatrix} \otimes \left\{\mathbf{I},\ \mathbf{R}_{\mathbf{i}_2}^{\pi}\right\}. \tag{25}$$

Applying operation $\mathbf{R}_{\mathbf{i}_2}^{\pi}$ on this pair of $(\mathbf{m},\ \mathbf{b})$, which corresponds to $\kappa = +1$ in Eqn. (15), results in another pair, which corresponds to $\kappa = -1$ in Eqn. (15). These two pairs of $(\mathbf{m},\ \mathbf{b})$ (denoted by $\left\{\mathbf{I},\ \mathbf{R}_{\mathbf{i}_2}^{\pi}\right\}$) correspond to the same phase transformation eigenstrain $\mathbf{C}$. There are 12 elements in left coset of $\mathcal{P}_2$ in $\mathcal{P}_{24}$. Applying these 12 operations on those two pairs of $(\mathbf{m},\ \mathbf{b})$ results in 24 pairs of $(\mathbf{m},\ \mathbf{b})$ (denoted by $\mathcal{P}_{24}$, corresponding to 24 habit planes). Finally, applying operations $\left\{\mathbf{I},\ \tilde{\mathbf{I}}\right\}$ on these 24 pairs of $(\mathbf{m},\ \mathbf{b})$ gives 48 different pairs of $(\mathbf{m},\ \mathbf{b})$. It is obvious that there are 12 different phase transformation eigenstrains in $18R$ martensite. Each element of left coset of $\mathcal{P}_2$ in $\mathcal{P}_{24}$ corresponds to one phase transformation eigenstrain $\mathbf{C}$.

4. SELF-ACCOMMODATION GROUP

Let

$$\mathcal{P}_4^* = \left\{\mathbf{I},\ \mathbf{R}_{\mathbf{i}_3}^{\pi}\right\} \otimes \left\{\mathbf{I},\ \mathbf{R}_{(\mathbf{i}_1-\mathbf{i}_2)/\sqrt{2}}^{\pi}\right\} = \left\{\mathbf{I},\ \mathbf{R}_{(\mathbf{i}_1-\mathbf{i}_2)/\sqrt{2}}^{\pi},\ \mathbf{R}_{\mathbf{i}_3}^{\pi},\ \mathbf{R}_{(\mathbf{i}_1+\mathbf{i}_2)/\sqrt{2}}^{\pi}\right\}. \tag{26}$$

It can be proved that $\mathcal{P}_4^*$ is a subgroup of $\mathcal{P}_{24}$. Take one pair of $(\mathbf{m},\ \mathbf{b})$ from $o-$basis, which corresponds to $\kappa = +1$ in Eqn. (15). Applying operations in $\mathcal{P}_4^*$ on this pair of $(\mathbf{m},\ \mathbf{b})$ results in four pairs of $(\mathbf{m},\ \mathbf{b})$. These four pairs of $(\mathbf{m},\ \mathbf{b})$, which correspond to four martensite variants, form a self-accommodation group. Four $\mathbf{m}$ correspond to $\left\{\mathbf{I},\ \mathbf{R}_{(\mathbf{i}_1-\mathbf{i}_2)/\sqrt{2}}^{\pi},\ \tilde{\mathbf{R}}_{\mathbf{i}_3}^{\pi},\ \tilde{\mathbf{R}}_{(\mathbf{i}_1+\mathbf{i}_2)/\sqrt{2}}^{\pi}\right\}$ which cluster along $(\mathbf{i}_1 - \mathbf{i}_2)/\sqrt{2}$ direction, and the other four $\mathbf{m}$ correspond to $\left\{\tilde{\mathbf{I}},\ \tilde{\mathbf{R}}_{(\mathbf{i}_1-\mathbf{i}_2)/\sqrt{2}}^{\pi},\ \mathbf{R}_{\mathbf{i}_3}^{\pi},\ \mathbf{R}_{(\mathbf{i}_1+\mathbf{i}_2)/\sqrt{2}}^{\pi}\right\}$ which cluster along $(\mathbf{i}_2 - \mathbf{i}_1)/\sqrt{2}$ direction. One group is inverse of the other. Traditionally, they are considered as one group. We may denote them as $\left\{\mathbf{I},\ \mathbf{R}_{(\mathbf{i}_1-\mathbf{i}_2)/\sqrt{2}}^{\pi},\ \mathbf{R}_{\mathbf{i}_3}^{\pi},\ \mathbf{R}_{(\mathbf{i}_1+\mathbf{i}_2)/\sqrt{2}}^{\pi}\right\}$ (or $\mathcal{P}_4^*$). Point group of austenite $\mathcal{P}_{24}$ may be decomposed into

$$\mathcal{P}_{24} = \{\mathbf{I}, \mathbf{R}_{\mathbf{i}_1}^{\pi/2}, \mathbf{R}_{\mathbf{i}_2}^{\pi/2}\} \otimes \{\mathbf{I}, \mathbf{R}_{\mathbf{i}_3}^{\pi/2}\} \otimes \mathcal{P}_4^*$$
$$= \{\mathbf{I}, \mathbf{R}_{\mathbf{i}_1}^{\pi/2}, \mathbf{R}_{\mathbf{i}_2}^{\pi/2}, \mathbf{R}_{\mathbf{i}_3}^{\pi/2}, \mathbf{R}_{(\mathbf{i}_1+\mathbf{i}_2+\mathbf{i}_3)/\sqrt{3}}^{2\pi/3}, \mathbf{R}_{(\mathbf{i}_1-\mathbf{i}_2+\mathbf{i}_3)/\sqrt{3}}^{2\pi/3}\} \otimes \mathcal{P}_4^* \tag{27}$$

Thus, there are six self-accommodation groups. Other five groups can be derived from one group by rotation $\mathcal{P}_4^*$. The rotation operation is taken from the left coset of $\mathcal{P}_4^*$ in $\mathcal{P}_{24}$. From Eqn. (27)

$$\tilde{\mathcal{P}}_{48} = \{\mathbf{I}, \mathbf{R}_{\mathbf{i}_1}^{\pi/2}, \mathbf{R}_{\mathbf{i}_2}^{\pi/2}, \mathbf{R}_{\mathbf{i}_3}^{\pi/2}, \mathbf{R}_{(\mathbf{i}_1+\mathbf{i}_2+\mathbf{i}_3)/\sqrt{3}}^{2\pi/3}, \mathbf{R}_{(\mathbf{i}_1-\mathbf{i}_2+\mathbf{i}_3)/\sqrt{3}}^{2\pi/3}\} \otimes \tilde{\mathcal{P}}_8^* \tag{28}$$

where

$$\tilde{\mathcal{P}}_8^* = \{\mathbf{I}, \tilde{\mathbf{I}}\} \otimes \mathcal{P}_4^*. \tag{29}$$

In one self-accommodation group, there are eight normal directions corresponding to four habit plane variants, which cluster along directions $\pm(\mathbf{i}_1 - \mathbf{i}_2)/\sqrt{2}$. There are 48 normal directions in total (corresponding to 24 martensite variants). They cluster along 12 (110) directions in austenite lattice coordinate system $[\mathbf{i}_1, \mathbf{i}_2, \mathbf{i}_3]$ (c-basis).

5. APPLICATION

The case investigated is cubic to monoclinic transformation ($DO_3 \rightarrow 6M$) in CuZnAl. Its lattice parameters were measured by Chakravorty and Wayman (1997). From Eqns. (15), and take $\kappa = +1$, we can obtain a pair of vectors $\mathbf{m}$ and $\mathbf{b}$. Operation $\mathcal{P}_4^*$ on this pair of $(\mathbf{m}, \mathbf{b})$ results in four pairs of $(\mathbf{m}, \mathbf{b})$, which form a self-accommodation group. According to Eqn.(27), the left coset of $\mathcal{P}_4^*$ in $\mathcal{P}_{24}$ may be written as

$$\{\mathbf{I}, \mathbf{R}_{\mathbf{i}_1}^{\pi/2}, \mathbf{R}_{\mathbf{i}_2}^{\pi/2}, \mathbf{R}_{\mathbf{i}_3}^{\pi/2}, \mathbf{R}_{(\mathbf{i}_1+\mathbf{i}_2+\mathbf{i}_3)/\sqrt{3}}^{2\pi/3}, \mathbf{R}_{(\mathbf{i}_1-\mathbf{i}_2+\mathbf{i}_3)/\sqrt{3}}^{2\pi/3}\}. \tag{30}$$

Applying operations in Eqn. (30) on this self-accommodation group yield six self-accommodation groups. There are four pairs of $(\mathbf{m}, \mathbf{b})$ in each group. We may denote them as $(1,2,3,4)$, $(5,6,7,8)$, □, $(21,22,23,24)$. The dark dots marked on the unit sphere in Figure 4 represent these 24 $\mathbf{m}$. Finally, applying $\tilde{\mathbf{I}}$ on these 24 pairs of $(\mathbf{m}, \mathbf{b})$ gives 24 pairs of $(\mathbf{m}, \mathbf{b})$ as marked by white spots in Figure 4. As we can see in Figure 4, $\{\mathbf{m}_1, \mathbf{m}_2, \tilde{\mathbf{m}}_3, \tilde{\mathbf{m}}_4\}$ cluster along $(\mathbf{i}_1 - \mathbf{i}_2)/\sqrt{2}$ direction; $\{\tilde{\mathbf{m}}_1, \tilde{\mathbf{m}}_2, \mathbf{m}_3, \mathbf{m}_4\}$ cluster along $(\mathbf{i}_2 - \mathbf{i}_1)/\sqrt{2}$ direction; $\{\tilde{\mathbf{m}}_{17}, \tilde{\mathbf{m}}_{18}, \mathbf{m}_{19}, \mathbf{m}_{20}\}$ cluster along $(\mathbf{i}_3 - \mathbf{i}_2)/\sqrt{2}$ direction; $\{\mathbf{m}_{17}, \mathbf{m}_{18}, \tilde{\mathbf{m}}_{19}, \tilde{\mathbf{m}}_{20}\}$ cluster along $(\mathbf{i}_2 - \mathbf{i}_3)/\sqrt{2}$ direction, and so on. A diamond, which represents one martensite self-accommodation group $\mathcal{P}_4^*$, is shown in Figure 5. It is surrounded by austenite with eight habit planes. The normal directions of these eight habit planes are $\{\mathbf{m}_1, \mathbf{m}_2, \tilde{\mathbf{m}}_3, \tilde{\mathbf{m}}_4\}$ and $\{\tilde{\mathbf{m}}_1, \tilde{\mathbf{m}}_2, \mathbf{m}_3, \mathbf{m}_4\}$. It is found that the interface between martensite variants (midrib) of this diamond may be classified into three different types, namely S_{12}, S_{13} and S_{14} (Zhang *et al.*, 1999b). The other interfaces maybe generated by rotation these three interfaces. According to group theory presented above, it is because that there are only three types of misorientation between two adjacent martensite variants in one self-accommodation group. For instance, S_{24} can be generated from S_{13} through $\mathbf{R}^{\pi}_{(\mathbf{i}_1-\mathbf{i}_2)/\sqrt{2}}$.

6. CONCLUSIONS

We present a clear and systematic description on the deformation from DO_3 austenite to $18R$ martensite in shape memory alloys using group theory. According to group theory, there are 24 elements in point group of austenite $\mathcal{P}_{24}$. Point group $\mathcal{P}_4$ is a subgroup of $\mathcal{P}_{24}$. Martensite point group $\mathcal{P}_2$ is a regular subgroup of $\mathcal{P}_{24}$. Phase transformation eigenstrain **C** is invariant for operations in $\mathcal{P}_2$. Point group $\mathcal{P}_4^*$ is a subgroup of $\mathcal{P}_{24}$, which stands for self-accommodation group in $18R$ martensite. We then apply them in CuZnAl shape memory alloy. This method has been successfully used in the estimation of the deformation energy in martensitic transformation (Zhu *et al.*, 2001). Currently we are working on applying this method in other shape memory alloys.

REFERENCES

Ball, J. M. and James, R. D. (1987), Fine phase mixtures as minimizers of energy, *Arch. Rat. Mech. Anal.*, 100 : 13–52.

Bhattacharya, K. (1992), Self-accommodation in martensite, *Arch. Rat. Mech. Anal.*, 120 : 201–244.

Chakravorty, S. and Wayman, C.M. (1997), Electron microscopy of internally faulted Cu-Zn-Al martensite, *Acta Metall.*, 25 : 989-1000.

Hane, K.F. (1999), Bulk and thin film microstructures in untwinned martesites, *J.Mech. Phys. Solids.*, 47:1917-1939.

Pitteri, M. and Zanzotto, G. (1998), Generic and non-generic cubic-tomonoclinic transformations and their twins, *Acta Mater.*, 46:225-237.

Saburi, T. and Nenno, S. (1981), The shape memory effect and related phenomena, in Aaronson, H. I., Laughlin, D. E., Sekerka, R. F., and Wayman, C. M., editors, *Proceedings of An International Conference on Solid to Solid Phase Transformations*, pp.1455–1479, The Metallurgical Society of AIME.

Saburi, T., Nenno, S., Kato, S. and Takata, K. (1976), Configuration of martensite variants in Cu-Zn-Ga, *J. Less Common Met.*, 50 : 223-236.

Saburi, T. and Wayman, C.M. (1979), Crystallographic similarities in shape memory martensite, *Acta Metall.*, 27 : 979-995.

Saburi, T., Wayman, C.M., Takata, K. and Nenno, S. (1980), The shape memory mechanics in 18*R* martensite alloys, *Acta Metall.*, 28 : 15-32.

Wechsler, M. S., Lieberman, D. S. and Read, T. A. (1953), On the theory of the formation of martensite, *J. Metals*, 197:1503–1515.

Zhang, J.X., Zheng, Y.F., Luo, Y.C. and Zhao, L.C. (1999a), Substructure and boundary structure of deformed 18*R* martensite in a Cu-Zn-Al alloy, *Acta Mater.*, 47:3497-3506.

Zhang, J.X., Zheng, Y.F. and Zhao, L.C. (1999b), The structure and mobility of the intervariant boundaries in 18*R* martensite in a Cu-Zn-Al alloy, *Acta Mater.*, 47:2125-2141.

Zhu, J.J., Huang, W.M. and Liew, K.M. (2001), Deformation energy in martensitic transformation (in preparation).

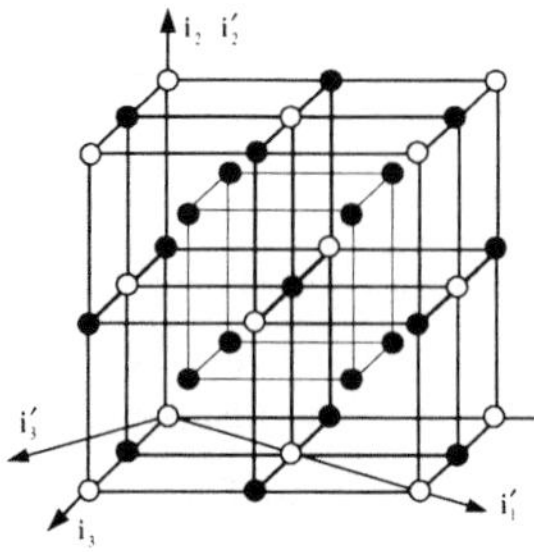

Figure 1. Lattice structure of DO_3 austenite.

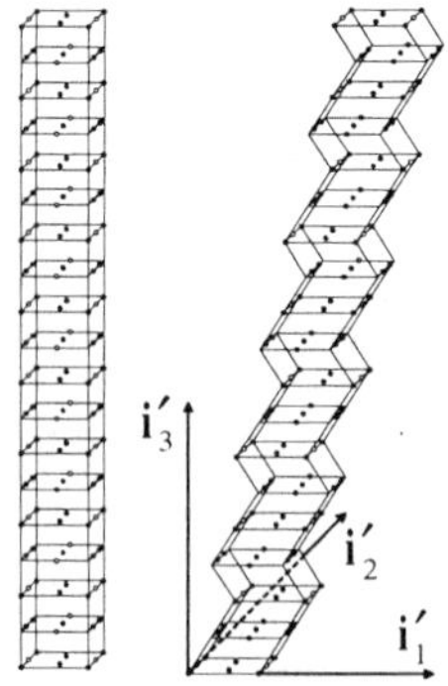

Figure 2. Deformation from austenite to martensite.

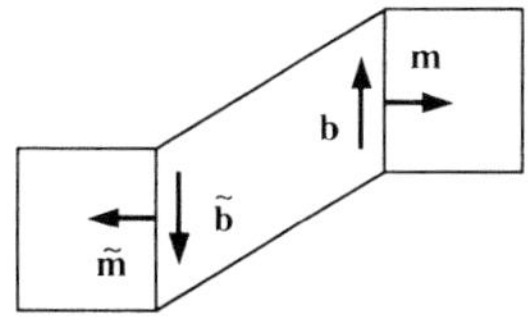

Figure 3. Two pairs of $(\mathbf{m}, \mathbf{b})$ in one martensite variant.

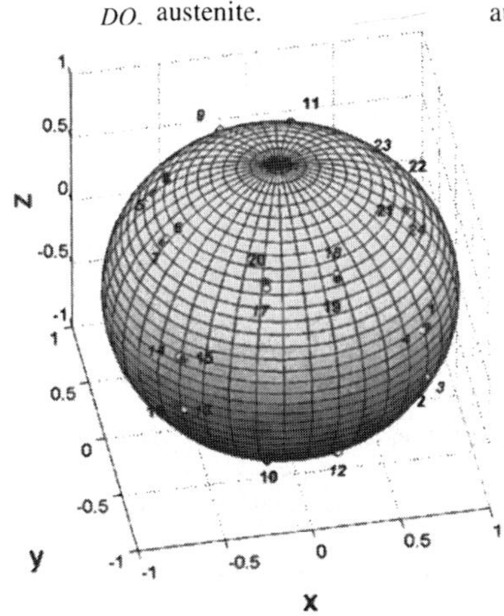

Figure 4. Distribution of 48 normal directions of martensite habit plane variants. (a) Overall view.

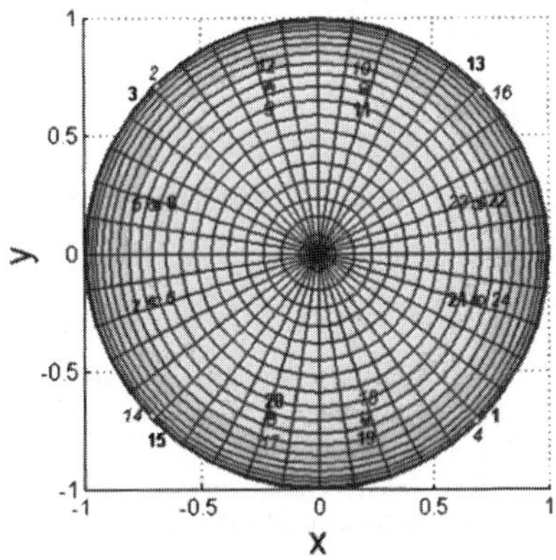

(b) Top view.

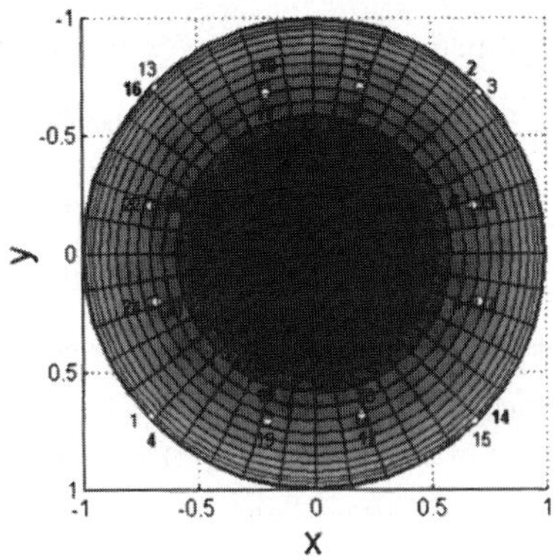

(c) Bottom view.

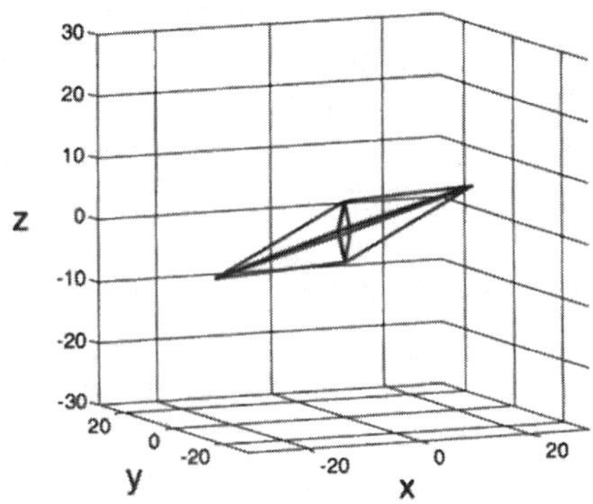

Figure 5. Self-accommodation diamond.

Pressurized Heterogeneous Thin Films of Martensitic Materials

Y.C. Shu
Institute of Applied Mechanics, National Taiwan University, Taipei, Taiwan, ROC

1 Introduction

The explosive growth of micro systems has created a great need for the development of microfluid devices. These devices including micropumps and microvalves have a wide range of applications in fields such as drug delivery, inkjet printing and cooling systems of electronic circuits. In such applications, micropumps with large pumping volume per cycle and high pumping pressure are essential to successfully control microfluid. They need high stroke to achieve large fluid displacement, and large actuation energy density to transmit high force. However, traditional MEMS-integrated micropumps are limited both in stroke and actuation force; as a result, the typical output pressure of these pumps is in the order of several tens of kPa.

Shape-memory alloy (SMA), on the other hand, is an excellent candidate in actuation due to extremely large recoverable strain and high work output per volume per cycle. In addition, the disadvantage of low response rate caused by cooling and heating bulk SMAs can be greatly improved at small scales because of the increase in the surface area to volume ratio. This makes these alloys in thin-film form ideal for use in MEMS-integrated microactuators and micropumps.

In this paper, we propose an analytic model to study the behavior of shape-memory micropumps. Our work was motivated by the most recent experiment on the fabrication of SMA actuated micropumps [5, 6]. They have reported that a TiNi micropump has been fabricated with pumping pressure up to several hundreds of kPa. The ratio of the maximum deflection to the radius of the TiNi diaphragm has been measured about 0.025. However, this ratio is smaller than that predicted by our theoretical model here. The discrepancy is not completely understood here, but we believe there should be plenty of room to improve this critical ratio to design large stroke micropumps. In particular, we show that (100) Cu-based

Q.P. Sun (ed.), IUTAM Symposium on Mechanics of Martensitic Phase Transformation in Solids, 171–178.

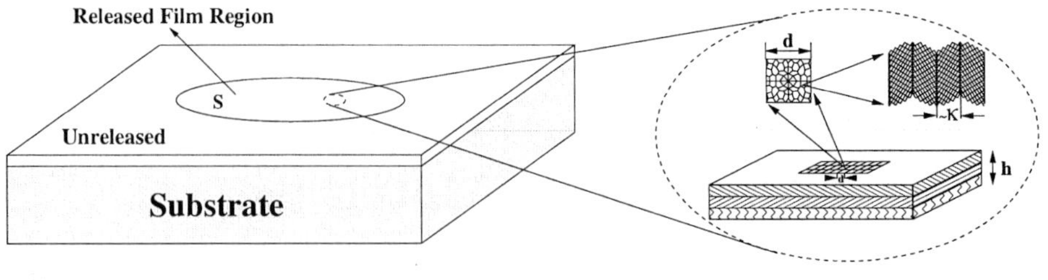

Figure 1: Prototype of a micropump using shape-memory material. The film is heterogeneous and contains three different length scales. It is released from the substrate in the chosen region S, but attached to it outside.

shape-memory thin films may have better behavior than TiNi film regards to the purpose of large deflection.

2 Rescaled Setting of the 3D Model

Consider a thin film released from the substrate in a well-defined region, but constrained on its lateral boundaries as shown in Figure 1. A pressure is applied from either above or below depending on actuation types including evacuation and pressurization types. We assume this film is heterogeneous and contains three relevant length scales (see Figure 1): the thickness of the film h, the typical grain size which is essentially d, and the typical length scale of the microstructure which is essentially κ. Since we anticipate large strains, we use the setting of finite deformations. Let $\mathbf{y}$ be the deformation of the film with plane S, and $\mathbf{x} = (x_1, x_2, x_3)$ be the material point relative to an orthonormal film basis $\{\mathbf{e}_1, \mathbf{e}_2, \mathbf{e}_3\}$. Suppose a hydrostatic pressure $p^{(h)}$ is applied on the lower surface of the film. The total energy of such a film is

$$e^{(h)}[\mathbf{y}] = \int_{S\times(0,h)} \left\{\kappa^2 \left|\nabla^2 \mathbf{y}\right|^2 + \varphi\left(\nabla \mathbf{y}, \frac{x_1}{d}, \frac{x_2}{d}, \frac{x_3}{h}\right)\right\} d\mathbf{x} - \frac{p^{(h)}}{3} \int_{S\times\{0\}} \mathbf{y}\cdot(\mathbf{y}_{,1} \wedge \mathbf{y}_{,2})\, dx_1 dx_2, \quad (2.1)$$

where the first term accounts for the "interfacial energy", the second term is the "elastic energy" with density φ, and the final term is interpreted as the energy of a fluid under the film with pressure $p^{(h)}$. For a martensitic material φ is nonconvex, with multi-well structure – one well for each phase or variant. To understand the behavior of this film, we need to minimize this energy amongst all possible deformations $\mathbf{y}$ at pressure $p^{(h)}$. This is a rather difficult problem. However, we know that the lateral extent (i.e., the in-plane dimensions) of the film is much larger than any of the length scales κ, d or h. So the macroscopic behavior of the film does not depend on every detail of the grains and multilayers, but only on some average features.

To make our analysis amenable to the Γ-convergence method, we consider the rescaled problem. We introduce the following change of variables:

$$\mathbf{z}_p = (z_1, z_2) = \mathbf{x}_p = (x_1, x_2), \quad z_3 = \frac{1}{h} x_3, \quad \mathbf{x} \in \Omega^h, \quad \mathbf{z} \in \Omega^1 = S \times (0,1). \tag{2.2}$$

To each deformation $\tilde{\mathbf{y}} : \Omega^h \to I\!R^3$ we associate a deformation $\mathbf{y} : \Omega^1 \to I\!R^3$ via $\mathbf{y}(\mathbf{z}(\mathbf{x})) = \tilde{\mathbf{y}}(\mathbf{x}), \ \mathbf{x} \in \Omega^h$. Let $\nabla_p = \mathbf{y}_{,1} \otimes \mathbf{e}_1 + \mathbf{y}_{,2} \otimes \mathbf{e}_2$ be the in-plane gradient. We consider the energy per unit film thickness $e_1^{(h)} = \frac{1}{h} e^{(h)}$. From (2.1) and (2.2),

$$\begin{aligned} e_1^{(h)}[\mathbf{y}] &= \int_{\Omega^1} \left\{ \kappa^2 \left(|\nabla_p^2 \mathbf{y}|^2 + \frac{2}{h^2} |\nabla_p \mathbf{y}_{,3}|^2 + \frac{1}{h^4} |\mathbf{y}_{,33}|^2 \right) + \varphi \left(\mathbf{y}_{,1} \mid \mathbf{y}_{,2} \mid \frac{1}{h} \mathbf{y}_{,3}, \frac{\mathbf{z}_p}{d}, z_3 \right) \right\} d\mathbf{z} \\ &\quad - \frac{1}{3} \frac{p^{(h)}}{h} \int_{S \times \{0\}} \mathbf{y} \cdot (\mathbf{y}_{,1} \wedge \mathbf{y}_{,2}) \, dz_1 dz_2. \end{aligned} \tag{2.3}$$

We have used the notation $\mathbf{F} = (\mathbf{f}_1 | \mathbf{f}_2 | \mathbf{f}_3) = \mathbf{f}_1 \otimes \mathbf{e}_1 + \mathbf{f}_2 \otimes \mathbf{e}_2 + \mathbf{f}_3 \otimes \mathbf{e}_3$ for $\mathbf{F} \in I\!M^{3\times 3}$ where $I\!M^{m\times n}$ is the set of all $m \times n$ matrices.

The energy density φ is assumed to be nonnegative, periodic in the in-plane variables, and satisfy some growth and coercivity conditions. From (2.3), it appears that the right order of magnitude is $p^{(h)} = h\,P$ where P is a constant independent of h.

We are interested in finding the limiting behavior of the film when all length scales κ, d and h tend to zero. Therefore, we take $\kappa = \kappa(h) > 0, \ d = d(h) > 0, \ \lim_{h\to 0} \kappa(h) = 0, \ \lim_{h \to 0} d(h) = 0$, and assume that they have fixed limiting ratios. We study the "variational limit" of (2.3) as h, d and κ tend to zero with fixed limiting ratios. The natural notion in this context is Γ-convergence which under a suitable technical hypothesis is nearly identical to that of convergence of minimizers. Using this notion, we show that, for any $\mathbf{y} \in W^{1,p}(\Omega^1, I\!R^3), \ p > 3$, $e_1^{(h)}$ Γ-converges to a functional $e_1^{(0)}$ of the form

$$e_1^{(0)}[\mathbf{y}] = \int_S \left\{ \bar{\varphi}(\nabla_p \mathbf{y}) - \frac{P}{3} \mathbf{y} \cdot (\mathbf{y}_{,1} \wedge \mathbf{y}_{,2}) \right\} d\mathbf{z}_p, \tag{2.4}$$

where $\bar{\varphi}$ is the effective energy density and will be determined explicitly later.

3 Effective Behavior

Let us consider the single crystal film first, i.e., $\varphi = \varphi(\mathbf{F})$.

Theorem 1 *Suppose* $\varphi = \varphi(\mathbf{F})$, $\kappa(h) \to 0$ *as* $h \to 0$. *Let* $\bar{\mathbf{F}} \in I\!M^{3\times 2}$ *and* $e_1^{(h)}$ *be defined by (2.3). Then,* $e_1^{(h)}$ Γ*-converges to the functional* $e_1^{(0)}$ *defined by (2.4) if*

$$\bar{\varphi}(\bar{\mathbf{F}}) = Q\varphi_0(\bar{\mathbf{F}}), \tag{3.1}$$

$$\varphi_0(\bar{\mathbf{F}}) = \inf_{\mathbf{b}\in I\!R^3} \varphi(\bar{\mathbf{F}}|\mathbf{b}). \tag{3.2}$$

where $Q\varphi_0$ *is the lower quasi-convex envelope of* φ_0.

James and Rizzoni [4] have considered a similar problem with fixed $\kappa > 0$ independent of h. As a result, their minimizing deformation converges in the class of deformation $\mathbf{y}$ belonging to the Sobolev space $W^{2,2}(\Omega^1, I\!R^3)$ while ours converges in $W^{1,p}(\Omega^1, I\!R^3), p > 3$. The proof of Theorem 1 is given by Shu [7, 8].

Now returning to heterogeneous thin films. The effective energy density $\bar{\varphi}$ depends qualitatively on the relative magnitudes of length scales or on the ratios of $\frac{\kappa}{d}$, $\frac{h}{d}$ and $\frac{\kappa}{h}$. A detailed characterization of $\bar{\varphi}$ has been studied by Shu [7, 8]. For the future use, we state an important result for columnar grains in which the grain size is much larger than the film thickness and the length scale of microstructure.

Theorem 2 *Suppose* $\varphi = \varphi(\mathbf{F}, \frac{\mathbf{z}_p}{d})$, $\frac{\kappa}{d} \to 0$ *and* $\frac{h}{d} \to 0$ *as* $h \to 0$. *Let* $\bar{\mathbf{F}} \in I\!M^{3\times 2}$, $Z = (0,1)^2$ *and* $e_1^{(h)}$ *be defined by (2.3). Then,* $e_1^{(h)}$ Γ*-converges to the functional* $e_1^{(0)}$ *defined by (2.4) if*

$$\bar{\varphi}(\bar{\mathbf{F}}) = \inf_{k\in I\!N} \inf_{\boldsymbol{\omega}\in W_0^{1,p}(kZ)} \fint_{kZ} \varphi_0(\bar{\mathbf{F}} + \nabla_p \boldsymbol{\omega}, \mathbf{z}_p) d\mathbf{z}_p, \tag{3.3}$$

$$\varphi_0(\bar{\mathbf{F}}^0, \mathbf{z}_p^0) = \inf_{\mathbf{b}\in I\!R^3} \varphi(\bar{\mathbf{F}}^0|\mathbf{b}, \mathbf{z}_p^0). \tag{3.4}$$

4 Application to Micropumps

4.1 Two-well Problem

We wish to determine the maximum deflection that a martensitic diaphragm can sustain under small or moderate pressure. Consider a model problem – the energy density φ_0 given by (3.2) has a two-well structure. Let

$$\bar{\mathbf{U}}_1 = \begin{pmatrix} \alpha & 0 \\ 0 & \beta \\ 0 & 0 \end{pmatrix}, \qquad \bar{\mathbf{U}}_2 = \begin{pmatrix} \beta & 0 \\ 0 & \alpha \\ 0 & 0 \end{pmatrix}$$

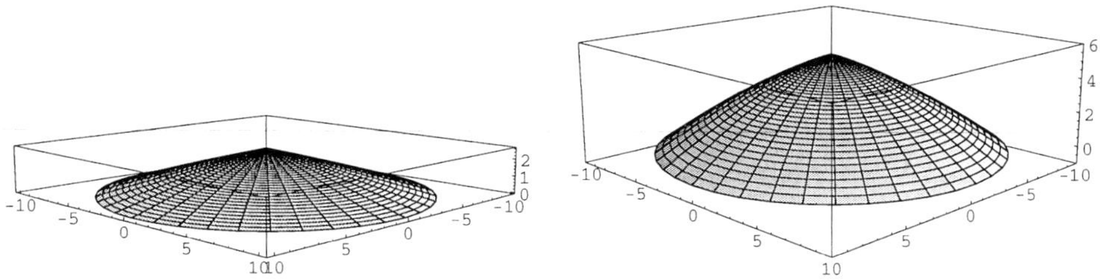

Figure 2: Deflection of Ni-Al film under small (left, $\frac{P}{a}, \frac{P}{b} << 1$) and moderate (right, $\frac{P}{a}, \frac{P}{b} \approx 1$) pressure.

for parameters $\beta > \alpha > 0$. For any $\bar{\mathbf{F}} \in I\!M^{3\times 2}$, define $\mathbf{D} \in I\!M^{2\times 2}_{\mathrm{sym}}$ by

$$\mathbf{D} = \begin{cases} \bar{\mathbf{F}}^T\bar{\mathbf{F}} - \bar{\mathbf{U}}_1^T\bar{\mathbf{U}}_1 & \text{if} \quad |\bar{\mathbf{F}}\mathbf{e}_1| - |\bar{\mathbf{F}}\mathbf{e}_2| \geq 0, \\ \bar{\mathbf{F}}^T\bar{\mathbf{F}} - \bar{\mathbf{U}}_2^T\bar{\mathbf{U}}_2 & \text{if} \quad |\bar{\mathbf{F}}\mathbf{e}_1| - |\bar{\mathbf{F}}\mathbf{e}_2| \leq 0. \end{cases}$$

Above $I\!M^{n\times n}_{\mathrm{sym}}$ is the set of all $n \times n$ symmetric matrices. Let

$$\lambda(\mathbf{D}) = \left(\frac{D_{11} + D_{22}}{2}\right) + \sqrt{\left(\frac{D_{11} + D_{22}}{2}\right)^2 + D_{12}^2}$$

be the larger of the two eigenvalues of $\mathbf{D}$ and $\mathcal{V} = \left\{\frac{1}{\sqrt{2}}(1,1)^T, \frac{1}{\sqrt{2}}(-1,1)^T\right\}$. The relaxed constitutive relation proposed by Bhattacharya and Dolzmann [1] is

$$\begin{aligned} Q\varphi_0(\bar{\mathbf{F}}) &= \frac{a}{2}\left(\det \bar{\mathbf{F}}^T\bar{\mathbf{F}} - \alpha^2\beta^2\right)_+^2 \\ &+ \frac{b}{2}\begin{cases} (\lambda(\mathbf{D}))_+^2 & \text{if } (|\bar{\mathbf{F}}\mathbf{e}_1| - |\bar{\mathbf{F}}\mathbf{e}_2|)(D_{11} - D_{22}) \geq 0, \\ \left(\max\limits_{\mathbf{v}\in\mathcal{V}} < \mathbf{v}, \mathbf{D}\mathbf{v} >\right)_+^2 & \text{otherwise} \end{cases} \end{aligned} \quad (4.1)$$

for $a, b > 0$, and $(A)_+ = \max\{A, 0\}$.

Assume S is a circular domain. Minimizing (2.4) using Theorem 1 and (4.1) gives the deflection of the Ni-Al film as shown in Figure 2 for small and moderate pressure. Note that the ratio of the maximum deflection to the radius of the film can be up to 0.25 even for small pressure.

4.2 Simplified Model

Frame-indifference implies the energy density has to satisfy

$$\varphi_0(\bar{\mathbf{F}}) = \varphi_0(\mathbf{R}\bar{\mathbf{F}}), \qquad \forall\, \bar{\mathbf{F}} \in I\!M^{3\times 2}, \qquad \forall\, \mathbf{R} \in \mathrm{SO}(3),$$

where φ_0 is given by (3.2). This also implies that there exists a function $W : \mathbb{M}^{2\times 2}_{\text{sym}} \to \mathbb{R}$ such that $\varphi_0(\bar{\mathbf{F}}) = W\left(\sqrt{\bar{\mathbf{C}}}\right)$ where $\bar{\mathbf{C}} = \bar{\mathbf{F}}^T\bar{\mathbf{F}}$ is positive semi-definite symmetric [3]. Let $u_1 = y_1 - z_1, u_2 = y_2 - z_2, u_3 = \eta$ be the in-plane displacements and out-of-plane deflection of the film. For the membrane fully constrained on its boundary ∂S with deflection η much larger than the film thickness h, we may assume $|u_{1,\alpha}u_{1,\beta}| << 1$, $|u_{2,\alpha}u_{2,\beta}| << 1$ for $\alpha, \beta = 1, 2$. It follows that the finite strain measure $\sqrt{\bar{\mathbf{C}}}$ can be approximated to

$$\sqrt{\bar{\mathbf{C}}} \approx \mathbf{E} = \boldsymbol{\epsilon}^1 + \boldsymbol{\epsilon}^2, \quad \boldsymbol{\epsilon}^1[\mathbf{u}] = \frac{1}{2}\left(\nabla_p \mathbf{u} + (\nabla_p \mathbf{u})^T\right), \quad \boldsymbol{\epsilon}^2[\eta] = \frac{1}{2}\nabla_p \eta \otimes \nabla_p \eta. \quad (4.2)$$

The energy density is then approximated to

$$\varphi_0(\bar{\mathbf{F}}) = W\left(\sqrt{\bar{\mathbf{C}}}\right) \approx W(\mathbf{E}). \quad (4.3)$$

For martensitic thin films, we assume that

$$W(\mathbf{E}) \begin{cases} = 0 & \mathbf{E} \in \mathbf{E}^{(1)} \cup \mathbf{E}^{(2)} \cup \cdots \cup \mathbf{E}^{(k)}, \\ > 0 & \text{otherwise,} \end{cases} \quad (4.4)$$

where

$$\mathbf{E}^{(i)} = \sqrt{\bar{\mathbf{U}}_i^T \bar{\mathbf{U}}_i} - \mathbf{I}_p, \; i = 1, \cdots, k,$$

$\mathbf{I}_p$ is the 2×2 identity matrix, and $\bar{\mathbf{U}}_i \in \mathbb{M}^{3\times 2}$ is the stress-free transformation strain such that $\varphi_0(\bar{\mathbf{U}}_i) = 0$ for $i = 1, \cdots, k$. We wish to determine the set of strains such that

$$\mathcal{S} = \{\mathbf{E} : \; QW(\mathbf{E}) = 0\}, \quad (4.5)$$

where QW is the relaxation of W. However, it is pretty difficult to determine the set $\mathcal{S}$ even for this simplified model. But we can estimate it in some special but common cases such as in martensites undergoing cubic to tetragonal, trigonal, orthorhombic, and some oriented thin films. In this situation, let $\mathcal{S}^c$ be the convex hull of transformation strains $\mathbf{E}^{(1)}, \cdots, \mathbf{E}^{(k)}$. Then, we may replace $\mathcal{S}$ by $\mathcal{S}^c$ for strains $\mathbf{E}$ with nonnegative eigenvalues (see Shu [8] for detailed explanation).

4.3 Single Crystal Micropumps

We now apply our simplified model to the design of shape-memory micropumps with large pumping volume per cycle. Consider a square domain S. We assume the deformed shape is a tent. Let m be the ratio of the central deflection of the film to the half side length of the square domain. To satisfy the fixed end boundary condition, we assume the macroscopic in-plane displacement $\mathbf{u} = \mathbf{0}$. For a tent deformation, the gradients of the deflection η in S are

$$\nabla_p \eta = \begin{pmatrix} m \\ 0 \end{pmatrix}, \begin{pmatrix} 0 \\ -m \end{pmatrix}, \begin{pmatrix} -m \\ 0 \end{pmatrix}, \begin{pmatrix} 0 \\ -m \end{pmatrix}. \tag{4.6}$$

Using (4.2), (4.3), (4.5) and (4.6) gives the condition for stress-free deformation

$$\begin{pmatrix} \frac{1}{2}m^2 & 0 \\ 0 & 0 \end{pmatrix}, \begin{pmatrix} 0 & 0 \\ 0 & \frac{1}{2}m^2 \end{pmatrix} \in \mathcal{S}. \tag{4.7}$$

Consider a 4-well problem which arises in the cubic to monoclinic-II transformation. Let the transformation strains be

$$\mathbf{E}^{(1)} = \begin{pmatrix} \alpha & \delta \\ \delta & \gamma \end{pmatrix}, \quad \mathbf{E}^{(2)} = \begin{pmatrix} \alpha & -\delta \\ -\delta & \gamma \end{pmatrix},$$
$$\mathbf{E}^{(3)} = \begin{pmatrix} \gamma & \delta \\ \delta & \alpha \end{pmatrix}, \quad \mathbf{E}^{(4)} = \begin{pmatrix} \gamma & -\delta \\ -\delta & \alpha \end{pmatrix},$$

where $\alpha > 0, \gamma > 0, \delta > 0$. In this case, we may replace $\mathcal{S}$ with $\mathcal{S}^c$ in (4.7) (see Shu [8]). However, it can be shown that (4.7) can not be satisfied. But if the crystal basis is rotated with some angle θ w.r.t. the film basis $\mathbf{e}_3$, we can have the membrane with tent shape. Indeed, choosing $\cos 2\theta = \frac{\alpha-\gamma}{\alpha+\gamma}$, we obtain

$$m = \sqrt{2(\alpha + \gamma)}. \tag{4.8}$$

For Cu-Zn-Al, $\alpha = 0.089, \delta = 0.025$, $\gamma = 0.007$, we have $m = 0.438$ and $\theta = 15.7°$. Note that if we use fully finite deformation kinematics and Bhattacharya-James thin film theory [2], $m = 0.45$ and $\theta = 15.6°$.

4.4 Polycrystalline Micropumps

We now turn to the practical cases – textured thin films such as (100), (110) or (111) films. In this situation, the properties of the film is basically transversely isotropic. So we consider a circular domain S and set m be the ratio of the maximum deflection of the diaphragm to the radius of the circular domain. Our analysis is based on Theorem 2 and the Taylor bound - assume each grain undergoes the identical deformation. Table 1 shows the results for different shape-memory films with some common textures.

Makino *et al* [5, 6] have successfully fabricated a TiNi actuated micropump with pumping pressure up to several hundreds of kPa. The measured ratio of deflection to the radius of the TiNi diaphragm is about 0.025. Clearly, from Table 1, our

Texture	m: maximum deflection/radius			
	Ni-Al	Cu-Al-Ni	Ti-Ni	Cu-Zn-Al
{100} film	0.18	0.19	0.15	0.20
{111} film	–	0.08	0.13	0.09
{110} sputtered film	–	0.10	0.15	0.09

Table 1: The predicted ratio of the maximum deflection to the radius of the circular diaphragm for various textures under small pressure.

theoretical prediction is larger than this value. The discrepancy is not completely understood here. But note that we have assumed there is no pre-stress in our film when the bias pressure vanishes. If the misfit tensile strain exists between film and substrate outside S, the values in Table 1 will be reduced significantly. Next, from the 4-well problem and Table 1, we see (100) Cu-based shape-memory film can have better behavior than TiNi film concerning the purpose of large deflection. Finally, we may apply our result to the design of pizeoelectricly actuated micropumps. For example, $PbTiO_3$ is a ferroelectric material with crystal structure undergoing cubic to tetragonal transformation. It follows that $m = 0.1$ for (100) $PbTiO_3$ textured film.

References

[1] K. Bhattacharya and G. Dolzmann. Relaxed Constitute Relations for Phase Transformation Materials. *J. Mech. Phys. Solids*, **48**:1493–1517, 2000.

[2] K. Bhattacharya and R. D. James. A Theory of Thin Films of Martensitic Materials with Applications to Microactuators. *J. Mech. Phys. Solids*, 47:531–576, 1999.

[3] H. Le Dret and A. Raoult. The Nonlinear Membrane Model as Variational Limit of Nonlinear Three-dimensional Elasticity. *J. Math. Pures. Appl.*, **74**:549–578, 1995.

[4] R. D. James and R. Rizzoni. Pressurized Shape Memory Thin Films. *J. Elasticity*, **59**:399–436, 2000.

[5] E. Makino, T. Shibata, and K. Kato. Dynamic Thermo-Mechanical Properties of Evaporated TiNi Shape Memory Thin Film. *Sensors and Actuators A*, **78**:163–167, 1999.

[6] E. Makino, T. Shibata, and K. Kato. Fabrication of TiNi Shape Memory Micropump. *Sensors and Actuators A*, **88**:256–262, 2001.

[7] Y. C. Shu. Heterogeneous Thin Films of Martensitic Materials. *Arch. Rational Mech. Anal.*, **153**:39–90, 2000.

[8] Y. C. Shu. In preparation. 2001.

LOAD PARTITIONING IN SHAPE MEMORY ALLOY POLYCRYSTALS STUDIED BY IN-SITU NEUTRON DIFFRACTION AND MODELING

P. ŠITTNER AND V. NOVÁK,
Institute of Physics, Academy of Sciences of the Czech Republic, Na Slovance 2, Prague 8, 18221, Czech Republic

AND

P. LUKÁŠ
Nuclear Physics Institute, Academy of Sciences of the Czech Republic, 250 68 Řež, Czech Republic

Abstract

A mechanism by which the macroscopic stress redistributes among oriented grains of CuAlZnMn shape memory alloy polycrystals transforming pseudoelastically in tension/compression has been studied experimentally by in-situ neutron diffraction techniques. Micromechanics model simulations for pseudoleastically deforming Cu-based alloys and NiTi are presented and the results are compared with the diffraction data. Key factors controlling the load partition are found to be the elastic anisotropy of the austenite phase, crystallographic type of the martensitic transformation and associated transformation anisotropy, grain size, texture and sense of the applied uniaxial load.

1. Introduction

Martensitic transformations /MT/ in shape memory alloy /SMA/ polycrystals occurring under applied stress are significantly affected by the requirement for cooperative deformation of the polycrystal grains. Since SMAs are commonly elastically anisotropic materials and stress induced martensitic transformations /SIMT/ are also strongly anisotropic processes, the continuity of strains at the grain boundaries necessitates heterogeneous distribution of the externally applied stress among variously oriented polycrystal grains - load partition. In this work, this load partition in pseudoelastically transforming Cu-based and NiTi SMA polycrystals is investigated by in-situ neutron diffraction experimental techniques (Šittner et al, 2001) and micromechanics modeling (Šittner and Novak, 2000). An idea that evolution of crystallographic orientation dependent stress heterogeneity in transforming polycrystals is partially responsible for limited recoverable strains of some SMAs in polycrystalline state is elaborated. The load partition mechanism of the CuAlZnMn polycrystal in tensile pseudoelastic test is discussed in detail, and simulation results for Cu-based and NiTi polycrystals in tension/compression pseudoelasticity are compared.

Q.P. Sun (ed.), IUTAM Symposium on Mechanics of Martensitic Phase Transformation in Solids, 179–188.

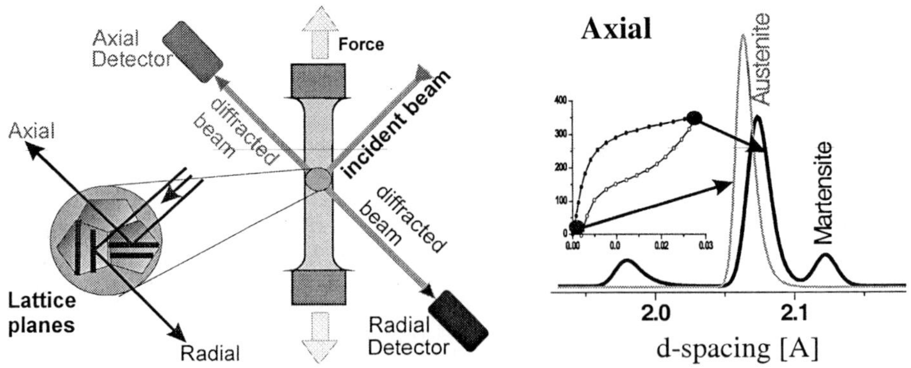

Figure 1. In situ neutron diffraction experiment on SMA polycrystal. Neutron diffraction spectra (right) recorded in selected stress-strain-temperature states during uniaxial thermomechanical load test on the SMA polycrystal are recorded in two geometrical configurations (left) with scaterring vector oriented either in parallel (axial) or perpendicularly (radial) to the load axis. Analysis of the austenite and martensite reflections in the diffraction spectra is carried out. Positions, integral intensities and possibly even shapes of the individual *hkl*-diffraction profiles are related to the macroscopic stress, strain or temperature and microstructure information on the transformation process in the polycrystal is deduced.

2. Neutron diffraction measurement of stresses, strains and phase fractions

Neutron diffraction (Fig.1), similarly to X-rays, can be used to probe martensitically transforming solids under thermomechanical loads. Due to the high penetration depth of thermal neutrons, the diffraction data are obtained from large specimen volume (~1cm^3). Martensite phase fractions ξ_{hkl} in SMA polycrystals can be evaluated from the peak integral intensities I_{hkl} of the austenite phase (Eq. 1).

$$\xi_{hkl} = (1 - \frac{I_{hkl}}{I_{0,hkl}}) \tag{1}$$

Internal elastic strains ε_{hkl} in individual phases can be calculated from the deviations of d_{hkl}-lattice spacings of particularly oriented crystal planes (Eq. 2) - i.e. from the relative shifts of the positions of diffraction peaks due to the applied stress.

$$\varepsilon_{hkl} = \frac{d_{hkl} - d_{0,hkl}}{d_{0,hkl}} = \frac{\Delta d_{hkl}}{d_{0,hkl}} = -\cot(\vartheta)\Delta\vartheta \tag{2}$$

ϑ is the diffraction angle of the measured *hkl*-reflection. $I_{0,hkl}$ and $d_{0,hkl}$ are the integral intensity and lattice spacing, respectively, measured at zero stress. They were used as initial values for the lattice strain and martensite phase fraction calculations.

With respect to the load partition mechanism which is the focus of this work, it is essential to keep in mind that the strains ε_{hkl}, called lattice strains further on, are not continuum strains in a common sense but: i) elastic strains, ii) phase specific strains-determined selectively only from the volume of the grain existing currently in a given phase - i.e. austenite or martensite in SMA, iii) crystal orientation specific strains - determined selectively only from those grains in the polycrystal specimen which are suitably oriented with respect to the scattering vector (see Fig. 1), iv) averaged strain values over those selected grains within the irradiated volume, and v) one component of the elastic strain tensor within the irradiated volume is generally obtained from a single measurement. Two geometric configurations of the diffraction experiment (Fig. 1) with scattering vector oriented either in parallel or perpendicularly to the tensile load axis

In the following, only simulation of the pseudoelastic deformation of Cu-based and NiTi polycrystals in tension/compression will be focussed. Simulated macroscopic stress-strain response of the NiTi aggregate is compared in figure 6a with the response of a single arbitrarily chosen [111] axial domain and with the [111]-axial component response. Material parameters were selected such that the NiTi alloy shows pseudoelasticity at room temperature. Notice that even if the individual domain deforms at constant stress level as assumed, the average σ^{AM}_{111}-e_{111} and σ_G-ε_G responses exhibit significant pseudoelastic "hardening", particularly in compression. Asymmetry and variety of σ^{AM}_{hkl}-e_{hkl} responses of variously oriented grains (Fig. 6b) derive from the transformation anisotropy of NiTi alloy presented in figures 5d,f. The stresses σ^{AM}_{hkl} in figure 6a,b are averages over both austenite and martensite phases in the *hkl*-oriented domains, in contrast to stresses σ_{hkl} which are calculated for the austenite phase only and are comparable to the neutron diffraction results. Based on figure 6, it can be easily rationalized why the <111> fiber texture modifies the macroscopic σ_G-ε_G curve of the NiTi as figure 7b shows. The simulated effects of common SMA wire drawing textures on the stress-strain response and asymmetry in tension/compression are compared in figure 7 for Cu-based (β_1-β'_1) and NiTi alloys. The texture effects on the stress-strain responses are predicted to be almost tension/compression symmetric for the Cu-based alloys but very asymmetric for the NiTi.

To achieve a sufficient statistics over the model variables in only a small subset (~200) of particularly oriented *hkl*-domains that fulfil the "diffraction condition" (Fig. 1), stresses strains and phase fractions in a very large number of domains (typically N~$2x10^5$) have to be tracked down during the simulation.. Figure 8 shows simulation results for 6 selected *hkl*-families of domains in the axial configuration of the textured CuAlZnMn polycrystal comparable with the neutron diffraction experimental results in figures 2d,3. Even if the CuAlZnMn polycrystal exhibits SIMT into a mixture of β'_1 and γ'_1 martensite phases, particularly in compression (Šittner and Novák, 2000), only the $\beta_1 \rightarrow \beta'_1$ MT has been considered here for simplicity. Simulated macroscopic response σ_G-ε_G is drawn together with σ_{hkl}-e_{hkl} component average responses in figure 8a. Figure 8b shows the lattice strain responses, ε_{hkl} -σ_G, comparable to those evaluated in the neutron diffraction experiment (Figs. 2d,3). Other figures demonstrate the simulated partition of austenite stress among variously oriented domains (Fig. 8c) and phase fraction evolution in these domains with macroscopic strain ε_G (Fig. 8d) .

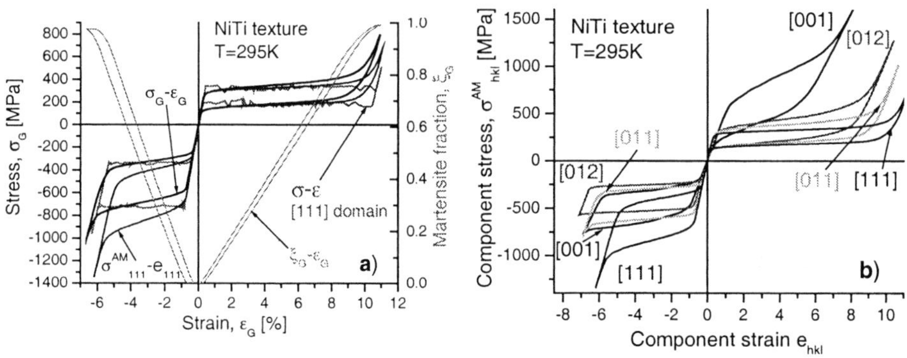

Figure 6. Simulated tension/compression pseudoelasticity of <111> fiber textured NiTi polycrystal.
a) macroscopic σ_G-ε_G response compared with the behavior of a single [111] oriented domain and with average component response σ^{AM}_{111}-e_{111}, of all domains oriented with [111] direction parallel to the load axis, b) average component responses of four *hkl*-families of domains. The [111] fiber texture (Fig. 7) characterizing preferential orientations of aggregate domains represents the experimental texture commonly encountered in NiTi drawn wires and bars.

5. Load partition in pseudoelastically transforming SMA polycrystals

Let us consider the simulated pseudoelastic deformation of the CuAlZnMn polycrystal in tension (Fig. 7a,8) with the characteristic elastic and transformation anisotropies illustrated in figure 5. In the early elastic range (σ_G<150MPa), the macroscopic stress is redistributed (Fig. 8c) in such a way that the elastically harder domains (close to [111]) share larger stress than the elastically softer domains (close to [001]) and the ε_{hkl}-σ_G responses are linear. Both the experimental (Fig.3) and simulation (Fig. 8b) results for elasticity (aggregate Young's modulus and diffraction elastic constants) correspond very well to the results of selfconsistent model calculations for anisotropic cubic metals (Daymond et al, 2000). Beyond the stress level σ_G>175 MPa, the SIMT starts occasionally in the domains in which the orientation dependent transformation stresses (Fig. 5c) have been firstly reached. However, the SIMT can not proceed continuously there since the local stresses in these domains decrease in a response to the transformation events forcing the SIMT to temporarily stop there. The load is taken over by the neighboring domains (Fig. 4). The SIMT hence acts as a kind of block hindering further increase of the local stress in the domains in which the transformation has already started. The load partition in the polycrystal hence becomes increasingly affected by the growing number of transformation events. This situation is clearly detected both in the simulation (Fig. 8b) and experiment (point 6 in Figs. 2b,d) as the deviation of the ε_{hkl}-σ_G curve from linearity. Stresses in most of the domains eventually approach the values close to the orientation specific σ^{tr} transformation stresses, which is a prerequisite for the onset of the continuously progressing SIMT in the polycrystal aggregate. At the beginning of the large scale SIMT, there is hence a characteristic stress heterogeneity in the aggregate - intergranular stresses - which can be basically inferred from figure 5c. Since the simulated lattice strains (Fig. 8b) and martensite volume fraction responses (Fig. 8d) agree even quantitatively well with the diffraction data (Figs. 2,3), we assume that the model describes properly the mechanism of the load partition in the CuAlZnMn polycrystal. The way the stresses, strains and martensite fractions are distributed among *hkl*-oriented axial domains in tension and evolve during the pseudoelastic test can be understood from figures 8a,c. The load partition mechanism of the CuAlZnMn polycrystal in tension necessitates that the macroscopic stress σ_G has to significantly increase to drive ahead the MT, even if the local stresses in the austenite phase of the transforming domains stay constant until these domains become fully transformed to the martensite phase. This is possible since the evermore increasing load becomes taken over by the martensite phase and by the yet untransformed austenite phase in the hardest near [111] oriented domains. The apparent

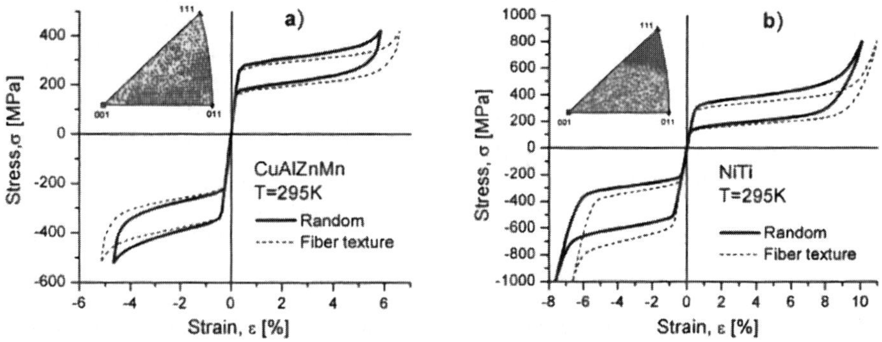

Figure 7. Simulated tension/compression pseudoelastic curves for randomly oriented and textured CuAlZnMn (a) and NiTi (b) alloys polycrystals. CuAlZnMn and NiTi drawn bar specimens exhibit three component <012>, <014> and <133> fiber texture (Neov et al, 2001) and <111> fiber texture, respectively.

pseudoelastic "hardening" of the CuAlZnMn SMA polycrystal in the inclined transformation "plateau" is thus solely due to this peculiar load sharing mechanism likely common for Cu-based SMAs transforming mainly via $\beta_1 \rightarrow \beta'_1$ cubic to monoclinic transformation. On unloading, after overcoming the stress hysteresis, the reverse transformation takes place, leading to an almost complete restoration of the original austenite intensity I_{hkl} and lattice strain ε_{hkl} (Fig. 2d). In contrast to the simulated lattice strains ε_{hkl} (Fig. 8b), however, the experimental lattice strains were not fully restored following the first pseudoelastic cycle. This needs a further work on the simulation side. Nevertheless, the variety of otherwise very complicated nonlinear hysteretic *hkl*-component experimental lattice strain and intensity responses in pseudoelastic cycle could be rationalized and interpreted using the simulations.

Magnitude of the recoverable strains exhibited by the SMAs derives from crystallographic transformation strain γ^{tr} characterizing the undergoing MT. The SMAs in polycrystaline state typically exhibit recoverable transformation strains much smaller than single crystals due to the anisotropy of transformation strains stemming from the structure symmetry changes associated with the MT (Bhattacharya and Kohn, 1996). This is fully confirmed by the present model. At the same time, moreover, since external loading is inevitable to induce macroscopic strains in SMA polycrystals, the arising intergranular stresses may further limit the recoverable strains of polycrystalline SMAs and the load partition mechanism is essential. In fact, what comes out of the present experiments and modeling, is that what further limits the recoverable strains is

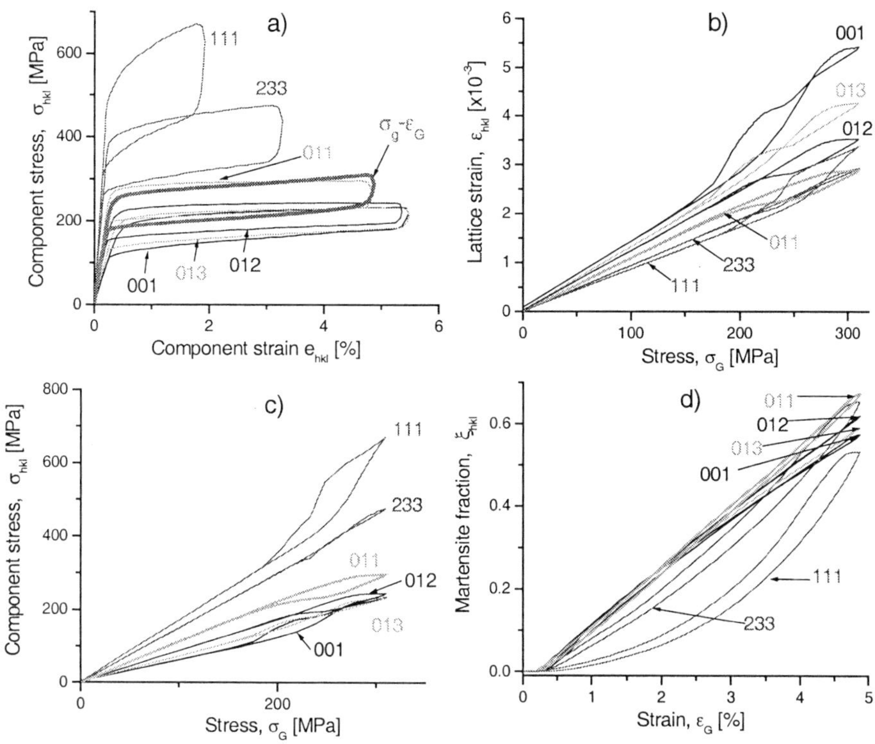

Figure 8: **Simulated pseudoelastic response of the textured CuAlZnMn polycrystal in tension** at T=295K in the axial configuration. σ_G, ε_G and ξ_G are, respectively, macroscopic stress, strain and martensite fraction volume. σ_{hkl}, ε_{hkl} e_{hkl} and ξ_{hkl} are, respectively, average stress, elastic strain, strain and martensite fraction volume in the families of *hkl*-oriented domains.

neither the elastic anisotropy of the austenite phase nor the transformation anisotropy of the undergoing MT alone, but the way these factors together affect the load partition mechanism. The NiTi polycrystals are essentially different from the Cu-based alloys in this respect. Not only the NiTi are less elastically anisotropic (Fig. 5a,b), but mainly the elastic and transformation anisotropies are not equally aligned as in the case of the Cu-based alloys (Fig. 5). In particular, the near [111] oriented grains in NiTi in tension sharing largest part of the load in the elastic range (Fig. 5b) require smallest stresses (Fig. 5d) to induce the martensitic transformation. The tendency of NiTi polycrystals towards the localized deformation mode via motion of few macroscopic interfaces commonly experienced in tensile experiments (Shaw and Kyriakides, 1997) might be related with these specific elastic and transformation anisotropies of NiTi in tension but macroscopic deformation instabilities can not be treated in frame of the present model. Since the transformation anisotropies of NiTi differ significantly between tension and compression, the NiTi alloys are predicted to show large asymmetry in their stress-strain responses and differing texture effects in tension and compression (Fig. 7b).

6. Conclusions

Mechanism of the load partition in CuAlZnMn SMA polycrystals transforming in tensile pseudoelastic cycle was studied experimentally by in-situ neutron diffraction techniques and theoretically by simulation using a recently proposed micromechanics model of SMA polycrystal transformation. It is found that neither the elastic nor the transformation anisotropy alone, but the way how these factors together affect the load partitioning among polycrystal grains during the martensitic transformation is impostant for good transformability of polycrystalline SMAs. The simulation results for Cu-based alloys and NiTi are discussed in this respect.

Acknowledgements-Support of the Grant Agency of the Academy of Sciences of the Czech Republic (contracts No. A1048107 and No. A1010909) and Grant Agency of the Czech Republic (contract 202/97/K038) are sincerely acknowledged. We further thank to Prof. G. M. Swallow from University of Loughborough, UK, Dr. M. Daymond from ISIS RAL Chilton, UK, and D. Neov from NPI Rez, Czech Republic for their great assistance in conducting the experiments.

References

Bhattacharya K., and Kohn R.V., (1996), Texture and recoverable strains in SMA polycrystals, *Acta materialia,* **44**, 529-543.

Daymond M.R., Tome C.N, Bourke M.A.M, (2000), Measured and predicted intergranular strains in textured austenitic steels, *Acta materialia*, **48** 553-564.

Neov D., Lukáš P., Černík M., Šittner P., (2001) Neutron diffraction analysis of solid phases transforming in CuAlZnMn shape memory alloy, in Proc. XVIII Conference on Applied Crystallography (CAC), 4-7 September 2000, Poland, in print.

Shaw, J.A. , Kyriakides, S. (1997) On the nucleation and propagation of phase transformation fronts in a NiTi alloy., *Acta Materialia,* **45**, 683-700.

Strunz P.(1997),Transformed Model Fitting- a Straightforward Approach to the Evaluation of Several Types of Neutron Diffraction Experiments, *Materials Structure in Chemistry, Biology, Physics and Technology* (Bulletin of the Czech and Slovak Crystallographic Association) **4**, 136-143.

Šittner P., Novák V., Lukáš P. and .Neov D. (2000), Characterization of stress induced martensitic transformation in shape memory alloys by in-situ neutron diffraction techniques, ASME ***AD-Vol.*** **60**, 25-32.

Šittner P. and Novák V. (2000) Anisotropy of martensitic transformations in modeling of shape memory alloy polycrystals, *International Journal of Plasticity* **16**,1243-1268.

Šittner P, Lukáš P., Neov D., Daymond M.R., Novák V. and Swallowe G.M., (2001) Stress induced martensitic transformation in CuAlZnMn polycrystal investigated by two in situ neutron diffraction techniques *Material Science and Engineering A,* in print

DYNAMIC CHARACTERISTICS OF Ti-Ni SMA THIN FILM MICROACTUATORS

S. MIYAZAKI, M. HIRANO AND T. YAMAMOTO
Institute of Materials Science, University of Tsukuba, Tsukuba
Ibaraki 305-8573, Japan

Abstract

Ti-Ni shape memory thin films are expected to be applied for microactuators due to the great advantages such as large shape recovery force and strain. In this study, a Ti-Ni thin film was deposited on a Si substrate with SiO_2 surface by using the radio-frequency (R.F.) magnetron sputtering method. A diaphragm microactuator was fabricated by applying the well-known Si photoetching method, and a heat-treatment was carried out in order to acquire shape memory effect. The dimension of the diaphragm was a square with the side of 560 μm, and the shape was convex in the martensite phase and flat in the parent phase. The shape changed repeatedly by thermal cycles of heating and cooling. In order to evaluate the dynamic characteristics of the diaphragm microactuator, three-dimensional shape analysis was carried out in terms of frequency, currents, and duty ratio. It is concluded that the dynamic properties of the Ti-Ni diaphragm are affected by the amplitude and frequency of electric current and heat transfer of atmospheric gas.

Keywords: shape memory alloys, Ti-Ni, microactuator, shape memory effect, martensitic transformation

1. Introduction

Ti-Ni shape memory alloy (SMA) thin films formed by sputter-deposition are one of the most promising candidates for microactuators due to their large shape recovery force reaching 600 MPa and strain exceeding 6% [1-4]. In recent years, many researches with respect to Ti-Ni microactuators have been done [5-9]. However, more understanding regarding the dynamic characteristics of microactuators is needed in order to optimize operational conditions. The SMA microactuator is driven by thermal cycling: deformation of the SMA thin film occurs due to phase transformation between high temperature (austenite) phase and low temperature (martensite) phase. It should be noted that the heating rate of the SMA microactuator can be easily increased by applying electric current, however, the cooling rate is often the rate-limiting factor for the phase transformation. The cooling rate is greatly improved by fabricating thinner films, since the natural cooling is accelerated with increasing surface/volume ratio. In this study, static and dynamic characteristics of a

Q.P. Sun (ed.), IUTAM Symposium on Mechanics of Martensitic Phase Transformation in Solids, 189–196.

diaphragm-type microactuator composed of a Ti-Ni SMA thin film and a SiO_2 film on a Si substrate were investigated by varying applied current, frequency and atmosphere.

2. Experimental Procedure

A Ti-Ni SMA thin film having a near-equiatomic composition was deposited on a SiO_2 layer formed on a (100) silicon substrate by the radio-frequency (R.F.) magnetron sputtering in an argon atmosphere. The thickness of the SMA thin film is about 1.5 μm. Chemical composition of the Ti-Ni SMA thin film was determined to be Ti-48.8at.%Ni by electron probe microanalysis (EPMA). The deposited Ti-Ni thin film was heat-treated in Ar atmosphere in order to crystallize and memorize the initial flat shape.

The fabrication of the microactuator was based on the Si semiconductor techniques. A negative photoresist was coated on the surfaces of both SiO_2 and Ti-Ni SMA film, and was pre-baked. Patterning was performed in the photoresist on the SiO_2. Then, exposed SiO_2 areas were etched away in a HF/NH_4F buffer solution. After the photoresist layer was removed, anisotropic etching of Si was carried out using an etchant containing ethylene diamine and pyrocatechol (EDP). Fig. 1 shows a schimatic illustration of the fabrication process used in this study. The dimension of the diaphragm was a square with the side of 560 μm.

Fig. 2 shows the cross-section of the diaphragm microactuator used in this study. It can be seen that the microactuator is convex in the martensite state due to an internal stress and the height at the central part decreases with heating above the reverse martensitic transformation start temperature (*As*). Then the microactuator becomes flat when heated above the reverse martensitic transformation finish temperature (*Af*). With cooling below the martensitic transformation start temperature (*Ms*), the flat shape becomes convex vice versa.

The movement of the diaphragm was examined by a three-dimensional shape analysis system equipped with a laser beam scanner (Fig. 3). The experiment was performed at room temperature (296-298 K). The temperature of the microactuator for the dynamic characterization was controlled by adjusting the frequency and amplitude of the applied current. The duty ratio, which was defined as the ratio of duration of current-on and -off, was fixed at 3:97. The microactuator was allowed to be cooled in the stream of the gas.

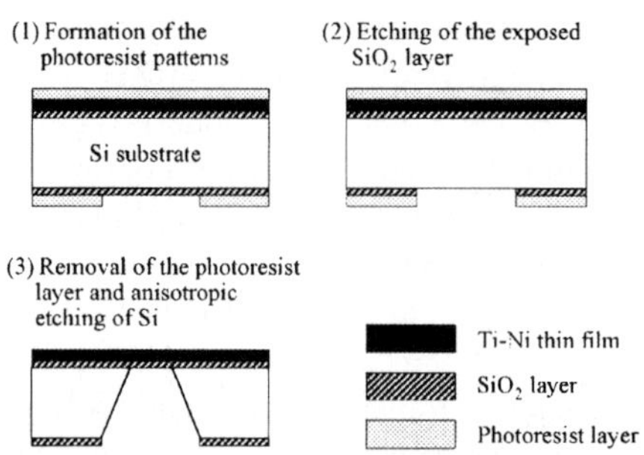

Fig. 1: Fabrication process of a diaphragm-type microactuator based on Ti-Ni SMA thin film.

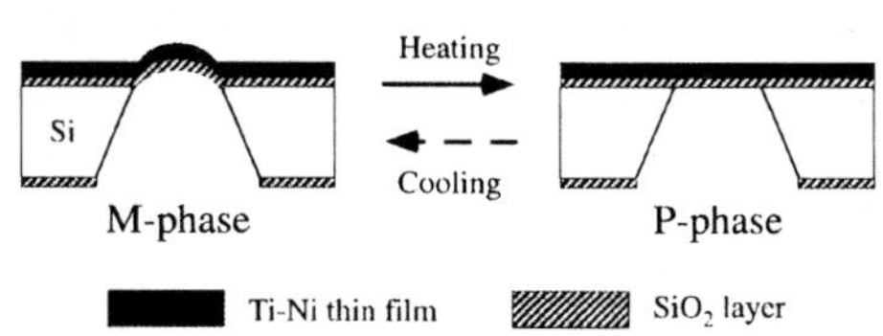

Fig. 2: Cross-sections of the microactuator in the P- and M-phases.

The temperature was continuously measured by a thermocouple attached to the Ti-Ni thin film part fixed to the Si substrate instead of to the moving actuator part. The dynamic properties of the diaphragm were investigated from the viewpoint of heat transfer to the atmosphere; air and helium gas were used. Either of these gases was flowed into the bottom side of the diaphragm under a constant and ambient pressure. It should be noted that the diaphragm was not effectively pressurized by applying these gases.

3. Results and Discussion

3.1. STATIC CHARACTERISTICS

In order to characterize the dynamic properties of the diaphragm microactuator, the static behavior of the diaphragm was measured at first. Some examples of shape measurements for the diaphragm are shown in Fig. 4 as contour diagrams. In order to achieve the equilibrium state at a constant temperature, each measurement was done after keeping the microactuator more than five minutes in a small furnace. No temperature change was recognized during each measurement. Time spent for each measurement was about two minutes. Thus, the actuation behavior measured in this way is called "static" characteristics in this study. It is clear that the microactuator is convex in the martensite state at room temperature and the height decreased with increasing temperature: 19.0 μm at 300 K, 13.6 μm at 357.5 K, 9.7 μm at 362.5 K and 0.3 μm at 403 K. At 403 K, the

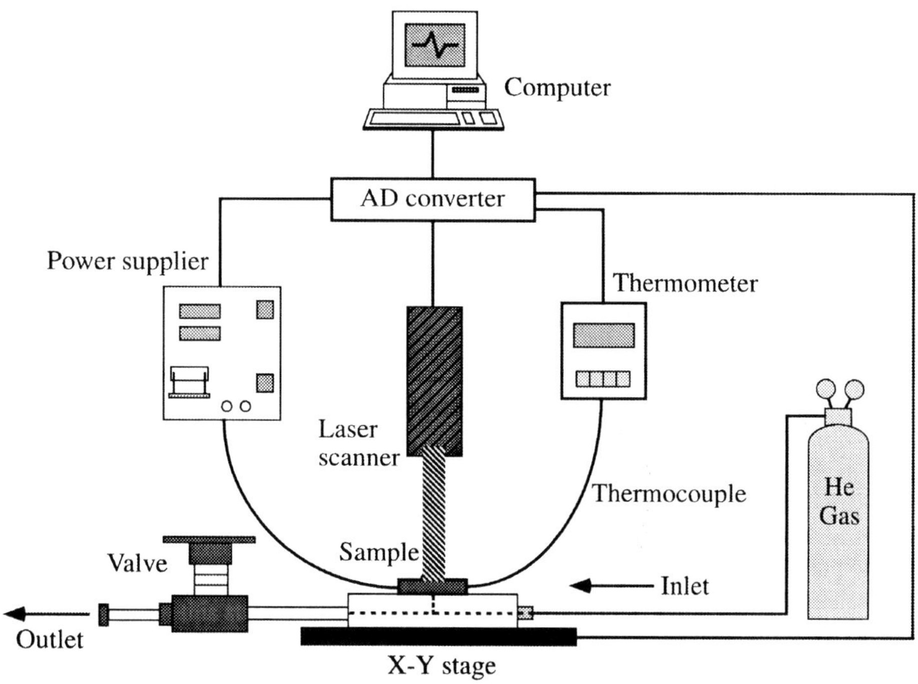

Fig. 3: A measuring system of three-dimensional shape analysis.

microactuator was almost flat indicationg that the temperature is judged to be above the reverse martensitic transformation finish temperature (*Af*). Alternatively, the flat shape becomes convex with decreasing temperature. Thus, the reversible shape change was confirmed by cooling and heating. According to these results, the maximum change in height of the microactuator is estimated at 19.0 μm.

3.2. DYNAMIC CHARACTERISTICS

It is clear in Fig. 4 that the maximum change in height of the diaphragm occurs at the center. Thus, in order to evaluate the dynamic properties, the actuation at the center of the diaphragm was measured as a representative. Then, the change in the height at the center part of the diaphragm (hereafter height) was investigated by changing current and frequency under the constant duty ratio of 3:97: the constant frequency was varried from 0.1 to 50 Hz. Hereafter, h^s_{max} and h^s_{min} are defined as the maximum and minimum heights obtained for the static condition described in section 3.1, and h_{max} and h_{min} are defined as the maximum and minimum heights measured after reaching a steady state in the actuation in the dynamic tests, respectively. In order to characterize the dynamic properties, the difference between h_{max} and h_{min} (= h_{max} - h_{min}) is used and called displacement.

Fig. 5 shows the variation of voltage, temperature and height of the diaphragm as a function of time under the condition of the applied constant current of 6.0 A and the frequency of 0.1 Hz. It is clear in Fig. 5 (b) that the temperature increased with increasing time but became of a steady state after 45 s. Consequently, the change in height became equilibrated: h_{max} and h_{min} were 18.1 μm and 0.8 μm, respectively. Thus, it is clear that the actuation occurs repeatedly by controlling the frequency of current. The displacement of the diaphragm in the dynamic actuation was estimated at 17.3 μm. Similarly, Fig. 6 shows the variation of voltage, temperature and height as a function of time under the condition of the applied current of 8.0 A and the frequency of 1 Hz. The steady state in this case was

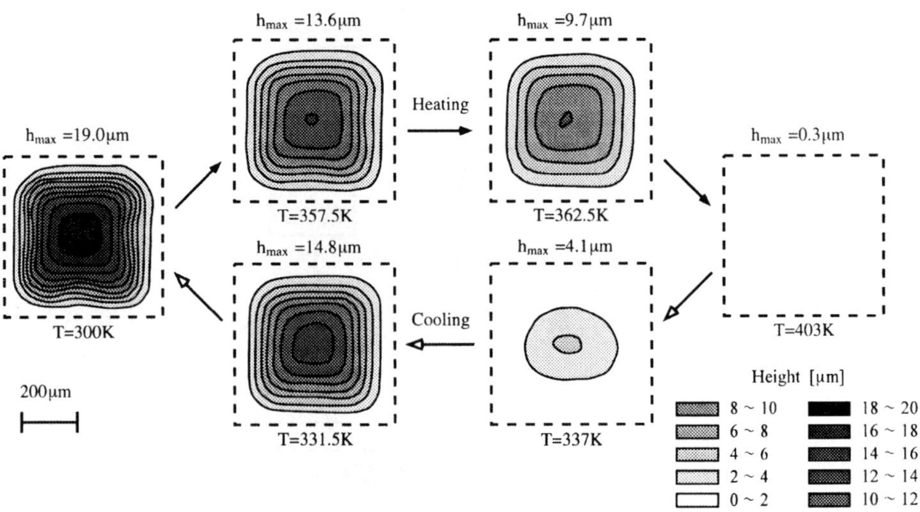

Fig. 4: Contour diagrams showing the shapes of a microactuator at various temperatures during heating and cooling. Darker part stands for higher position.

achieved after 30 s as seen in Fig. 6 (b) and (c). The frequency of applied current is higher in Fig. 6 (1 Hz) than in Fig. 5 (0.1 Hz). Accordingly the variation in temperature during measurements become smaller in Fig. 6 (about 6 K) than in Fig. 5 (about 30 K). The temperature in Fig. 5 seems to be close to the real temperature at the actuating part, because the frequency is low enough. However, the temperature in Fig. 6 with a higher frequency condition does not represent the real temperature. It is also found in Fig. 6 that h_{max} and h_{min} were 13.2 μm and 2.9 μm, respectively. The displacement in this case is 10.3 μm, which is smaller than that (19.0 μm) in Fig. 5.

3.3. INFLUENCE OF FREQUENCY ON THE RELATION BETWEEN CURRENT AND DISPLACEMENT

Fig. 7 shows the displacement expressed as a function of current with the constant

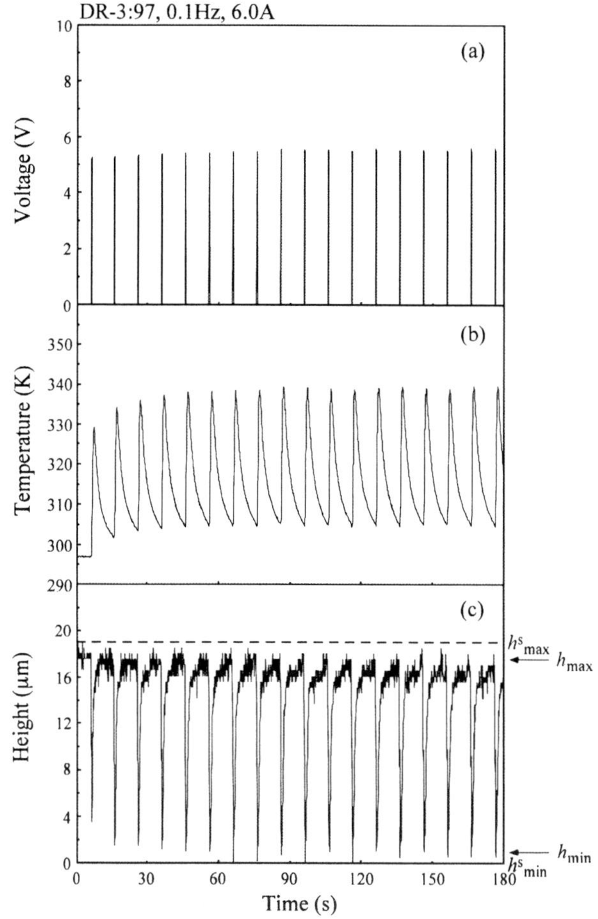

Fig. 5: (a) Voltage, (b) temperature and (c) height of the actuator as a function of time under the condition of duty ratio of 3:97, current of 6.0 A and frequency of 0.1 Hz.

duty ratio of 3:97 for various constant frequencies. It is clear that the actuator worked in a wide range of current and frequency. Thus, it can be said that the diaphragm type microactuator based on Ti-Ni SMA thin film exhibits satisfactory actuation. When the frequency was 0.1Hz, the displacement increased up to near h^s_{max} with increasing current. However, the actuation under higher frequency requires higher current. In case of higher frequency, the displacement increased with increasing current reaching a maximum, then decreased with further increasing current. The maximum displacement decreased with increasing frequency. Besides, the current corresponding to the maximum displacement increased with increasing frequency. Therefore, the displacement vs. current curve shifts to the higher current region with increasing frequency. This is because the heat transfer becomes smaller with increasing frequency. The largest displacement is obtained when cooling and heating are balanced. It is also concluded that the dynamic properties of the microactuator based on SMA are affected by the current, frequency and duty ratio.

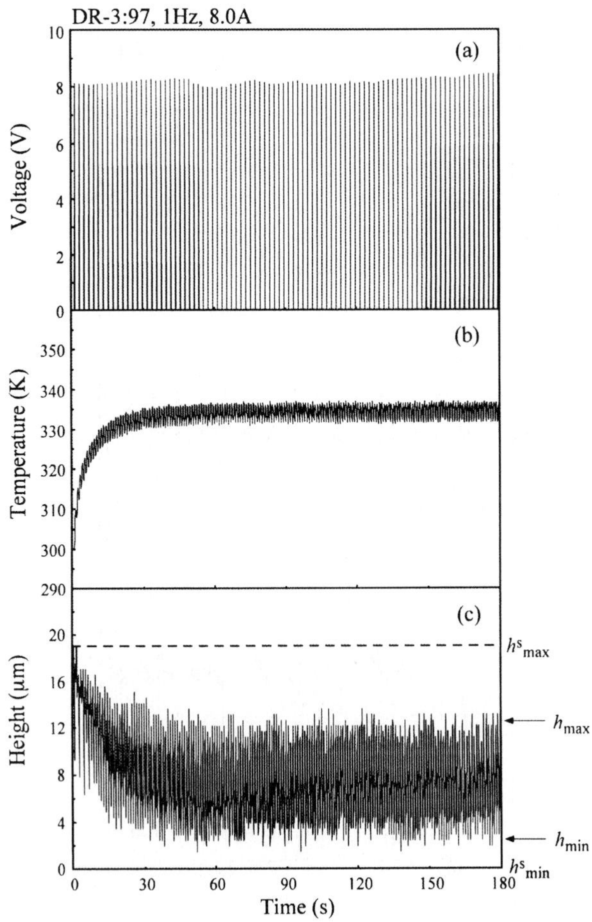

Fig. 6: (a) Voltage, (b) temperature and (c) height of the actuator as a function of time under the condition of duty ratio of 3:97, current of 8.0 A and frequency of 1.0 Hz.

3.4. EFFECT OF THERMAL CONDUCTIVITY OF GASES

Fig. 8 shows the displacement in air and helium gas as a function of current at various frequencies. Since the actuation of the actuator was confirmed in a wide range of current and frequency, it can be said that the diaphragm-type microactuator based on the Ti-Ni SMA thin film was successfully fabricated. The increase in displacement close to h^s_{max} was obtained with increasing current at the frequency of 0.1 Hz. When the frequency was increased, the displacement gradually increased up to the maximum value with increasing current, then decreased with further increasing current. It is also clear that a larger current was needed for the actuation when higher frequency was used. It should be noted that the maximum displacement decreased and the necessary corresponding current for each maximum displacement increased with increasing frequency. Therefore, the displacement vs. current curve shifted to the larger current region with increasing frequency, that is, the

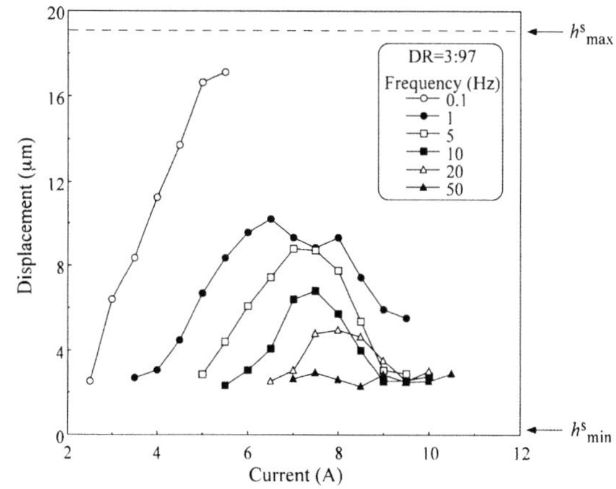

Fig. 7: Displacement as a function of current under various frequencies.

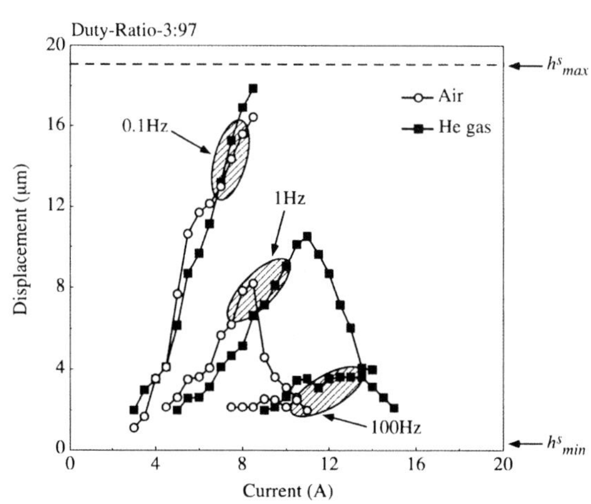

Fig. 8: Current dependence of displacement at various frequencies measured in air and helium gas, respectively.

curve shifted to the right side in Fig. 5. This is because a quantity of heat is insufficient for achieving a large displacement due to a shorter period of heating at a higher frequency, thus a higher current is needed with increasing frequency. The largest displacement is obtained when cooling and heating are balanced.

When the frequency was 0.1 Hz, the similar displacement was obtained irrespective of the gas used. Thus, no difference in the maximum displacement was recognized by changing the gas when the driving frequency was low enough. On the other hand, when the frequency was 1 Hz and 100 Hz, the displacement achieved using helium gas was larger than that in air. The larger displacement can be obtained under an atmosphere having a larger thermal conductivity, because the cooling rate increases with increasing thermal conductivity. It is clarified that the dynamic properties of the microactuators based on SMA are affected by the applied current, frequency and thermal conductivity of atmosphere.

4. Conclusions

[1] A diaphragm type microactuator based on Ti-Ni SMA thin film exhibited satisfactory actuation.
[2] Reversible shape change between convex and flat shapes occurred repeatedly by heating and cooling, and the actuation behavior was affected by frequency and current.
[3] hen the frequency is less than 1 Hz, the displacement of the diaphragm increased with increasing current.
[4] When the frequency is more than 1 Hz, the displacement increased with increasing current reacheing a maximum, then decreased with further increasing current.
[5] Response can be improved by increasing thermal conductivity of atmospheric gas.

Acknowledgements

This work was partially supported by Grants-in-Aid for Scientific Research(B: 10555210 and A: 11305043) from the Ministry of Education, Science, Sports and Culture.

References

[1] Miyazaki, S. and Ishida, A. (1999) *Mater. Science Eng.*, **A273-275**, 106-133.
[2] Miyazaki, S., Ohmi, Y., Otsuka, K. and Suzuki, Y. (1982) *J. Phy.*, **43**, C4-255-260.
[3] Miyazaki, S. (1990) in Duering, T.W., Melton, K.N., Stökel, D. and Wayman, C.M. (eds.), *Engineering Aspects of Shape Memory Alloys*, Butterworth-Heinemann, Guildford, UK, pp.394-413.
[4] Miyazaki, S. and Otsuka, K. (1989) *ISIJ Int.*, **29**, 353-377.
[5] Walker, J. and Gabriel, K. (1990) *Sensors and Actuators*, **A21-A23**, 243-246.
[6] Wolf, R.H. and Heuer, A.H. (1995) *J. Microelectromech. Syst.*, **4**,206-212.
[7] Quandt, E., Halene, C., Holleck, H., Feit, K., Kohl, M., Schloßmacher, P., Skokan, A. and Skrobanek, K. (1996) *Sensors and Actuators*, **A53**, 434-439.
[8] Makino, E., Kato, K. and Shibata, T. (1999) *Sensors and Actuators*, **A75**, 156-161.
[9] Makino, E., Mitsuya, T. and Shibata, T. (2000) *Sensors and Actuators*, **A79**, 128-135.

ON THE DESCRIPTION OF MARTENSITIC PHASE TRANSFORMATIONS USING TENSORIAL TRANSFORMATION KINETICS

O.T. BRUHNS AND C. OBERSTE-BRANDENBURG
Institute of Mechanics, Ruhr-University Bochum,
D-44780 Bochum, Germany

1. Introduction

In steels undergoing an austenite martensite phase transition, plastic deformation can occur during the transformation even if the load level is well below the yield stress of the coexisting phases (TRIP) (Fischer *et al.*, 1996). A significant influence of the orientation and the magnitude of the applied stress on the transformation kinetics, the microstructure and the mechanical behavior can be observed experimentally (Gautier and Patoor, 1997; Videau *et al.*, 1996).

Various authors (e.g. contributions of M. Berveiller, E. Patoor etc. in this volume) have tried to account for this orientation dependence by including the fully crystallographic complexity of the transformation. This leads to extremely difficult problems. The alternative of using a scalar description (where the driving force is the Gibbs energy) is unsatisfactory as it misses the orientation dependence. This paper presents a new approach which retains the orientation dependence by the introduction of tensorial driving force (and conjugate flow), but is simpler than the crystallographic approach. This description is used to develop a macroscopic material law which describes the material behavior during an austenite martensite transformation in TRIP-steels. The model is implemented into a finite element package. The ability to simulate large thermodynamically coupled problems is demonstrated.

2. Micromechanical Foundation of the Macroscopic Tensors

In this section, a local examination of the phase boundary is done. This consideration leads to the introduction of the tensorial variables describing the phase transition kinetics. Homogenization techniques will be used to constitute a formal definition of the macroscopic tensors used in the macroscopic material law.

Q.P. Sun (ed.), IUTAM Symposium on Mechanics of Martensitic Phase Transformation in Solids, 197–204.

2.1. LOCAL DESCRIPTION AT THE PHASE BOUNDARY

The local examination of the phase boundary is done in the referential configuration defined by the deformation gradient $\mathbf{F} = \partial\chi/\partial\mathbf{X}$ with the position of a material point in the actual configuration $\mathbf{x} = \chi(\mathbf{X}, t)$. This allows to distinguish between the descriptions of the movement of the phase boundary, and of the deformation of the body. The jump of a, not necessarily scalar, function f is denoted by the brackets $[f] := f^{+} - f^{-}$, where "+" denotes the value at the side of the parent phase and the "–" in the new phase. The velocity of the phase boundary orthogonal to itself in the referential description will be denoted by U_N. The balance laws at the phase boundary can be expressed as

$$\rho_0 U_N [u] + [\mathbf{P} \cdot \dot{\mathbf{x}} - \mathbf{Q}] \cdot \mathbf{N} = 0 \tag{1}$$

for the energy and

$$\rho_0 U_N [s] - \left[\frac{\mathbf{Q}}{T}\right] \cdot \mathbf{N} = -\dot{s}_\Lambda \leq 0 \tag{2}$$

for the entropy, where ρ_0, u, $\mathbf{P}$, $\mathbf{Q}$, s, T and $\dot{s}_\Lambda$ denote, in the reference configuration, the mass density, the internal energy, the first Piola-Kirchhoff stress, the heat flux, the entropy, the absolute temperature, and an entropy production at the interface due to the movement of the interface, respectively.

The procedure outlined by Heidug and Lehner (1985) will be followed here as it is formulated in the referential description and relies only on the balance equations at the phase boundary. Grinfeld (1991) gives an overview of developments that led to the introduction of the chemical potential tensor. Under the assumption of a coherent boundary $[\mathbf{x}] = \mathbf{0}$, the absence of a jump in the temperature, with the balance of momentum for the quasistatic case $[\mathbf{P} \cdot \mathbf{N}] = \mathbf{0}$, and neglecting interfacial energy, one can identify the term

$$T\dot{s}_\Lambda = \rho_0 U_N \mathbf{N} \cdot \left[\boldsymbol{\mu}^R\right] \cdot \mathbf{N} \geq 0, \quad \boldsymbol{\mu}^R := \varphi \mathbf{1}^R - \frac{1}{\rho_0} \mathbf{C} \cdot \mathbf{S} \tag{3}$$

as a description of the local dissipation caused by the movement of the phase boundary. Here, the (specific) chemical potential tensor $\boldsymbol{\mu}^R$ has been introduced, where φ, $\mathbf{C}$ and $\mathbf{S}$ denote the specific free Helmholtz energy, the right Cauchy-Green tensor and the second Piola-Kirchhoff stress.

In Eq. $(3)_1$, the local thermodynamic driving force can be identified to be $\mathbf{N} \cdot \left[\boldsymbol{\mu}^R\right] \cdot \mathbf{N}$, depending on the orientation of the phase boundary. As a description independent of the local orientation of the interface is desired, the reasoning by Maugin (1993) or Gurtin (1995) (there for the isothermal Eshelby-tensor) is followed, who formulated a balance law using the Eshelby-tensor. Thus, $\boldsymbol{\mu}^R$ can be seen as the extension of the scalar Gibbs energy to the case of a non-hydrostatically stressed solid.

The use of a tensorial driving force implies, following the procedure of the Legendre transformation, the use of a tensorial thermodynamic flux. Based on the expression describing the local dissipation at the phase boundary (3), it will be assumed that the dissipation induced by the movement of the phase boundary can be expressed as

$$T\dot{s}_\Lambda = \left[\boldsymbol{\mu}^R\right] : \dot{\mathbf{m}}^R \geq 0\,, \quad \dot{\mathbf{m}}^R := \rho_0\, U_N\, \mathbf{N} \otimes \mathbf{N}\,, \tag{4}$$

where $\dot{\mathbf{m}}^R$ is the thermodynamic flux.

2.2. HOMOGENIZATION

For simplicity, the growth of a single new phase in a matrix will be considered. The extension to a description with more inclusions of different phases is straightforward. To use these tensorial values in a material law, they have to fit into the classical thermodynamic frame. By integration over the whole domain of the heterogeneous body $\mathcal{R}_0$ and using the jump conditions at the interface Λ_0, the first law of thermodynamics can be expressed as

$$\begin{aligned} \dot{U} = \frac{\mathrm{D}}{\mathrm{D}t} \int_{\mathcal{R}_0} \rho_0 u \,\mathrm{d}V_0 &= \int_{\partial\mathcal{R}_0} (\mathbf{v} \cdot \mathbf{P} - \mathbf{Q}) \cdot \mathbf{N} \,\mathrm{d}A_0 \\ &+ \int_{\mathcal{R}_0} \rho_0 \left(\mathbf{k} \cdot \mathbf{v} + r\right) \mathrm{d}V_0 - \int_{\Lambda_0} [\mathbf{v} \cdot \mathbf{P} - \mathbf{Q}] \cdot \mathbf{N} \,\mathrm{d}A_0\,, \end{aligned} \tag{5}$$

where $\mathbf{k}$ and r are volumetric body forces, and energy sources. The first two terms on the right hand side are terms of a homogeneous body and the third term is due to phase transition. The second law of thermodynamics then demands

$$\int_{\mathcal{R}_0} \rho_0 \left(T\dot{s} - \dot{u}\right) \mathrm{d}V_0 + \int_{\partial\mathcal{R}_0} \mathbf{N} \cdot \mathbf{P} \cdot \mathbf{v} \,\mathrm{d}A_0 + \int_{\Lambda_0} \left[\boldsymbol{\mu}^R\right] : \dot{\mathbf{m}}^R \,\mathrm{d}A_0 \geq 0\,. \tag{6}$$

Again, the first two terms are due to the description of a homogeneous body and the third term is due to the description of a moving phase boundary.

The macroscopic kinetic variable is defined by integrating of the local variable over the phase boundary, i.e. $\dot{\mathbf{M}}^R := \int_{\Lambda_0} \dot{\mathbf{m}}^R \,\mathrm{d}A_0$. A relative measure $\dot{\boldsymbol{\xi}}^R := \dot{\mathbf{M}}^R / M^{tot}$ will be used, where M^{tot} is the constant total mass of the body under consideration. The mean value of the driving force is defined

$$\Delta\boldsymbol{\mu}^R := A_\Lambda^{-1} \int_{\Lambda_0} \left[\boldsymbol{\mu}^R\right] \mathrm{d}A_0\,, \tag{7}$$

where A_Λ is the area of the phase interface, and the tensorial thermodynamic flux can be connected to the classical thermodynamic flux, the (specific) mass-change, by

$$\operatorname{tr}(\dot{\mathbf{M}}^R) = \int\limits_{\Lambda_0} \rho_0\, U_N \operatorname{tr}(\mathbf{N} \otimes \mathbf{N})\, \mathrm{d}\mathbf{A_0} = \dot{\mathbf{M}}\,, \tag{8}$$

where $\dot{M}$ denotes the mass change of the phase under consideration. This relation can be used to fulfill the requirement that the total mass of the heterogeneous body shall not change. The condition in the scalar case, can be expressed as $\sum_\alpha \dot{\xi}^\alpha = 0$, with ξ^α the mass fraction of the phase α in the reference configuration, and can be extended to the tensorial case using $\dot{\xi}^\alpha = \operatorname{tr}(\dot{\boldsymbol{\xi}}^{R,\alpha})$

$$\sum_\alpha \operatorname{tr}(\dot{\boldsymbol{\xi}}^{R,\alpha}) = 0\,, \quad \dot{\boldsymbol{\xi}}^{R,\alpha} = \dot{\boldsymbol{\xi}}^R|_\alpha\,. \tag{9}$$

This, together with the fact that for a hydrostatic stress state $\boldsymbol{\sigma} = p\,\mathbf{1}$ the chemical potential tensor collapses to $\boldsymbol{\mu}^R = \mathbf{1}\ (\varphi - p/\rho) = \mathbf{1}g$, where g denotes the Gibbs energy, emphasizes and supports the proposition that these tensorial values are the natural extension of the scalar concept.

3. Macroscopic Description of the Material Behavior

The tensorial variables are embedded into a material law describing the TRIP-effect during a phase transformation. In order to develop a model which can be used to calculate real structures with momentarily available computing power, a compromise had to be found between a description based on the relatively well describable phenomena on the mesoscale, and a computationally efficient phenomenological description on the macroscale. It has been chosen in the way that the material behavior of the phases are modeled independently, but assumed to be homogeneous.

To simplify the notation, $\boldsymbol{\mu}$ represents the chemical potential tensor describing a transformation from austenite to martensite and $\dot{\boldsymbol{\xi}}$ is the kinetic variable describing the growth of the martensite inclusions.

The mesoscopic material behavior of both phases is modeled using a linear elastic-viscoplastic law of overstress type (Bruhns and Anding, 1999). The model was adapted for the use in the referential description. The yield stress depends on temperature, whereas the tangent modulus for the asymptotic linear hardening behavior is the same for all temperatures. The elastic behavior is modeled with different elastic moduli, and the behavior during plastic deformation differs concerning different yield stresses of the phases.

It is assumed that the thermal properties of the phases do not differ significantly. Thus, the effect of phase transition enters the thermal problem only via the latent heat of the transformation. Therefore, the thermal description does not distinguish the phases, and as constitutive law, Maxwell-Cattaneo's law in the form

$$\tau_q\, \dot{\mathbf{Q}} + \mathbf{Q} = -\lambda_q \nabla_{\!X} T \tag{10}$$

with the material parameters τ_q and λ_q is used.

As the homogenization procedure for the stresses, a volumetric averaging has been applied to lead to

$$\mathbf{S} = \xi\,\mathbf{S}^M + (1-\xi)\,\mathbf{S}^A, \quad \xi = \mathrm{tr}\,(\boldsymbol{\xi}), \tag{11}$$

the global stress $\mathbf{S}$, and the stresses of the phases $\mathbf{S}^M$ and $\mathbf{S}^A$. For the strains, a kinematical reference model (Raniecki and Bruhns, 1991) has been applied, leading to the approximation of the global strain rate $\dot{\mathbf{E}}$ to be

$$\dot{\mathbf{E}} = \dot{\mathbf{E}}^{AM} + \dot{\mathbf{E}}^{PT}, \tag{12}$$

where $\dot{\mathbf{E}}^{AM}$ is the strain rate in the mesoscopic phases, and assumed to be the same. The form for $\dot{\mathbf{E}}^{PT}$ is an adaption of an approach often used in phenomenological material laws (e.g. Leblond *et al.*, 1989; Fischer *et al.*, 2000). These phenomenological laws do not explicitly distinguish the stress states of the phases, therefore this approach was adapted to this two phase description and was chosen

$$\dot{\mathbf{E}}^{PT} = \frac{3}{2}K^* \frac{\mathbf{S}'^A + \mathbf{S}'^M}{2} \frac{\partial f^\xi(\xi)}{\partial \xi}\,\dot{\xi}, \quad f^\xi(\xi) = (2-\xi)\xi\,. \tag{13}$$

A further development of this expression, based on the the kinetic variable $\boldsymbol{\xi}$ and the stresses of the phases may be desirable in the future. Furthermore, the inclusion of the volumetric change due to the phase transformation is necessary.
The approach to develop an evolution law for $\boldsymbol{\xi}$ is based on two assumptions:

- The dissipation induced by the transformation reaches a maximum for the process under consideration, i.e. $\boldsymbol{\mu} : \dot{\boldsymbol{\xi}} \rightarrow \mathrm{Max}$,
- The transformation starts only after reaching a certain threshold. A barrier F^ξ similar to the yield function in plasticity has been introduced as

$$F^\xi = (\gamma^\xi \cdot \boldsymbol{\mu} - \boldsymbol{\alpha}^\xi) : (\gamma^\xi \cdot \boldsymbol{\mu} - \boldsymbol{\alpha}^\xi) - g^\xi(\kappa^\xi) = 0\,. \tag{14}$$

  Similar concepts based on the maximization of the dissipation rate can be found e.g. in Levitas and Stein (1997) or in Fischer *et al.* (1998). The internal variables $\boldsymbol{\alpha}^\xi$ and κ^ξ are similar to the kinematic and isotropic hardening variables in plasticity. In the current context, only the transformation from austenite to martensite, but not the reverse transformation, will be considered. In analogous way as in plasticity, where also only one of the two variables can be identified, if reverse loading is excluded, we here neglect the influence of $\boldsymbol{\alpha}^\xi$. The internal variable γ^ξ was introduced to satisfy the condition of the global mass balance.

These two conditions lead to a constraint optimization problem which is solved by introducing the Lagrange multiplier λ^ξ and arriving at the Lagrange function

$\mathcal{L}^\xi$ with $\mathcal{L}^\xi = \boldsymbol{\mu} : \dot{\boldsymbol{\xi}} - \lambda^\xi F^\xi$. The evolution law for $\dot{\boldsymbol{\gamma}}^\xi$ has been chosen to be $\dot{\boldsymbol{\gamma}}^\xi = -\boldsymbol{\mu}^{-1} \cdot \dot{\boldsymbol{\mu}} \cdot \boldsymbol{\gamma}^\xi \operatorname{tr}(\boldsymbol{\xi})$ to fulfill the condition $(\boldsymbol{\mu} \cdot \boldsymbol{\gamma}^\xi)^\cdot = \mathbf{0}$ for $\operatorname{tr}(\boldsymbol{\xi}) = 1$. This condition assures that the transformation stops after reaching $\operatorname{tr}(\boldsymbol{\xi}) = 1$. As the initial value, $\boldsymbol{\gamma}|_{t=0} = \mathbf{1}$ must be chosen. The law for $\dot{\kappa}^\xi$ has been chosen to be $\dot{\kappa}^\xi = (\boldsymbol{\gamma}^\xi \cdot \boldsymbol{\mu}) : \dot{\boldsymbol{\xi}}$, with the initial condition $\kappa^\xi|_{t=0} = 0$. Using the flow rule derived from $\mathcal{L}^\xi$ by partial differentiation with respect to $\boldsymbol{\mu}$, i.e. with $\dot{\boldsymbol{\xi}} = \lambda^\xi \partial F^\xi / \partial \boldsymbol{\mu}$, and the consistency condition $\dot{F}^\xi = 0$ leads to the evolution law for $\boldsymbol{\xi}$:

$$\dot{\boldsymbol{\xi}} = \boldsymbol{\gamma}^\xi \cdot (\boldsymbol{\gamma}^\xi \cdot \boldsymbol{\mu}) \left\langle \frac{\boldsymbol{\gamma}^\xi \cdot (\boldsymbol{\gamma}^\xi \cdot \boldsymbol{\mu}) : \dot{\boldsymbol{\mu}} - \boldsymbol{\mu} \cdot (\boldsymbol{\gamma}^\xi \cdot \boldsymbol{\mu}) : (\boldsymbol{\mu}^{-1} \cdot \dot{\boldsymbol{\mu}} \cdot \boldsymbol{\gamma}^\xi) \operatorname{tr}(\boldsymbol{\xi})}{\frac{1}{2} \frac{\partial g^\xi}{\partial \kappa^\xi} (\boldsymbol{\gamma}^\xi \cdot \boldsymbol{\mu}) : \boldsymbol{\gamma}^\xi \cdot (\boldsymbol{\gamma}^\xi \cdot \boldsymbol{\mu})} \right\rangle^\xi \quad (15)$$

where the argument of the McCauley brackets is non-zero only if $\partial F^\xi / \partial \boldsymbol{\mu} : \dot{\boldsymbol{\mu}} \geq 0$ and $F^\xi = 0$.

For function g^ξ, a phenomenological approach has been chosen which allows an adaption of the model to the transformation kinetics observed during experiments. To compare the tensorial form presented here with scalar descriptions, a hydrostatic stress state may be assumed. Setting $\boldsymbol{\mu} = \mathbf{1}\,\mu$, $\boldsymbol{\gamma}^\xi = \mathbf{1}\,\gamma^\xi$, and $\dot{\boldsymbol{\xi}} = \frac{1}{3}\mathbf{1}\,\dot{\xi}$, Eq. (14) becomes $F^\xi = 3(\gamma^\xi \mu)^2 - g^\xi$ and Eq. (15) $\dot{\xi} = 6\gamma^\xi(1-\xi)\dot{\mu}/[\partial g^\xi / \partial \kappa^\xi]^{-1}$. The evolution laws for the internal variables are $\dot{\gamma}^\xi = -\mu^{-1}\dot{\mu}\,\gamma^\xi\,\xi$ and $\dot{\kappa}^\xi = \gamma^\xi \mu \dot{\xi}$. The law describing the transformation kinetics is of the type where an evolution law for the martensite fraction is formulated in dependence on thermomechanical properties (see e.g. Oberaigner *et al.*, 1993). A detailed comparison will be subject of a more extensive report.

4. Multiaxial Simulations

Multiaxial simulations of the material model can be performed either by use of the FEM-implementation or by direct integration of the system of differential equations. Since the material model does not allow to solve the equations for the strain increments, an iterative algorithm of the modified Newton-Raphson type has been implemented to solve for a desired stress increment. It turned out that using the elastic tangent matrix as an approximation of the consistent tangent matrix is sufficient concerning the convergence.

4.1. MARTENSITE-START TEMPERATURE

As one example to demonstrate the influence of the parameters on the results, the dependence of the martensite-start temperature on a variation of two parameters is presented. A test on a thin walled tube under axial and torsional load has been modeled. The mechanical load has been applied to the tube at a temperature well above the martensite start temperature. To initiate the transformation, the specimen has been cooled down by applying a constant thermal flux on its boundary. In Figure 1, the relative change of the martensite start temperature related to the

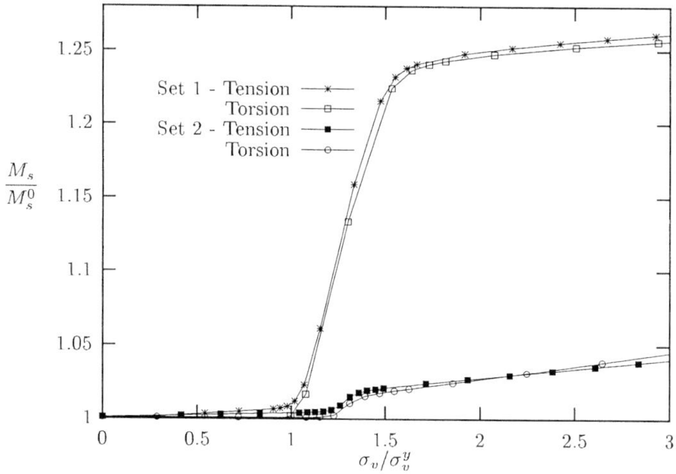

Figure 1. Martesite start temperature vs applied stress.

applied stress is illustrated. The results for two parameter sets are shown: they differ concerning the modeling of the stress independent part of the approximation of the Helmholtz free energy and the threshold value of F^{ξ}. For both sets, an axial and a torsional load have been applied. For stresses below the yield stress of the austenite σ_v^y, a linear dependence of the martensite start temperature on the stress is observed.

If the stress reaches a level just below the yield stress, the choice of the parameter becomes the dominant effect, a large jump can be observed for the first parameter set whereas the jump for the second set is rather small.

4.2. ORIENTATION EFFECT

As noted above, tensor $\boldsymbol{\xi}$ is used as a variable to describe the dependence of the material behavior on the applied stress. Therefore, it is necessary that this variable can reflect the influence of the orientation and the level of the applied load during the transformation, appropriately. In Figure 2, the relative evolution of the 11-component of $\boldsymbol{\xi}$ is shown during a transformation under tension. The orientation is relatively small for loads below the yield stress but reaches a higher level for loads above. This effect can be observed during experiments, cf. Gautier and Patoor (1997).

5. Summary

A new pair of thermomechanical variables describing the transformation kinetics in a non-hydrostatically stressed solid has been presented. Based on these variables, a material law has been developed to describe the material behavior of a TRIP-steel during phase transformation. The ability of the model to calculate

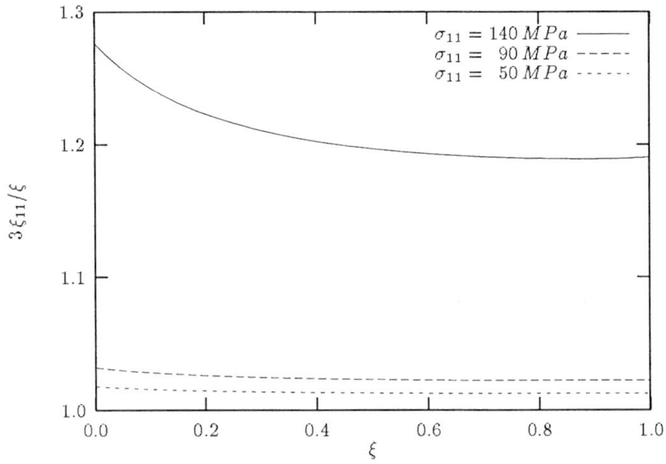

Figure 2. Relative evolution of ξ_{11} during a transformation.

large structures, e.g. in a quenching process, has been demonstrated. As the next steps, the adaption of the model to experimental data is necessary. A frame for the identification and analysis of the parameters has been established.

References

Bruhns, O. and D. Anding: 1999, 'On the simultaneous estimation of model parameters used in constitutive laws for inelastic material behaviour'. *Int. J. Plasticity* **15**, 1311–1340.

Fischer, F., G. Reisner, E. Werner, K. Tanaka, G. Cailletaud, and T. Antretter: 2000, 'A new view on transformation induced plasticity'. *Int. J. Plasticity* **16**, 723–748.

Fischer, F., Q.-P. Sun, and K. Tanaka: 1996, 'Transformation-induced plasticity (TRIP)'. *Appl. Mech. Rev.* **49**, 317–364.

Fischer, F. D., E. D. Oberaigner, K. Tanaka, and F. Nishimura: 1998, 'Transformation induced plasticity revised an updated formulation '. *Int. J. Solids Struct.* **35**, 2209–2227.

Gautier, E. and E. Patoor: 1997, 'Experimental Observations for Shape Memory Alloys and Transformation Induced Plasticity Phenomena'. In: M. Berveiller and F. Fischer (eds.): *Mechanics of Solids with Phase Changes*, Vol. 368 of *CISM Courses and Lectures*. Springer-Verlag, pp. 69–103.

Grinfeld, M.: 1991, *Thermodynamic methods in the theory of heterogenous systems*. Longman Scientific & Technical.

Gurtin, M.: 1995, 'The Nature of Configurational Forces'. *Arch. Rat. Mech. Anal.* **131**, 67–100.

Heidug, W. and F. Lehner: 1985, 'Thermodynamics of Coherent Phase Tranformations in Nonhydrostatically Stressed Solids'. *Pure Appl. Geophys.* **123**, 91–98.

Leblond, J., J. Devaux, and J. Devaux: 1989, 'Mathematical Modelling of Transformation Plasticity in Steels I: Case of Ideal-Plastic Phases'. *Int. J. Plasticity* **5**, 551–572.

Levitas, V. and E. Stein: 1997, 'Simple micromechanical Model of thermoelastic martensitic Transformations'. *Mech. Res. Comm.* **24**, 309–318.

Maugin, G.: 1993, *Material Inhomogeneities in Elasticity*. Chapman & Hall, London.

Oberaigner, E., F. Fischer, and K. Tanaka: 1993, 'A new micromechanical formulation of martensite kinetics driven by temparature and/or stress'. *Arch. Appl. Mech.* **63**, 522–533.

Raniecki, B. and O. Bruhns: 1991, 'Thermodynamic reference model for elastic-plastic solids undergoing phase transformations'. *Arch. Mech.* **43**, 343–376.

Videau, J.-C., G. Cailletaud, and A. Pineau: 1996, 'Experimental Study of the Transformation-Induced Plasticity in a Cr-Ni-Mo-Al-Ti Steel'. *Journal de Physique* **III**(6), C1–465–474.

MECHANICAL INSTABILITY OF NiTi IN TENSION, COMPRESSION AND SHEAR

D. FAVIER[1], Y. LIU[2], L. ORGEAS[1], G. RIO[3]

[1] *Laboratoire Sols, Solides Structures, CNRS-Université Joseph Fourier-Ins. Nat. Polytechnique, B.P.53, 38041 Grenoble Cedex 9, France*

[2] *Department of Mechanical and Materials Engineering, University of Western Australia, Crawley, WA 6009, Australia*

[3] *Laboratoire de Génie Mécanique et Matériaux, Université de Bretagne Sud, 2 Rue le Coat Saint Haouens, 56325 Lorient, France*

1. Introduction

Shape memory alloys (SMAs) are known to exhibit a range of novel thermomechanical properties due to thermoelastic martensitic transformations. The superelasticity is a quasi-elastic deformation far beyond the conventional elastic limit of the material when deformed at certain temperatures above A_f, the finishing temperature of the reverse martensite-to-austenite transformation. This phenomenon is associated with a stress-induced martensitic (SIM) transformation. If the temperature is lowered to below M_f, the finishing temperature of the forward austenite-to-martensite transformation, the deformation proceeds by martensite reorientation. This deformation mode is known as ferroelasticity, in recognition of its phenomenological similarity to ferromagnetism.

Numerous experimental works have been performed aiming at characterising the thermo-mechanical behaviour of SMAs in order to provide experimental evidences for the establishment of constitutive models. These studies have been devoted to both single-crystal and polycrystal specimens, mainly using Cu-based and NiTi SMAs. Most mechanical testing has been carried out in tension using wire or strip specimens. Superelastic tensile tests in most NiTi alloys exhibit Lüders type engineering stress plateaus with an upper yield stress during loading and an inverse stress peak during unloading. This localised deformation has been studied by *in-situ* optical observations [1] and more recently by *in-situ* full-field temperature monitoring and by local deformation measurement using multiple miniature extensometers [2,3]. Shaw and Kyriakides have proposed to approximate the NiTi behaviour as that of a finitely deforming elasto-plastic solid exhibiting a trilinear up-down-up nominal stress-strain response in tension [4]. Two models have been tested. The first one supposed that the true stress-logarithmic strain behaviour in tension was always rising although the nominal stress-strain curves of the intermediate branch had a negative slope. The middle branch of the true stress-strain response for the second model had a negative slope. The two models have been implemented in a finite element code in order to simulate experimental inhomogeneous tensile tests [4] neglecting the thermo-mechanical

Q.P. Sun (ed.), IUTAM Symposium on Mechanics of Martensitic Phase Transformation in Solids, 205–212.

coupling of the tensile sample with its environment. For the two models, Lüders-like tensile behaviour was well simulated, at least qualitatively. However, for the first model, the localisation was interpreted more as a geometrical instability than a material instability. In the second model, the onset and the propagation of the localised deformation was interpreted as being associated with nucleation and propagation of the martensitic transformation. Shaw *et al* rejected the first model arguing that the second model seemed to better fit experimental data. Their conclusions did not take into account (1) thermomechanical coupling, known to have a strong influence on the localisation [5,6], and (2) experimental results obtained under other stress conditions than tension. Thermo-mechanical coupling has been later studied by Shaw [7], but only for the second model.

This paper presents experimental results, physical arguments and numerical simulation aimed at clarifying the instability phenomena observed for NiTi polycrystalline alloys. In the first part, experimental results on equiatomic NiTi involving tension and other stress states are presented. The second part offers a discussion of the nucleation-propagation hypothesis for the martensitic transformation of NiTi polycristalline alloys from a metallurgical perspective. The last part presents numerical simulation of tension, compression and shear tests.

2. Experimental Results in Tension, Compression and Shear of NiTi

A commercial polycrystalline equiatomic NiTi alloy was used. As-received 3.2 mm and 1.3 mm thick sheets were homogenized at 1173 K for 3.6 ks, cold-rolled with an 18 % thickness reduction and annealed at 703 K for 1.8 ks. The treated material showed a two-step A→R→M transformation on cooling ($R_s \approx 321$ K, $R_f \approx 308$ K, $M_s \approx 283$ K, $M_f \approx 226$ K) and a one-step transformation M→A on heating ($A_s \approx 301$ K, $A_f \approx 328$ K).

The specimens for the tension-compression tests were bone-shaped with a deformation gauge section of 40 mm in length, 5.6 mm in width and 2.7 mm in thickness [8]. To enable compressive loading on such specimens, a specially designed anti-buckling gripping system was designed. The specimens for the shear tests were rectangular in shape with a shear gauge section of $30 \times 3 \times 1.05$ mm^3. Isothermal testing conditions were achieved using an oil bath for temperature control with an accuracy of ± 1 K. Deformation was carried out using a slow strain rate of 10^{-3} s^{-1} for tension-compression and 1.75 10^{-3} s^{-1} for shear. For the tension-compression tests, both local strain (ε %, LM) and global strain (ε %, GM) were measured. LM was measured using an extensometer of a 10 mm gauge length that was mounted on the specimen in the oil bath, whereas GM was calculated from the total length variation of the specimen [9]. The shear strain γ was calculated from the measurement of the relative displacement of the mobile grip of the shear testing apparatus with respect to the stationary grip, using a local extensometer mounted in the oil bath.

Three tension-compression loops are shown in Figure 1 for the same sample tested at 333, 338 and 343 K sequentially. The stress was estimated as the engineering stress F/S_0 (F and S_0 being respectively the current applied force and the initial cross-section) whereas the strain ε was estimated as the engineering strain $\Delta l/l_0$ (l_0 and Δl being respectively the initial gauge length and its variation). Superelastic behaviour was observed in tension and compression for the three temperatures ($> A_f$). Transformation stresses were higher in compression than in tension. Conversely, transformation strains were higher in tension than in compression. Lüders-like deformation behaviour, as characterized by a distinctive stress plateau on an engineering stress-strain curve, was observed in tension during both the forward and reverse transformations, but not in compression. Two virgin samples were also tested separately in tension and in compression at 333 K, as shown in Figure 2. The occurrence of Lüders-like deformation under the condition of monotonic loading conforms to the observations established for the alternative tension-compression loading conditions. The heterogeneous nature of the Lüders-like deformation is illustrated in Figure 3, where the measurement of the global strain (GM) is plotted as a function of the local strain (LM) for the tensile test shown in Figure 2. The corresponding points are labelled on the curves shown in Figures 2 and 3.

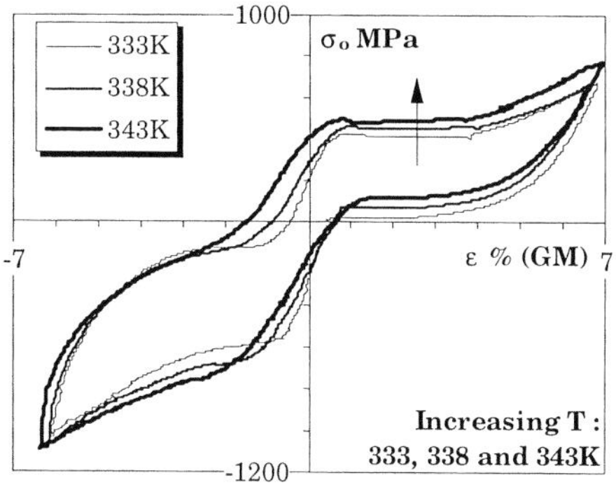

Figure 1: Isothermal tension-compression loops at $T = 333$, 338 and 343 K.

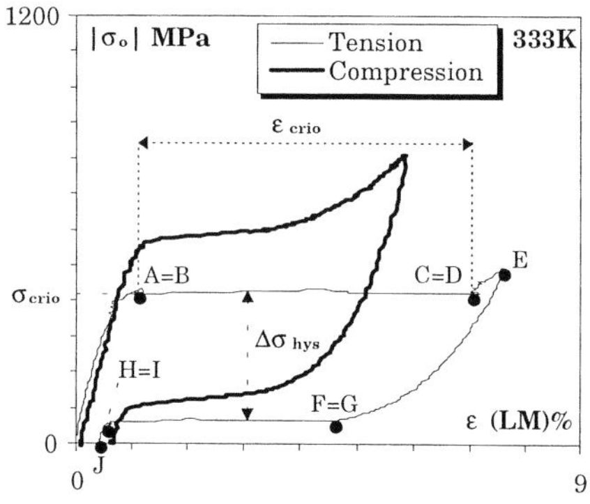

Figure 2: Tension and compression tests on two virgin samples at $T = 333$ K.

Figure 4 shows a specimen deformed in shear at temperatures above A_f. The stress-strain curves demonstrate that the deformation was mostly superelastic and that no Lüders-like behaviour occurred.

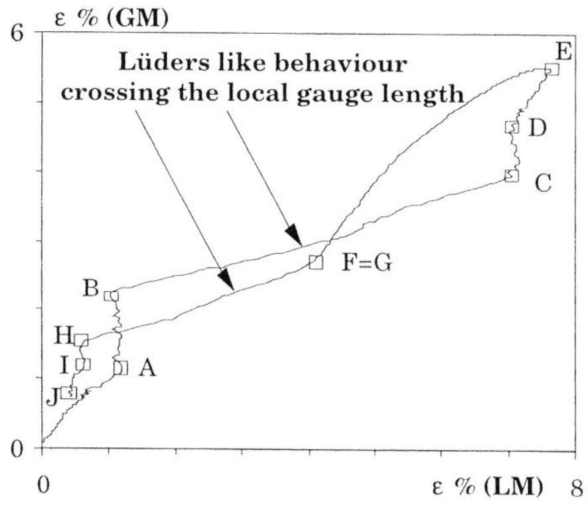

Figure 3: Global strain versus local strain measurement during the tensile test of Fig.2.

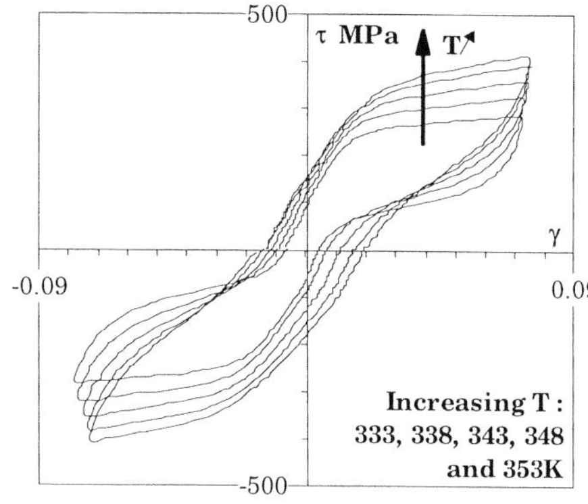

Figure 4: Isothermal shear loops at $T = 333$, 338, 343, 348 and 353 K.

3. Localised Deformation and Nucleation-propagation Hypothesis

3.1. REMARKS ON LOCALISATION OF DEFORMATION

Figure 5 shows the deformation of a Ti-50.0at%Ni wire specimen in tension via martensite reorientation, as measured by three LM measurements [10]. The three engineering tensile stress-strain curves are characterized by an initial stage of deformation prior to the stress drop at the onset of a stress plateau, which terminated at 6% strain. The localisation of deformation during this Lüders-like behaviour is emphasised by plotting the measurement of the three local strains as a function of the time elapsed during deformation, as shown in Figure 6.

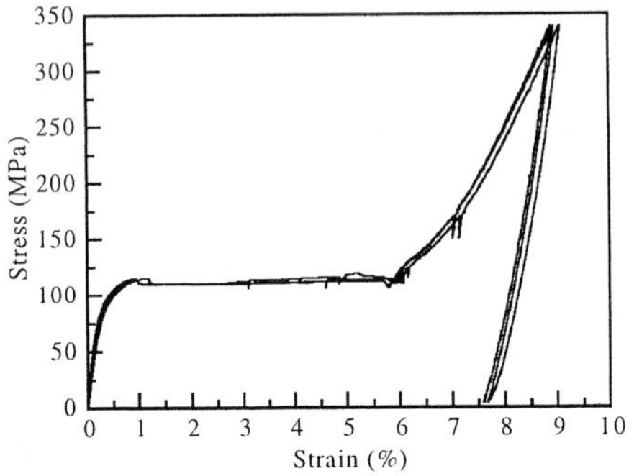

Figure 5: Tension test of martensite in NiTi.

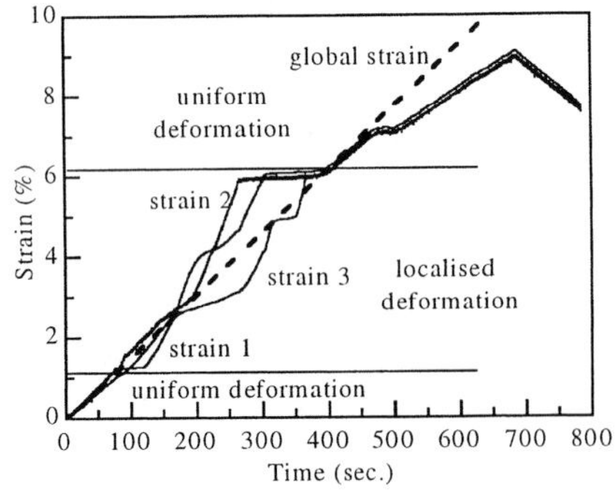

Figure 6: Measurement of local strains as a function of deformation time.

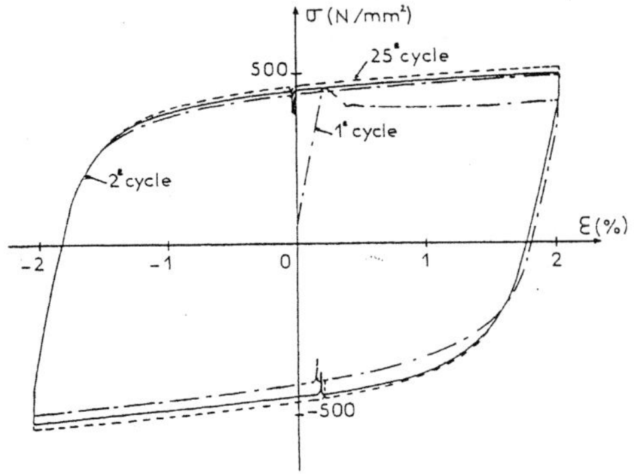

Figure 7: Tension-compression of mild steel.

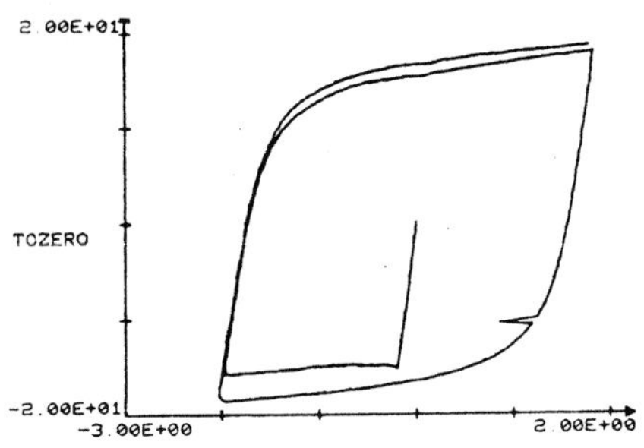

Figure 8: Torsion of mild steel.

It is well-known that Lüders deformation is often observed in mild steel tensile tests, as shown in Figure 7 [11]. For this type of materials, Lüders deformation is observed for other stress states, as seen in Figure 8, which shows a shear stress-strain curve deduced from a torsion test using a thin-tube specimen. It has been suggested that there exist some similarities in macroscopic events during tensile tests between superelastic NiTi alloys and fine-grained steel strips or wires [3]. This suggestion does not seem to prevail for other stress states, as demonstrated in the comparison between shear of NiTi sheets, where no Lüders bands were observed (Figure 4), and torsion of mild steel thin tubes (Figure 8).

3.2. THE NUCLEATION-PROPAGATION HYPOTHESIS

The stress plateau exhibited by NiTi deformed in tension (Figures 2 and 3) has been recognised as being associated with stress-induced martensitic transformation [1]. To explain the localization observed during this plateau, the theory postulates that the transformation requires a higher stress for initial nucleation and a lower stress for transformation propagation. This theory justifies the use of constitutive models which incorporate material instabilities, like the second model used by Shaw et al [4]. However, this postulate for the stress-induced martensitic transformation in NiTi is questionable on several accounts [12]:

(1) Whereas Lüders-like behaviour has been widely observed in NiTi during tensile deformation, there has been no convincing experimental evidence in the literature showing the occurrence of similar deformation behaviour for stress-induced transformation under other deformation modes, where nucleation is also a prerequisite for transformation. Instead, there has been experimental evidence of the absence of Lüders-like behaviour during compression (Fig. 2) or shear (Fig. 4) of samples that show Lüders-like behaviour in tension.
(2) Lüders-like behaviour is also observed during tensile deformation by martensite reorientation, in which no phase transformation and thus no nucleation is involved (Fig. 5).
(3) It has been frequently observed that the mechanical instability may start at tensile strain larger than 1% well beyond the linear elastic range. It has also been well established that recoverable strain continues to increase with deformation beyond the end of the instability [13,14]. These observations suggest that the stress plateau is only a manifestation of localised deformation and does not indicate either the start or the finish of the transformation.
(4) From a metallurgical point of view, whereas a deformation wave can propagate across grain boundaries from grain to grain in a polycrystalline matrix, a phase transformation cannot propagate across grain boundaries without nucleation in adjacent grains.

These pieces of evidence suggest that the mechanical instability of NiTi during tensile deformation is due to reasons other than the transformation nucleation-propagation hypothesis. The actual mechanism responsible for this instability has yet to be identified.

4. Numerical Simulations

This section revisits conclusions that Shaw and Kyriakides drew from consideration only of localisation observed during tensile superelastic tests of NiTi alloys. These authors have performed two sets of numerical simulations of localisation using two models, both being based on traditional rate-independent, finitely deforming J_2 flow theory with isotropic hardening [4]. The first set assumed that the true stress-strain response in tension maintained its stability during the plateau whereas in the second set the intermediate branch of the true stress-strain response had a negative slope. The only argument supporting the second model was that it gave a larger initiation peak in tension, in better agreement with experimental results. However, these conclusions were drawn from simulations performed neglecting thermomechanical coupling. In more recent simulations using only the second type of constitutive equation, it has been shown that this coupling is very important [7].

Our simulations are restricted to isothermal cases and thus do not address the strong effect of thermomechanical coupling. The only purpose is to show that simulation using constitutive equations similar to those of the first model proposed by Shaw in [4] agrees well not only with experimental Lüders-like tensile tests, but also with compression tests for which homogeneous deformation was experimentally observed.

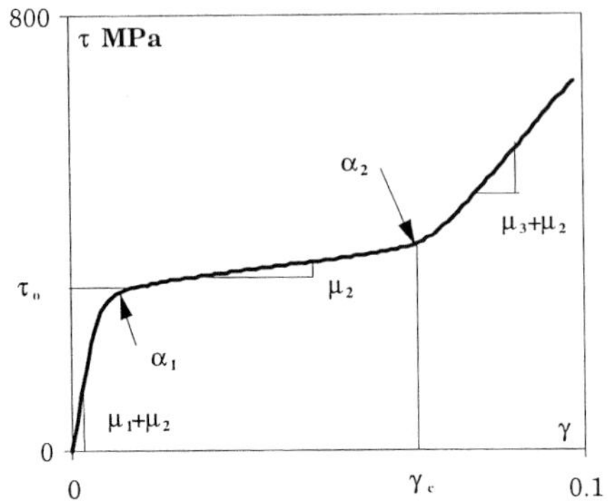

Figure 9: Definition of the parameters of the hyperelastic model in shear.

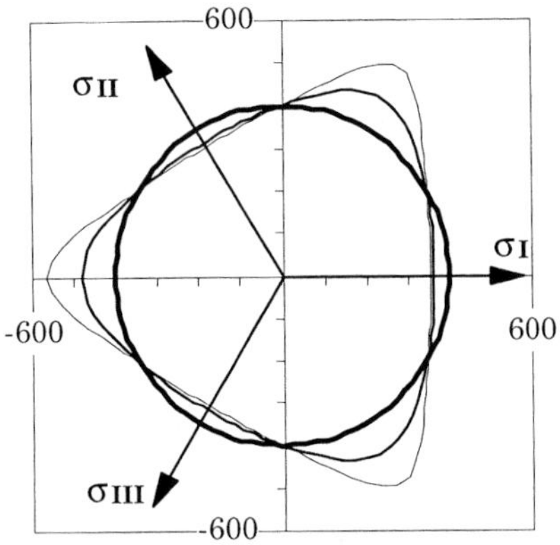

Figure 10: Transformation criteria plotted in the deviatoric stress plane.

Numerical results presented in this section use a simple isotropic hyperelastic scheme and account for large deformations. The stresses are derived from the total strains using a strain energy function which is a scalar function of the volume change and of the second and third invariants of the Almansi strain tensor. The set of constitutive equations were extensively described in [15]. As an example, prediction of the model in a pure shear loading is illustrated with the stress-strain curve plotted in Figure 9. As evident in this figure, the model prediction is similar to the tri-linear stress-strain curve used by Shaw in [4], except that the three linear stages are connected by arcs. To run the simulation, the hyperelastic scheme requires firstly the determination of 7 constitutive parameters:

- a transformation stress, *i.e.* τ_0, and a transformation strain, *i.e.* γ_c,
- shear stress-strain moduli, for the initial, middle and final stages observed during superelasticity, *i.e.* $(\mu_1 + \mu_2)$, μ_2 and $(\mu_3 + \mu_2)$, respectively,
- two parameters, *i.e.* α_1 and α_2. α_1 allows a continuous transition between the initial stage and the middle stage associated with the transformation stage. α_2 characterizes gradual transition between the middle stage and the final stage.

To account for the tension-compression asymmetry (Figures 1 and 2), the transformation criterion is not of Von-Mises type; the latter would correspond to the thick circle of Figure 10 which represents several transformation criteria plotted in the deviatoric stress plane. The thin curves are defined by three parameters, the shear transformation stress τ_0 (which is a linear function of the temperature, obeying a

Clausius-Clapeyron relation) and two additional parameters γ_0 and n [16]. Such thin curves can easily be fitted from tension-compression and shear superelastic tests.

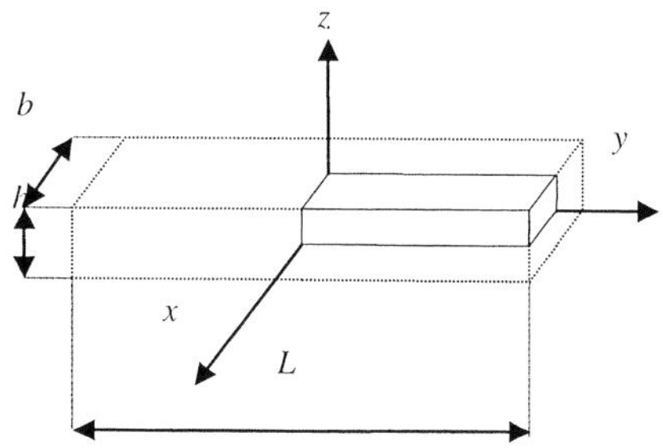

Figure 11: Geometry of the strip.

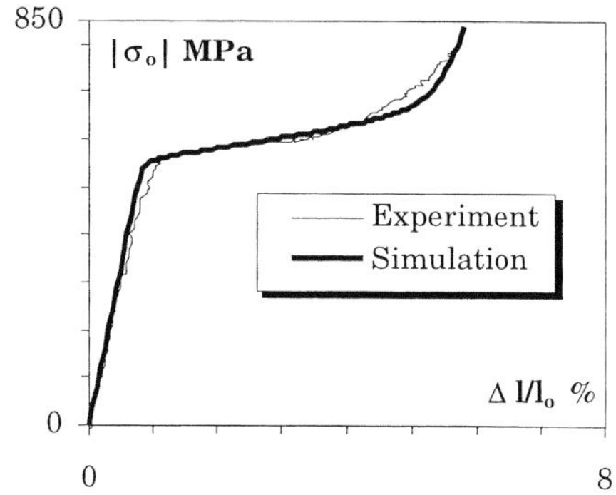

Figure 12: Modelling of the compression test

Models similar to Shaw's two models are obtained depending on the value of μ_2. Our simulation was performed with $\mu_2 = 245$ MPa, resulting in a positive slope for the true stress-strain function for shear, tension and compression stress state. Values of the other parameters together with the expressions of the curves defining the stress-strain response (Figure 9) and transformation criteria (Figure 10) are given in [15].

A strip of length $L = 50$ mm, thickness $h = 1$ mm and width b = 5 mm is considered, as depicted in Figure 11. Due to symmetry, only one eighth of the strip was meshed. The finite elements were 3D eight-node brick continuum elements. The results presented below have been obtained with a gauge section of the meshed model consisting of 1 element through the half-thickness $h/2$, 8 elements through the half-width $b/2$ and 80 elements through the half-length $L/2$. The influence of the meshing was tested previously, showing that the mesh described was satisfactory for the accuracy of results and for non-costly computing [15].

To model uniaxial tension-compression loading, the following set of boundary conditions were prescribed on the strip schematised in Figure 11: $U_y(y = 0) = U_x(x = 0) = U_z\ (z = 0) = 0$ and $U_y\ (y = L/2)$ = a linear function of time. To induce Lüders-like deformation, a geometric imperfection was introduced in the strip by considering that the width of the strip was not perfectly constant but was a function of y:

$$b = b_0 + \Delta b \sin(\pi y/L) \qquad \text{with} \qquad b_0 = 5 \text{ mm}.$$

In this work, Δb was equal to 10^{-2} mm. The results of the simulations were qualitatively similar for values of Δb ranging between 10^{-3} mm and 10^{-1} mm [15].

Figure 12 shows experimental and modelled engineering stress-strain curves for a compression test performed at 333 K. The simulation displays homogeneous deformation and the comparison between experiment and numerical simulation is fairly good. Similar results are plotted in Figure 13 in the case of a tensile test performed at the same temperature. The force per unit of initial area is plotted as a function of the total length variation, both for experiment and numerical simulation. These two curves agree almost perfectly. Two strain measurements have been calculated from the results of the simulation. The first one, *i.e.* $\Delta l/l$, is based on the displacement of the point P_1 initially located at $x = z = 0$ and $y = 7.5$ mm, whereas the second one, *i.e.* $\Delta L/L$, is based on the displacement of the top of the strip ($y = 25$ mm). Figure 14 shows the comparison of experimental and simulated results concerning the localisation of the deformation. The thick curve shows the simulated "global strain $\Delta L/L$" plotted as a function of the simulated "local strain $\Delta l/l$" in thick stroke. The thin curve represents the experimental global strain measurement ε % GM as a function of the experimental local measurement ε % LM, as shown in Figure 3 during the loading OABCD.

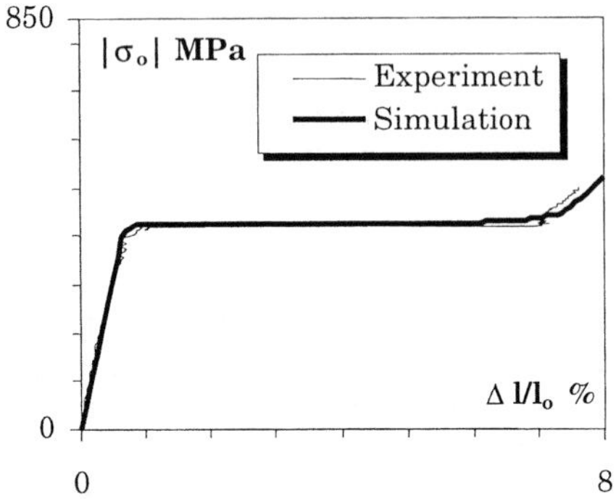

Figure 13: Modelling of the tensile test.

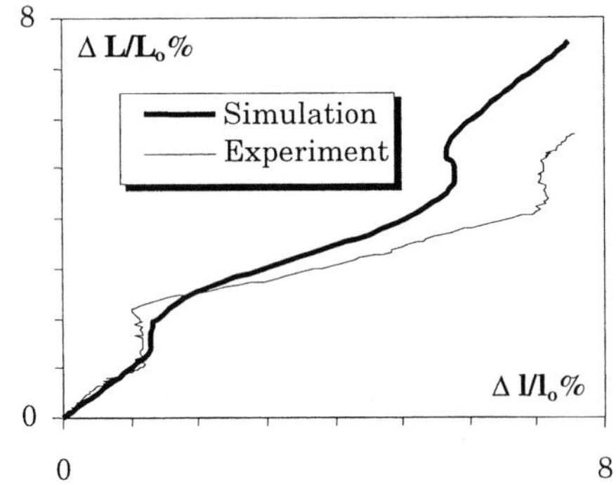

Figure 14: Modelling of the localisation during tensile test.

5. Conclusions

Several theories and constitutive equations have been proposed on the basis of experimental results of superelastic tensile tests exhibiting Lüders-like behaviour. This behaviour is not observed for other stress conditions like compression and shear, which negates the common association of the inhomogeneous deformation with phase transformation. The present paper shows that inhomogeneous tensile tests together with homogeneous compression tests can be well simulated using usual stable constitutive equations.

REFERENCES

1. Miyazaki S., Imai T., Otsuka K., Suzuki Y., *Scripta Met,* **15** (1981), 853-856.
2. Shaw J.A. and Kyriakides S., *J. of Mech. Phys. Solids,* **43, 8** (1995), 1243-1281.
3. Shaw J.A. and Kyriakides S., *Acta Materialia,* **45, 2** (1997), 683-700.
4. Shaw J.A. and Kyriakides S., *Int. J. of Plasticity,* **13, 10** (1998), 837-871.
5. McCormick P.G. , Liu Y., Miyazaki S., *Materials Science and Engineering, A,* **A167** (1993), 51-56.
6. Balandraud X., Ph. D thesis, University of Montpellier, France (2000).
7. Shaw J.A. , *Int. J. of Plasticity,* **16** (2000), 541-562.
8. Orgéas L., Favier D., *J. de Physique IV,* **5** C5 (1995), 605-611.
9. Orgéas L., Favier D., *Acta Materialia,* **46, 15**, (1998), 5579-5591.
10. Liu Yinong, Liu Yong, Van Humbeeck J., *J. Scripta Mat.,* **39, 8** (1998), 1047-1055.
11. Favier D., Guélin P., *Arch. Mech,* **37, 3** (1985), 201-219.
12. Liu Y., *Mat. Sci and Eng.,* **A271,** (1999), 506-508.
13. Miyazaki S., Otsuka K., Suzuki Y., *Scripta metal.,* **15** (1981), 287-292.
14. Liu Y., Liu Y., Van Humbeeck J., *Acta Materialia,* **47** (1999), 199-209.
15. Orgéas L., Favier D., Rio G., *Revue Européenne des Eléments Finis,* **7, 8** (1998), 111-136.
16. Orgéas L., Favier D., *J. de Physique IV,* (2001), 67-74.

HETEROGENEOUS NUCLEATION IN A 2-D MODEL OF MARTENSITIC TRANSFORMATION WITH VOLUME CHANGES

WENWU CAO AND RAJEEV AHLUWALIA
Materials Research Laboratory, The Pennsylvania State University, University Park, Pennsylvania 16802

We study domain pattern formation in a 2-D square to rectangle martensitic transformation using a time-dependent Ginzburg-Landau model. The model is based on a free energy in terms of the deviatoric, bulk and shear components of the strain tensor. In this work, we include a coupling between the deviatoric and bulk strains that allows for volume changes of the unit cell. An effective free energy in terms of deviatoric strains is constructed by eliminating the bulk and shear strains subject to the elastic compatibility relations. Heterogeneous nucleation is studied by considering randomly distributed stress centers that represent crystal defects. Computer simulations of the model reveal that the stress centers nucleate product phase domains that grow only in finite size. Hence, there is always some retained parent phase at the end of the transformation.

1. Introduction

The microstructure of martensitic materials exhibits fascinating domain patterns. Typically, the product phase domains grow anisotropically and have low aspect ratio. Usually, the parent and product phases are found to coexist with a pattern of crisscross domains that are separated by intermediate austenite regions that failed to transform[1]. The martensite domains may themselves be internally twinned. Clearly, there is a need for theoretical models to explain the complex microstructure observed in real materials. Recently, the microstructure of these materials has been studied by simulation models based on the Landau-Ginzburg approach [2-4]. This is a very useful technique to study domain patterns, as one does not need to make *a priori* assumptions about the domain wall orientations. The microstructure is a consequence of the appropriate symmetry built into the free energy.

In this work, we are particularly interested in the "Proper Ferroelastic Martensitic Transformations". Here, the strain itself is the order parameter as the unit cell undergoes a shape change. A 3-D Landau-Ginzburg free energy for proper ferroelastic transformations was first proposed by Barsch and Krumhansl [5]. Recently, Shenoy and coworkers made a 2-D simulation for the square to rectangle transition and showed that the microstructure in these materials was a consequence of elastic compatibility requirement[3]. Here we use the elastic compatibility approach to study a square to rectangle transition, in which a coupling between bulk and deviatoric strain explicitly occurs in the free energy. This coupling corresponds to the volume changes associated with the unit cell. We show below that the simple coupling discussed above drastically influences the simulated microstructures.

Q.P. Sun (ed.), IUTAM Symposium on Mechanics of Martensitic Phase Transformation in Solids, 213–220.

2. Model

The model is based on a free energy functional in terms of the bulk strain $\phi_1 = (\eta_{xx} + \eta_{yy})/\sqrt{2}$, the deviatoric strain $\phi_2 = (\eta_{xx} - \eta_{yy})/\sqrt{2}$ and the shear strain $\phi_3 = \eta_{xy}$, where η_{ij} represents the component of the linearized strain tensor. The free-energy is given as

$$F = F_d[\phi_2] + F_h[\phi_1, \phi_3] + F_c[\phi_2, \phi_1], \tag{1}$$

where $F_d[\phi_2]$ is the free-energy contribution due to the dominant devatoric strains. The term $F_h[\phi_1, \phi_3]$ is the harmonic contribution due to bulk and shear strains and $F_c[\phi_1, \phi_2]$ is the coupling between bulk and deviatoric strain. The deviatoric contribution is given by

$$F_d[\phi_2] = \int d\vec{r}[f(\phi_2) + \frac{\gamma_1}{2}(\vec{\nabla}\phi_2)^2 + \frac{\gamma_2}{2}(\nabla^2\phi_2)^2], \tag{2}$$

where strain gradient energy contribution has also been included. Since the deviatoric mode is the dominant mode for the transition, a sixth order polynomial is chosen for the $f(\phi_2)$ to describe a first order phase transition. The harmonic free-energy due to bulk and shear strain is given as

$$F_h[\phi_1, \phi_3] = \int d\vec{r}[\frac{A_1}{2}\phi_1^2 + \frac{A_3}{2}\phi_3^2], \tag{3}$$

where, A_1 and A_3 are constants that can be obtained from the elastic constants of the material. The coupling term between bulk and deviatoric strain is given as

$$F_c[\phi_1, \phi_2] = \int d\vec{r}[A_{12}\phi_1\phi_2^2]. \tag{4}$$

We should also consider the fact that the strains ϕ_1, ϕ_2 and ϕ_3 are not independent but are linked by the elastic compatibility relations [6]. In terms of the strain variables, these relations are expressed as

$$\nabla^2\phi_1 - (\frac{\partial^2}{\partial x^2} - \frac{\partial^2}{\partial y^2})\phi_2 - \frac{\partial^2}{\partial x \partial y}\phi_3 = 0. \tag{5}$$

We now make use of an elimination procedure similar to the one first proposed by Kartha et. al [6]. In this approach, the bulk and shear strains are eliminated subject to the constraint expressed in Eqn. (5). For our model, the elimination procedure results in an effective long-range interaction that replaces the terms $F_h[\phi_1,\phi_3]+F_c[\phi_1,\phi_2]$. The effective interaction is conveniently expressed in Fourier space as

$$F_{eff}=\int d\vec{k}C_1(\vec{k})\,|\,\phi_2(\vec{k})\,|^2+\int d\vec{k}C_2(\vec{k})\,|\,\Gamma(\vec{k})\,|^2+\int d\vec{k}C_3(\vec{k})\phi_2(-\vec{k})\Gamma(\vec{k}) \qquad (6)$$

where $\phi_2(\vec{k})$ and $\Gamma(\vec{k})$ are respectively the Fourier transforms of $\phi_2(\vec{r})$ and $\phi_2(\vec{r})^2$ and the kernels

$$C_1(\vec{k})=(1/2)(k_x{}^2-k_y{}^2)^2/[k^4/A_1+8k_x{}^2k_y{}^2/A_3] \qquad (7)$$

$$C_2(\vec{k})=-(A_{12}{}^2/2A_1A_3)(8k_x{}^2k_y{}^2)/[k^4/A_1+8k_x{}^2k_y{}^2/A_3] \qquad (8)$$

$$C_3(\vec{k})=(A_{12}/A_1)(k_x{}^2-k_y{}^2)(k_x{}^2+k_y{}^2)/[k^4/A_1+8k_x{}^2k_y{}^2/A_2] \qquad (9)$$

where, $k^4=(k_x{}^2+k_y{}^2)^2$.

After eliminating the bulk and the shear strain, we have a free energy only in terms of the deviatoric strain $\phi_2(\vec{r})$. To study the domain pattern formation, we study the time evolution of $\phi_2(\vec{r})$ using the time dependent Ginzburg-Landau formalism. The equation of motion is given as

$$\frac{\partial\phi_2(\vec{r})}{\partial t}=-L\frac{\delta F[\phi_2(\vec{r})]}{\delta\phi_2(\vec{r})} \qquad (10)$$

where L is kinetic coefficient and $F=F_d[\phi_2]+F_{eff}[\phi_2]$. We adopt a rescaling procedure of the free energy for the simulations, similar to the one discussed in [7] and accordingly use the polynomial

$$f(\phi_2)=\frac{\tau}{2}\phi_2(\vec{r})^2-\phi_2(\vec{r})^4+\frac{1}{2}\phi_2(\vec{r})^6-\sigma(\vec{r})\phi_2(\vec{r}) \qquad (11)$$

where τ is a rescaled temperature. All other variables in the effective free energy, such as A_1, A_3 and A_{12}, are also assumed to be rescaled. The term $\sigma(\vec{r})$ represents an inhomogeneous stress field that represents localized crystal defects [7]. We consider a configuration of randomly distributed defects with stress field given as

$$\sigma(\vec{r})=\sum_{j=1}^{N_d}\sigma_0(\vec{r}_j)\exp(-|\,\vec{r}-\vec{r}_j\,|^2). \qquad (12)$$

3. Simulations

We now describe our simulations of domain pattern formation using the above described model. The constants chosen for the local free energy are: $\tau = -0.1, \gamma_1 = 0.5, \gamma_2 = 0.01$. The amplitude of the local stress field $\sigma_0(\vec{r}_j)$ is chosen as being uniformly distributed in the interval [-0.6,0.6] and the total number of defects Nd is 288, which is 2% of the total grids. The parameters $A_1 = 1$ and $A_3 = 2$. For the coupling constant, we consider two cases below to illustrate the associated microstructure pattern formation:

Case A: This corresponds to $A_{12} = 1.2.$, where the transformed regions undergo a decrease in volume at the level of homogeneous elasticity. We discretize equation (10) on a 128×128 grid. Starting from small amplitude initial conditions, corresponding to an austenite phase, equation (10) is solved at each discrete grid point at each time step. Figure 1. shows the time evolution of the domain patterns obtained by simulating equation (10) for t = 40 (the time is assumed to be rescaled by the kinetic coefficient L) steps. In the snapshot corresponding to t=3, we can observe nucleation of product phase domains from the localized defects. Here the gray background corresponds to the Austenite phase while the dark and light regions represent the two variants of the martensitic phase. Notice that initially the domains grow very fast along [11] or $[\bar{1}1]$ directions with little sidewise growth. The onward growth is usually stopped by collision with other growing domains. In the very late stages, there is more sidewise motion and eventually the final pattern consists of crisscross martensite domains that partition the austenite matrix. Thus, there is retained austenite at the end of the transformation. These domain patterns are similar to the ones simulated in the phenomenological model in [4]. However, the martensite domains are not internally twinned due to the over damped model used by us. Inclusion of inertial effects in this model may induce internal twinning [8].

Case B: Here we consider the case $A_{12} = -1.2$, which corresponds to the situation where the volume of the unit cell increases at the level of homogeneous free energy. Figure 2 shows the domain pattern evolution for this case. This case appears to be different than the one shown in figure 1. Here the nuclei give rise to islands of the product phase that grow along the 45° directions. The growing domains may coalesce with adjacent domains. The final microstructure can be seen in the snapshot corresponding to t=40. It is clear that the nature of the simulated domain patterns is quite different for this case. However, even for this case, there is retained austenite at the end of the transformation. This difference in patterns is likely due the long-range kernel $C_3(\vec{k})$ that depends linearly on A_{12}. In figure 3, we display the time dependence of the volume fraction of the martensite phase for both the cases discussed above. We can clear see that after a small incubation period, there is a sudden increase in the volume fraction. This corresponds to the initial nucleation stage. Thereafter, the volume fraction appears to saturate. Notice that the saturation value is lower for the case with negative A_{12}.

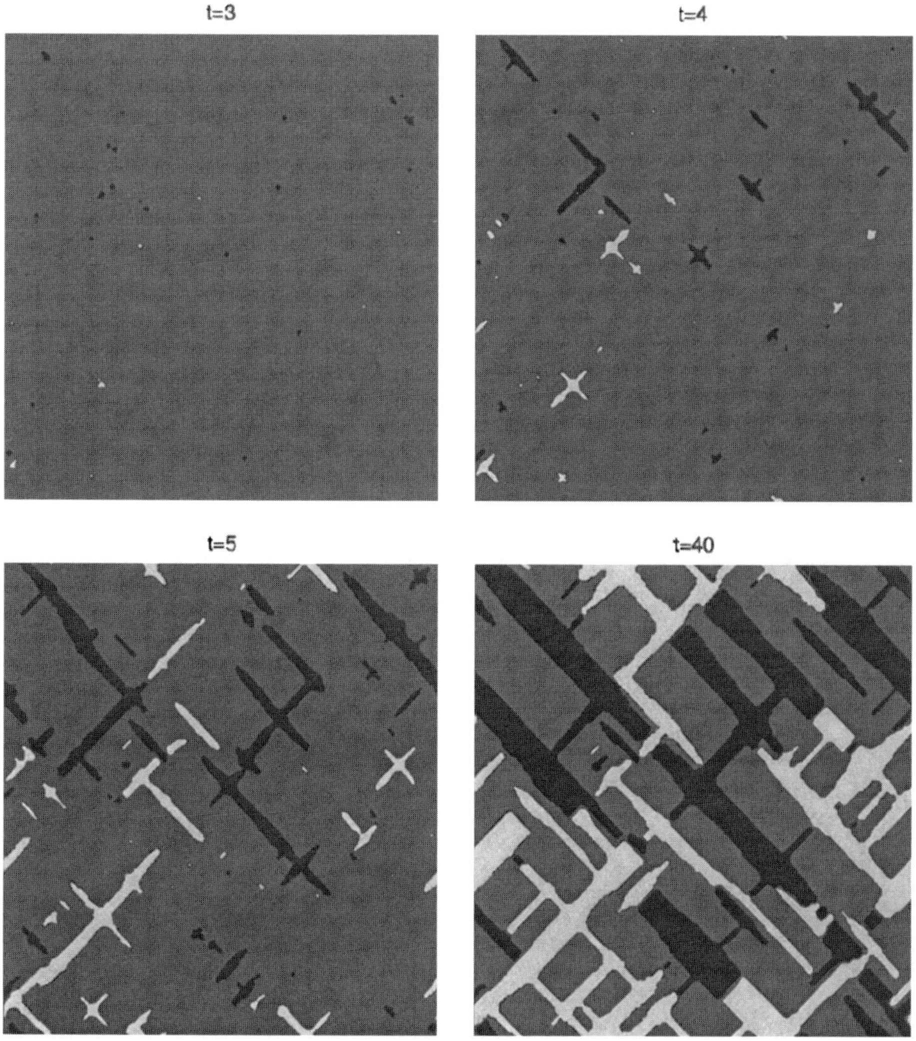

Figure 1. Domain pattern evolution for $A_{12} = 1.2.$ case. The light and dark colors represent the two variants of the transformed martensite phase while the gray background represents the austenite phase.

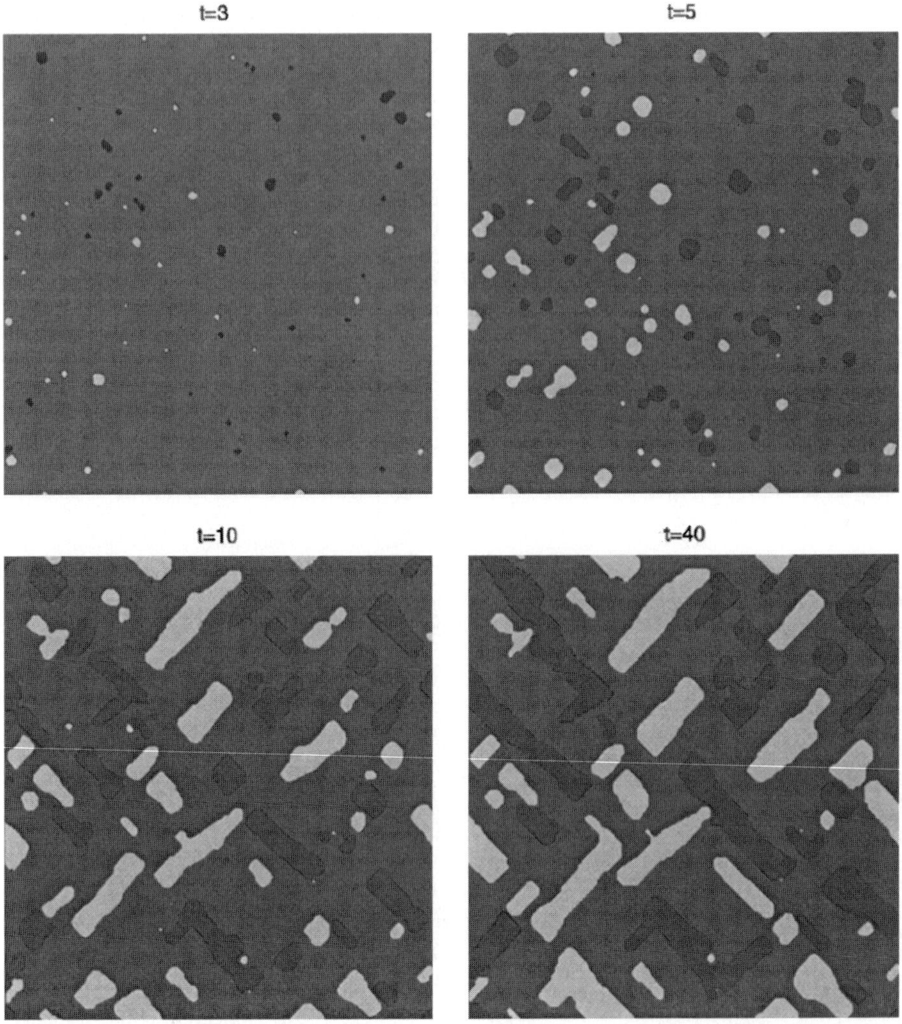

Figure 2. Pattern evolution for the case $A_{12} = -1.2.$. Color scheme is the same as that of figure 1.

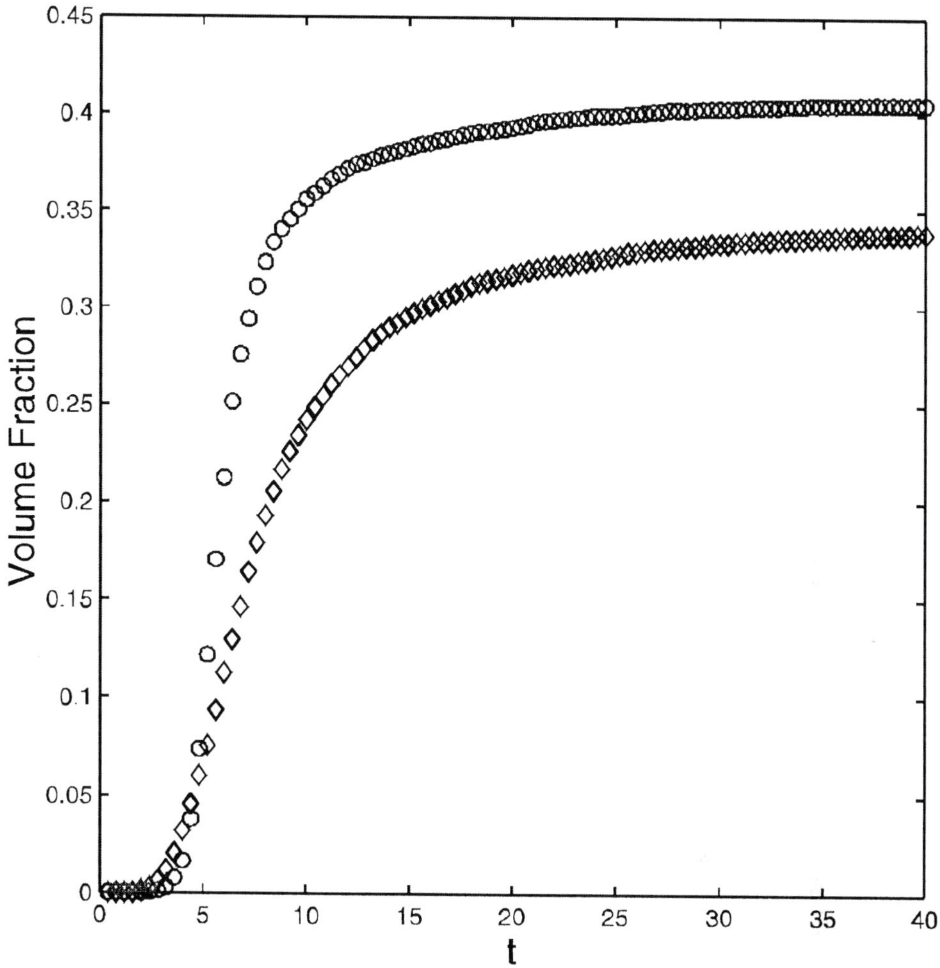

Figure 3. Time evolution of the volume fraction of martensite phase. Circles represent $A_{12} = 1.2$. and diamonds represent $A_{12} = -1.2$.

4. Discussion

We have made a computer simulation study of domain pattern formation in martensites based on a time-dependent Ginzburg-Landau model. It is found that inclusion of volume changes in the theory influences the domain patterns considerably. By including the coupling between bulk and deviatoric strains, we have shown that there is always some untransformed austenite left at the end of the transformation. This should be contrasted with case studied in [3] corresponding to $A_{12} = 0$ where twin patterns were observed without any residual Austenite. Of course, even for $A_{12} \neq 0$, at low enough temperature we expect that there would be no more austenite since the driving force would be high enough.

5. References

1. Olson G. B. and Owen W. S., Editors, (1992) *Martensite,* ASM International.
2. Wang, Y. and Khachaturyan, A.G. (1997) Three-dimensional field model and computer modeling of martensitic transformations, *Acta Materialia* **45,**759-773.
3. Shenoy, S.R., Lookman ,T., Saxena A. and Bishop, A.R. (1999) Martensitic textures: Multiscale consequences of Elastic Compatibility, *Phys.Rev. B* **60**, R12537-R12541.
4. Ahluwalia ,R. and Ananthakrishna G. (2001) Power-Law Statistics of Avalanches in a Martensitic Transformation, *Phys. Rev. Lett.* **86,** 4076-4079.
5. Barsch, G.R. and Krumhansl, J. A. (1984) Twin Boundaries in Ferroelastic Media without Interface Dislocations, *Phys. Rev. Lett* **53**, 1069-1072
6. Kartha ,S., Krumansl, J.A., Sethna, J.P., Wickham,L. (1995) Disorder driven Pretransitional Tweed Patterns in Martensitic Transformations, *Phys. Rev. B* **52,**803-822.
7. Cao, W.W. and Krumhansl, J. A. (1990) Defect-Induced Heterogeneous Transformations and Thermal Growth in Athermal Martensite, *Phys. Rev. B* **41,** 11319-11327.
8. Bales, G.S. and Gooding, R.J. (1991) Interfacial Dynamics at a First-Order Phase Transition involving Strain-Dynamic Twin formation, *Phys. Rev. Lett.* **67**,3412-3415.

Phase Transformation of Thin Wires in Tension

P. Tong, D. C. C. Lam and Q. P. Sun
Department of Mechanical Engineering
Hong Kong University of Science and Technology, Hong Kong, China

Abstract

A phase transformation constitutive relation including the effect of strain gradient is introduced to examine the extension of a thin wire. The resulting equilibrium equation in terms of strain is a nonlinear (even for small strain deformation) third order differential equation while that of the conventional elasticity theory is only first order and linear for small strain. Additional boundary conditions over those of the conventional theory are needed to make the solution unique and can be derived from a variational functional. The solutions show necking, bulging and periodic striation. We have established the closed form solution for phase transformation with trilinear stress-strain relation and examined the phase transformation behavior in details.

1. Introduction

If a wire of superelastic NiTi shape memory alloy [1] is stretched quasi-statically at a steady rate, the deformation starts as a uniform extension in the austenite state and gradually evolves to a non-uniform deformation, in which the wire thins down in one or more sections along the wire length and the net axial force actually decreases. This is a necking phenomenon during phase transformation in the wire where part of austenite is transforming into a mixture of austenite and martensite. As stretching continues, the neck(s) deepens and lengthens until the necks coalesce or a neck reaches an end of the wire as the transformation to martensite is completed over part of the wire. This process continues until all necks sweep over the entire wire, which becomes all martensite. These features were studied and identified in [2] and [3] and observed recently [4].

Phase transformation introduces nonlinearity (even at small strain) and strain gradient along the wire. The strain gradient, in turn, introduces higher order derivatives making the equilibrium equation a third order differential equation of strain. For conventional elasticity theory, the equilibrium equation is first order and linear if the strain is small. Additional boundary conditions, derived from the minimum of potential energy, beyond those of the conventional theory are needed to make the solution unique. We obtained a closed-form solution for a trilinear stress-strain relation [2]. The analytic solution helps us identify the characteristics of zones and instability.

2. One Dimensional Phase Transformation of a Thin Wire under Tension

2.1 Basic Equations and Assumptions

Consider a polycrystalline fine-grained Ti-Ni shape memory alloy wire of length $2l_0$ under tension [5]. Let l and z refer to the current and undeformed axial coordinates and ε be the axial strain. For small strain, no distinction among different strain measures is necessary, we have

(1) $\varepsilon = dl/dz - 1 = du/dz$,

where u is the axial displacement. We assume the stress strain relation in the form

Q.P. Sun (ed.), IUTAM Symposium on Mechanics of Martensitic Phase Transformation in Solids, 221–232.

(2) $$T = \tau(\varepsilon) + \frac{1}{2}\frac{d\gamma}{d\varepsilon}\left(\frac{d\varepsilon}{dz}\right)^2 - \frac{d}{dz}\left[\gamma(\varepsilon)\frac{d\varepsilon}{dz}\right] = \tau(\varepsilon) - \frac{1}{2}\frac{d\gamma}{d\varepsilon}\left(\frac{d\varepsilon}{dz}\right)^2 - \gamma(\varepsilon)\frac{d^2\varepsilon}{dz^2},$$

where $\tau(\varepsilon)$ is a nonlinear function of ε, which depends on the phase transformation characteristics of the material. Figure 1 illustrates a trilinear $\tau(\varepsilon)$. The second and third terms in the right most side of Eq. (2) specify the effect of strain gradient where $\gamma(\varepsilon)\,(>0)$ has the dimension of force and is proportional to the square of a characteristic length related to the size of transition zone. The problem is nonlinear if $\tau(\varepsilon)$ is so and/or if $\gamma(\varepsilon)$ is a function of ε. Equation (2) is a second order differential equation in terms of the axial strain. Thus the equilibrium equation, being the gradient of T, is third order. In terms of displacement, the equilibrium equation is fourth order as opposed to a second order differential equation for the conventional elasticity theory. As a result, additional boundary conditions (a total of 3 in terms of strain, or 4 in terms of displacement) are needed to determine the solution uniquely.

To derive the additional boundary conditions, we define the potential energy as

(3) $$\Pi = A\{\int_{-l_0}^{l_0}[U(\varepsilon) + \frac{1}{2}\gamma(\varepsilon)\left(\frac{d\varepsilon}{dz}\right)^2]dz - \alpha\bar{T}[u(l_0) - u(-l_0)]\},$$

where A is area, $2l_0$ is the length of the wire, $\bar{T}$ is the applied stress at the ends of the wire, α is a constant that

(4) $\alpha = 0$ for displacement-controlled loading,
$= 1$ for force-controlled loading,

and $U(\varepsilon)$ is the strain energy density such that

(5) $\tau = dU/d\varepsilon$.

We assume that the minimum of the potential energy corresponds to the solution of the phase transformation problem. Without loss of generality, we assume $A = 1$.

The first variation of Π is

$$\delta\Pi = \int_{-l_0}^{l_0}[\tau\,\delta\varepsilon + \frac{1}{2}\frac{d\gamma}{d\varepsilon}\left(\frac{d\varepsilon}{dz}\right)^2\delta\varepsilon + \gamma(\varepsilon)\frac{d\varepsilon}{dz}\frac{d\delta\varepsilon}{dz}]dz - \alpha\bar{T}[\delta u(l_0) - \delta u(-l_0)].$$

The stationary condition with respect of u requires that

(6) $$\frac{dT}{dz} = 0 \qquad \text{for} \qquad -l_0 \le z \le l_0,$$

(7) $$\frac{d\varepsilon}{dz} = 0 \qquad \text{at} \qquad z = \pm l_0,$$

and

(8) $u(\pm l_0) = \bar{u}^{\pm}$ for displacement-controlled loading,
$T = \bar{T}$ for force-controlled loading,

where $\bar{u}^{\pm}$ are the prescribed values of u at $z = \pm l_0$ and T is defined in Eq. (2).

From Eq. (6), T is constant. Then, we can rewrite Eq. (2) as

(9) $$\left(\frac{d\varepsilon}{dz}\right)^2 = 2\frac{W(\varepsilon)}{\gamma(\varepsilon)},$$

or

(10) $$\frac{d\varepsilon}{dz} = \mp\left[2\frac{W(\varepsilon)}{\gamma(\varepsilon)}\right]^{1/2},$$

where the '$\mp$' sign are for decreasing and increasing strains, respectively, and

(11) $$W(\varepsilon,\varepsilon_1) = \int_{\varepsilon_1}^{\varepsilon}[\tau(\varepsilon) - T]d\varepsilon$$

with ε_1 being a constant, which satisfies the equation (Fig. 1),

(12) $$\tau(\varepsilon_1) = T.$$

Equation (10) is a first order nonlinear ordinary differential equation, which can be solved by numerical integration. Consider the case of negative right hand side in Eq. (10). One first finds the root $\varepsilon_2 (> \varepsilon_1)$ of

$$W(\varepsilon_2,\varepsilon_1) = 0.$$

Then, one starts the integration with the initial condition $\varepsilon = \varepsilon_2$ until ε reaches ε_1. Thus, the solution satisfies $d\varepsilon/dz = 0$ at $\varepsilon = \varepsilon_1$. For negative right hand side in Eq. (10), the strain decreases monotonically from ε_2 to ε_1 as z increases. The wire elongation is

(13) $$\Delta l = \int_{-l_0}^{l_0}\varepsilon(z)dz = 2(l - l_0)$$

where $\varepsilon(z)$ is the solution of Eq. (10).

2.2 Closed form solutions for tri-linear transformation $\tau = \tau(\varepsilon)$

We consider a tri-linear stress/strain relation $\tau(\varepsilon)$ (Fig. 1) for homogeneous transformation as shown below:

(14) $$\begin{aligned} &\tau(\varepsilon) = E\varepsilon, && \text{for } \varepsilon \le \varepsilon_{m1}, \\ &\tau(\varepsilon) = \tau_{m1} - E_1(\varepsilon - \varepsilon_{m1}), && \text{for } \varepsilon_{m1} < \varepsilon \le \varepsilon_{m2}, \\ &\tau(\varepsilon) = \tau_{m2} + E_3(\varepsilon - \varepsilon_{m2}), && \text{for } \varepsilon_{m2} < \varepsilon, \end{aligned}$$

where

(15) $$\tau_{m1} = E\varepsilon_{m1}, \qquad \tau_{m2} = \tau_{m1} - E_1(\varepsilon_{m2} - \varepsilon_{m1}),$$

are the local maximum and minimum of $\tau(\varepsilon)$ with the corresponding strains ε_{m1}, ε_{m2}. Phase transformation starts at ε_{m1}. The function $W(\varepsilon)$ and its roots for trilinear $\tau(\varepsilon)$ are given in Appendix A. For constant γ, we can solve the problem in closed form.

Consider the symmetric solution only. The right half domain is defined as $-l_3 \le z \le l_1 + l_2$ with $l_0 = l_1 + l_2 + l_3$. Due to symmetry, $\varepsilon(-z - l_3) = \varepsilon(z + l_3)$ and

$$d\varepsilon/dz = 0$$

at the center of the wire ($z = -l_3$) and at the end of the wire ($z = l_1 + l_2$). One can divide the domain in three zones with Eq. (2) rewritten as

(16) $$\gamma\frac{d^2\varepsilon}{dz^2} + T = \begin{cases} -E_1(\varepsilon - \varepsilon_{m1}) + \tau_{m1} & \text{for } \varepsilon_{m1} < \varepsilon < \varepsilon_{m2} \text{ in } 0 < z < l_1 \quad \text{(zone 1)} \\ E\varepsilon & \text{for } \varepsilon \le \varepsilon_{m1} \text{ in } l_1 \le z \le l_1 + l_2 \quad \text{(zone 2)} \\ E_3(\varepsilon - \varepsilon_{m2}) + \tau_{m2} & \text{for } \varepsilon_{m2} \le \varepsilon \quad \text{in } -l_3 \le z \le 0 \quad \text{(zone 3)} \end{cases}$$

in which T is a constant. The strain and strain gradient of two adjacent zones are continuous. Depending on the wire length and the total elongation, the solution can have 1, 2 (which corresponds to some of l's being zero) or 3 zones. Zone 1, also called the transition zone, is adjacent to zone 2 or zone 3, or is between zones 2 and 3. The material is in martensite and austenite phases in zones 2 and 3, respectively, and is a mixture of martensite and austenite in zone 1. Appendix B gives the closed form solution of Eq. (16). The results are summarized below:

- The 1-zone solution has constant strain, which is obtained from Eq. (16) by setting $d^2\varepsilon/dz^2$ and two of l's to zero. The solution is zone-1 only if $\varepsilon_{m1} < \Delta l/l_0 \le \varepsilon_{m2}$, where $\Delta l/l_0$ is the elongation ratio. The zone-1 solution is stable only if the wire is short ($l_0 = l_1 \le \pi\sqrt{\gamma/E_1}$) under displacement-controlled loading. For load-controlled loading, the zone-1 solution is always unstable.
- For $\varepsilon_{m1} < \Delta l/l_0 \le \varepsilon_{m2}$ and $l_0 \ge \pi\sqrt{\gamma/E_1}$, the solution has multiple zones. In this case the wire is under phase transformation. The axial stress T varies from τ_{m1} to τ_{m2} and can be greater or less than the Maxwell stress. When $\Delta l/l_0$ first exceeds ε_{m1}, the solution has two zones, namely zones 1 and 2; T is *linearly proportional to* Δl and the sizes of zones 1 and 2 remain fixed [(B.9)] as T or Δl varies. As elongation increases from ε_{m1}, T reduces from τ_{m1} to T_b given by (B.11). It can be shown that

  (17) $$(\tau_{m1}+\tau_{m2})/2 \le T_b \le (\tau_{m1}\sqrt{E_1+E}+\tau_{m2}\sqrt{E})/(\sqrt{E}+\sqrt{E_1+E})$$

  for different wire lengths. The lower limit above is for short wires with $l_0 = \pi\sqrt{\gamma/E_1}$, the lower limit of l_0 for the existence of 2-zone solution. The upper limit is larger than the Maxwell stress and is for long wires corresponding to $\tan L_1 = -\sqrt{E_1/E}$.
- After T reaches the lower limit given by (B.11), continuing stretch will produce a 3-zone solution. In this case, the axial stress T decreases as Δl increases but they are *no longer linearly proportional to* each other. The size of zone 3 increases while that of zone 2 decreases as stretching increases. The size of zone 1 also varies slightly.
- Eventually zone 2 disappears and only zones 1 and 3 remain. Further stretching decreases T from T_{b2} given by (B.22) to τ_{m2} when $\Delta l/l_0 = \varepsilon_{m2}$. It can be shown that T_{b2} is bounded by

  (18) $$(\tau_{m1}\sqrt{E_3}+\tau_{m2}\sqrt{E_1+E_3})/(\sqrt{E_3}+\sqrt{E_1+E_3}) \le T_{b2} \le (\tau_{m1}+\tau_{m2})/2,$$

  The upper bound above is for short wires with $l_0 = \pi\sqrt{\gamma/E_1}$, the lower limit of l_0 for the existence of 2-zone solution. The lower bound is less than the Maxwell stress and is for long wires corresponding to $\tan L_1 = -\sqrt{E_1/E_3}$. For the zone 1-3 solution, like the zone 1-2 solution, T is *linearly proportional to* Δl and the sizes of zones 1 and 3 are constant [(B.19)]. When $\Delta l/l_0 \ge \varepsilon_{m2}$, all austenite has been transformed to martensite, the solution will again have only one zone, namely zone 3.

- The solution can be simplified for long wires by approximating $\tanh L_2, \tanh L_3$ as 1, if the sizes of zones 2 and 3 are large ($L_2, L_3 >> 1$). Then, the size of the transition zone (zone 1) or thickness of the austenite-martensite interface becomes

$$L_1 = \pi - \tan^{-1}\sqrt{E_1 / E} \qquad \text{[2-zone solution with zones 1 and 2, (B.13)]},$$

$$L_1 = \pi - \tan^{-1}\sqrt{E_1 / E_3} \qquad \text{[2-zone solution with zones 1 and 2, (B.24)]},$$

$$L_1 = \pi - \tan^{-1}[\sqrt{E_1}(\sqrt{E_3} + \sqrt{E})/(\sqrt{E}\sqrt{E_3} - E_1)] \quad \text{[3-zone solution, (B.34)]},$$

which are all constant. For the 3-zone solution, the axial stress equals the Maxwell stress [(B.33)]

2.3 Example.

We consider a wire with $E = E_3 = 52700$ MPa, $E_1 = 6060$ MPa, $\varepsilon_{m1} = 0.011$, $\varepsilon_{m2} = 0.046$, where E, E_3, E_1 are Young's moduli of austenite, martensite and the mixture of martensite and austenite. Thus $\tau_{m1} = 580$ MPa, $\tau_{m2} = 378$ MPa, $\tau_{mx} = 474$ MPa. Figure 2 plots T *vs.* $\Delta l / l_0$ for different wire lengths ($l_0\sqrt{E/\gamma} = 9$, 12, 30, ∞). The bounds of Eq. (17) for the zone 1-2 solution are 476.54 MPa for long wires and 473.65 MPa for $l_0\sqrt{E/\gamma} = 9.2644$. The bounds of Eq. (18) for the zone 1-3 solution are 470.76 MPa for long wires to 473.65 MPa for $l_0\sqrt{E/\gamma} = 9.2644$.

In addition to the 2- and 3-zone solutions, other solutions exist with appropriate constant strain zones or in the form of repeated pattern of 2- or 3-zone solution for long wires (Appendix B). These solutions do not have global minimum strain energy for a given total elongation. Strictly, they are not a genuine solution. However, the strain energy is often very close to the global minimum value. Due to the variation in material properties and/or wire radius, those deformations may actually occur.

As an example, for $l_0\sqrt{E/\gamma} = 15$, $l_2'\sqrt{E/\gamma} = 6$ and $(\Delta l_1 + \Delta l_2)/l_0 = 0.018$, the total non-dimensional strain energy for half of the wire is 0.00262704592 based on the zone 1-2 solution plus an appropriate constant strain zone over $l_1 \le z \le l_1 + l_2 + l_2' = l_0 + l_2' = l_0'$ with $\Delta l' / l_0' = 0.0156019368$. In the mean time, the corresponding total non-dimensional strain energy for the zone 1-2 solution of the wire of half-length $l_0'\sqrt{E/\gamma} = 21$ is 0.00262704589, which is very slightly less than that of the solution with an additional constant strain zone. Many digits are carried in the above example to illustrate the small difference in the strain energies of different solutions.

The solution with zones 1, 2, 2, 1 is a solution for the wire of semi-length $l_0' = 2(l_1 + l_2)$. This solution has strain energy higher than that of the 2-zone solution of the semi-wire length $2(l_1 + l_2)$ for the same total elongation. For the same problem considered before, the total strain energy of the zone-1-2-2-1 solution for the half wire length $l_0'\sqrt{E/\gamma} = 30$ is 0.00398. The corresponding strain energy of the zone 1-2 solution is 0.00376, which is slightly less. In practice, since the differences in the strain energy among the different solutions are small, the solution with added constant strain zones and/or repeated zones may exist for a long wire due to variations in material properties and/or wire radius. The solution with repeated zones 1-2-2-1 is also a solution. This solution is in a form of periodic striation.

The deformation process can be summarized as follows: In stretching a Ti-Ni shape memory alloy wire at a steady rate under displacement control, the deformation is always uniform if the wire is sufficiently short. We call it a 1-zone solution. For a load-controlled case, the solution is unstable when the stretching reaches certain critical value, regardless wire length. For long wires, the deformation starts as uniform extension and gradually evolves to non-uniform deformation. The solution is stable under displacement-controlled loading. Initially the solution has 2 zones: zones 1 and 2. The wire thins down in zone 1, called necking, and the net force actually decreases as stretching increases. In zone 1, phase transformation takes place that part of austenite transforms into martensite. The size of zone 1 remains a constant as it continues to thin down. Zone 2 remains in austenite phase. Further stretching produces a third zone, namely zone 3, in which the transformation from austenite to martensite has completed.

Continuing stretching enlarges zone 3 and reduces zone 2. During this time, zone 1 varies according to (B.31) for trilinear phase transformation. The axial stress continues to decrease as stretching increases, but the change is small. When zone 2 reduces to zero, again the solution has only 2 zones, i.e., zones 1 and 3. Continuing stretching eventually turns all austenite in zone 1 into martensite and the solution has only 1 zone as zone 1 disappears. The axial stress T starts to increase again for further stretching.

In unloading from a zone-3 state, the process is reverse. First zone 1 appears as a bulge in the wire. Then zone 2 appears and then followed by disappearing of zone 3 and subsequently zone 1 as bulging sweeps over the entire wire.

Concluding Remarks

Phase transformation introduces nonlinearity and higher order derivatives in the equilibrium equations. Additional boundary conditions beyond those of the conventional theory are needed to make the solution unique and can be derived by minimizing the potential energy. For loading, the deformation starts as a simple uniform extension and gradually evolves to a non-uniform deformation with the appearance of necking. The neck sweep across the entire domain as the material transforms from austenite to martensite. The process reverses in unloading with bulging evolves and sweeps over the entire wire. The process is nonlinear.

For phase transformation with trilinear stress strain relation, we obtained the closed form solution, which provides details of phase transformation in stretching and unloading. Depending on the amount of stretching and the wire length, the solution can have 1, 2, or 3 zones. Under phase transformation, the axial stress can be greater or less than the Maxwell stress. For short wires, only 1-zone solution exists. The deformation is always uniform and the solution is stable for displacement-controlled loading.

For $\varepsilon_{m1} \leq \Delta l / l_0 \leq \varepsilon_{m2}$, part of the wire is under phase transformation. The solution is no longer uniform and can have 2 or 3 zones if $l_0 > \pi\sqrt{\gamma / E_1}$. For 2-zone solution, the size of the transition zone (zone 1) is constant [(B.9), (B.19)] and T is linearly proportional to the elongation ratio, $\Delta l / l_0$. For long wires, T for solution with zones 1 and 2 and that with zones 1 and 3 are above and below the Maxwell stress, respectively.

For three-zone solutions, the zone sizes are no longer constant and T is not linearly proportional to $\Delta l / l_0$. However, T has values close to $(\tau_{m1} + \tau_{m2})/2$ as E_1 is usually much less than E and E_3. For longer wires with large zones 2 and 3, the transition zone size approaches a constant [Eq. (B.34)] asymptotically and the axial stress T becomes a constant equal to the Maxwell stress [Eq. (B.33)].

In addition to the two- and three-zone solutions, solutions with extra constant strain zones or in the form of repeated pattern of 2- or 3-zone solution exist for long wires. These solutions do not give the global minimum strain energy for a fixed total elongation. Strictly, they are not a genuine solution. However, the strain energy is often very close to the global minimum value. Due to the variation in material properties and/or wire radius, those deformations may actually occur in practice.

Appendix A. $W(\varepsilon)$ for constant γ and trilinear $\tau(\varepsilon)$

The Maxwell stress is the stress level dividing the phase transformation curve to make areas ABC and CDE equal as shown in Fig. 1. The Maxwell stress, for trilinear $\tau(\varepsilon)$, is

$$\text{(A.1)} \qquad \tau_{mx} = \frac{\tau_{m1} + \tau_{m2}}{2} + \frac{\sqrt{E_3}\sqrt{E + E_1} - \sqrt{E}\sqrt{E_3 + E_1}}{\sqrt{E_3}\sqrt{E + E_1} + \sqrt{E}\sqrt{E_3 + E_1}} \frac{\tau_{m1} - \tau_{m2}}{2}$$

which reduces to $(\tau_{m1} + \tau_{m2})/2$ when $E = E_3$. We can easily determine W from Eq. (11). For $E_3 = E$, we have

$$W(\varepsilon) = E(\varepsilon - \varepsilon_1)^2 / 2, \qquad \text{for } \varepsilon \le \varepsilon_{m1},$$

$$\text{(A.2)} \qquad W(\varepsilon) = -\frac{E_1 \varepsilon^2}{2} + [(\varepsilon_{m1} - \varepsilon_1)E + E_1 \varepsilon_{m1}]\varepsilon + \frac{E\varepsilon_1^2 - (E + E_1)\varepsilon_{m1}^2}{2}, \quad \text{for } \varepsilon_{m1} < \varepsilon \le \varepsilon_{m2}$$

$$W(\varepsilon) = \frac{E\varepsilon^2}{2} - [(\varepsilon_{m2} - \varepsilon_{m1})(E + E_1) + E\varepsilon_1]\varepsilon + \frac{E + E_1}{2}(\varepsilon_{m2}^2 - \varepsilon_{m1}^2) + \frac{E\varepsilon_1^2}{2}, \quad \text{for } \varepsilon_{m2} < \varepsilon$$

in which the relation given in Eq. (15) and $T = E\varepsilon_1$ have been used. The function $W(\varepsilon)$ is zero at ε_1. The other root of $W(\varepsilon) = 0$ is

$$\text{(A.3)} \qquad \varepsilon_2 = \varepsilon_1 + (\varepsilon_{m1} - \varepsilon_1)\left[1 + \frac{E}{E_1} + \sqrt{\frac{E}{E_1}(1 + \frac{E}{E_1})}\right], \qquad \text{for} \qquad \varepsilon_{m1} < \varepsilon_2 \le \varepsilon_{m2},$$

$$\text{(A.4)} \qquad \varepsilon_2 = \varepsilon_1 + (\varepsilon_{m1} - \varepsilon_1)\left[1 + \frac{E_1}{E} - \sqrt{(1 + \frac{E_1}{E})(\frac{E_1}{E} - 2\frac{\varepsilon_{m1} - \varepsilon_1}{\varepsilon_{m2} - \varepsilon_{m1}})}\right], \quad \text{for} \qquad \varepsilon_{m2} < \varepsilon_2 .$$

Note that $W(\varepsilon)$ is positive for $\varepsilon \le \varepsilon_{m1}$.

Appendix B. Closed form solution for constant γ and trilinear $\tau(\varepsilon)$

Depending on the wire length and elongation, we have 1-, 2- or 3-zone solution. Let

$$L_1 = l_1\sqrt{E_1/\gamma}, \qquad L_2 = l_2\sqrt{E/\gamma}, \qquad L_3 = l_3\sqrt{E_3/\gamma}$$

be the non-dimensional sizes of zones 1, 2 and 3. We examine the solution below:

One-zone solution. The 1-zone solution is simply

(B.1) $\varepsilon_0 = \Delta l / l_0 = T / E$, for $\varepsilon_0 \le \varepsilon_{m1}$, $T \le \tau_{m1}$

(B.2) $\varepsilon_0 = \Delta l / l_0 = \varepsilon_{m2} + (T - \tau_{m2}) / E_3$, for $\varepsilon_{m2} \le \varepsilon_0$, $\tau_{m2} \le T$

(B.3) $\varepsilon_0 = \dfrac{\Delta l}{l_0} = \varepsilon_{m1} + \dfrac{\tau_{m1} - T}{E_1}$, for $\varepsilon_{m1} < \varepsilon_0 < \varepsilon_{m2}$, $\tau_{m2} \le T \le \tau_{m1}$, $l_0\sqrt{E_1/\gamma} \le \pi$.

Solution with zones 1 and 2. For $l_0\sqrt{E_1/\gamma}$ larger than π and $\varepsilon_{m1} \le \Delta l / l_0 \le \varepsilon_{m2}$, the solution have multiple zones. Consider the case that the solution has two regions, i.e., a zone of mixed martensite and austenite with size l_1 (zone 1) and an austenite region of size l_2 (zone 2) with $l_3 = 0$. We have $0 \le z \le l_1 + l_2$ and

(B.4) $l_1 + l_2 = l_0$.

The solution can be written in the following form: In zone 1, $0 \le z \le l_1$,

(B.5) $\varepsilon = \varepsilon_{m1} + [1 - \cos(z\sqrt{E_1/\gamma}) / \cos L_1](\tau_{m1} - T) / E_1$,

(B.6) $\dfrac{\Delta l_1}{l_1} = \varepsilon_{m1} + \dfrac{\tau_{m1} - T}{E_1}(1 - \dfrac{\tan L_1}{L_1})$, $E_{n1} = \dfrac{1}{2}\varepsilon_{m1}^2 L_1\sqrt{\dfrac{E}{E_1}} + \dfrac{1}{2}\dfrac{\tau_{m1}^2 - T^2}{E_1^2}\sqrt{\dfrac{E_1}{E}}(L_1 - \tan L_1)$

where $\varepsilon_{m1} \le \Delta l_1 / l_1 \le \varepsilon_{m2}$ and E_{n1} is the non-dimensional strain energy of zone 1. Equation (B.5) gives $\varepsilon = \varepsilon_{m1}$ at $z = l_1$ and dε/dz = 0 at z = 0. In zone 2, $l_1 \le z \le l_1 + l_2$,

(B.7) $\varepsilon = T / E + [(\tau_{m1} - T) / E]\cosh[\sqrt{E/\gamma}(z - l_1 - l_2)] / \cosh L_2$,

(B.8) $\dfrac{\Delta l_2}{l_2} = \varepsilon_{m1} + \dfrac{\tau_{m1} - T}{E}(\dfrac{\tanh L_2}{L_2} - 1)$, $E_{n2} = \dfrac{1}{2}\varepsilon_{m1}^2 L_2 + \dfrac{1}{2}\dfrac{\tau_{m1}^2 - T^2}{E^2}(\tanh L_2 - L_2)$,

where $\Delta l_2 / l_2 \le \varepsilon_{m1}$ is the elongation ratio and E_{n2} is the strain energy of zone 2. Equation (B.7) gives $\varepsilon = \varepsilon_{m1}$ at $z = l_1$ and dε/dz = 0 at $z = l_1 + l_2$. To insure continuity of the strain gradient at $z = l_1$, we require

(B.9) $\tanh L_2 = -\sqrt{E/E_1}\tan L_1$ or $L_1 = \pi - \tan^{-1}(\sqrt{E_1/E}\tanh L_2)$,

with $\pi - \tan^{-1}\sqrt{E_1/E} < L_1 \le \pi$. The total elongation ratio and strain energy are

(B.10) $\Delta l / l_0 = \Delta l_1 / l_0 + \Delta l_2 / l_0$,

$$E_n = E_{n1} + E_{n2} = \frac{1}{2}\varepsilon_{m1}^2 L + \frac{1}{2}[\frac{E}{E_1}\sqrt{\frac{E}{E_1}}(L_1 - \tan L_1) + \tanh L_2 - L_2]\frac{\tau_{m1}^2 - T^2}{E^2}.$$

To determine the solution for given E, E_1 and wire length l_0, we first solve for L_1, L_2 or l_1, l_2 from (B.4) and (B.9). We can then determine Δl in terms of T or vice versa from (B.10). The ranges of Δl or T are limited by the condition $\varepsilon_{m1} \le \varepsilon \le \varepsilon_{m2}$ in zone 1 and $\varepsilon \le \varepsilon_{m1}$ in zone 2. Note that Δl is linearly proportional to T.

(B.9), independent of Δl or T, implies constant size for zones 1 and 2. No non-zero solution for l_2 exists if $l_0\sqrt{E/\gamma} < \pi\sqrt{E/E_1}$ [(B.4),(B.9)]. Evaluating (B.5) at z=0 gives

$$T = \tau_{m1} - E_1(\varepsilon_0 - \varepsilon_{m1})(\cos L_1)/(\cos L_1 - 1),$$

where ε_0 is the strain at z = 0. Since $\varepsilon_{m1} \le \varepsilon_0 \le \varepsilon_{m2}$ and $\cos L_1/(\cos L_1 - 1)$ is non-negative for $\pi - \tan^{-1}(\sqrt{E_1/E}) \le L_1 \le \pi$, we have

(B.11) $T_b = \tau_{m1} - E_1(\varepsilon_{m2} - \varepsilon_{m1})(\cos L_1)/(\cos L_1 - 1) \le T \le \tau_{m1}$.

The bounds of T_b are given by Eq.(17). One can use (B.10) to determine the range of Δl.

We can greatly simplify the solution for long wires by approximating the strain and the zone size [(B.7) and (B.9)] of zone 2 as

(B.12) $\varepsilon = T/E + \exp[-\sqrt{E/\gamma}(z - l_1)](\tau_{m1} - T)/E$,

(B.13) $\tan L_1 = -\sqrt{E_1/E}$ or $L_1 = \pi - \tan^{-1}\sqrt{E_1/E}$.

Appropriate constant strain zones can be added to the solution above to give solutions for longer wires. For example, for a wire of half length

$$l_0' = l_1 + l_2 + l_2'$$

over the domain $0 \le z \le l_1 + l_2 + l_2'$. Consider the solution, which is the same as the zone 1-2 solution over $0 \le z \le l_1 + l_2$, and

$$\varepsilon = \varepsilon_{22} = T/E + (\tau_{m1} - T)/(E\cosh L_2)$$

for $l_1 + l_2 \le z \le l_1 + l_2 + l_2'$. The total elongation ratio is

$$\Delta l'/l_0' = (\Delta l_1 + \Delta l_2 + \varepsilon_{22} l_2')/l_0'.$$

This solution has higher strain energy than that of the two-zone solution of same wire length and elongation and thus is not a true solution. Based on the calculated results, the difference from the strain energy of the 2-zone solution is very small for long wires.

Solution with zones 1 and 3. The zone 1-3 solution consists of zones 1 and 3 over $-l_3 \le z \le l_1$ with $l_2 = 0$, $0 \le z \le l_1$ for zone 1 and $-l_3 \le z \le 0$ for zone 3. We have

(B.14) $l_1 + l_3 = l_0$,

The solution is similar to that with zones 1 and 2. In zone 1,

(B.15) $\varepsilon = \varepsilon_{m2} + \{1 - \cos[\sqrt{E_1/\gamma}(z - l_1)]/\cos L_1\}(\tau_{m2} - T)/E_1$,

(B.16) $\Delta l_1/l_1 = \varepsilon_{m2} + [1 - (\tan L_1)/L_1](\tau_{m2} - T)/E_1$,

$$E_{n1} = \frac{L_1\sqrt{E}}{2\sqrt{E_1}}[\varepsilon_{m1}(2\varepsilon_{m2} - \varepsilon_{m1}) - \frac{E_1}{E}(\varepsilon_{m2} - \varepsilon_{m1})^2] + \frac{1}{2}(\tan L_1 - L_1)\frac{T^2 - \tau_{m2}^2}{E_1\sqrt{E_1}\sqrt{E}}$$

and in zone 3,

(B.17) $\varepsilon = \varepsilon_{m2} + \{1 - \cosh[\sqrt{E_3/\gamma}(z + l_3)]/\cosh L_3\}(T - \tau_{m2})/E_3$,

(B.18) $\Delta l_3/l_3 = \varepsilon_{m2} + [1 - (\tanh L_3)/L_3](T - \tau_{m2})/E_3$,

$$E_{n3} = \frac{L_3\sqrt{E}}{2\sqrt{E_3}}[\varepsilon_{m1}(2\varepsilon_{m2} - \varepsilon_{m1}) - \frac{E_1}{E}(\varepsilon_{m2} - \varepsilon_{m1})^2] + \frac{1}{2}(L_3 - \tanh L_3)\frac{T^2 - \tau_{m2}^2}{E_3\sqrt{E_3}\sqrt{E}}$$

Equation (B.15) gives $\varepsilon = \varepsilon_{m2}$ at z = 0 and dε/dz = 0 at $z = -l_3$, whereas (B.17) gives $\varepsilon = \varepsilon_{m2}$ at z = 0 and dε/dz = 0 at $z = l_1$. Continuity of $d\varepsilon/dz$ at z = 0 gives

(B.19) $\tanh L_3 = -\sqrt{E_3 / E_1} \tan L_1$ or $L_1 = \pi - \tan^{-1}[\sqrt{E_1 / E_3} \tanh L_3]$

with $\pi - \tan^{-1}\sqrt{E_1 / E_3} \le L_1 \le \pi$ at z = 0. The total elongation ratio and strain energy are

(B.20) $\Delta l / l_0 = (\Delta l_1 + \Delta l_3) / l_0$,

(B.21) $E_n = E_{n1} + E_{n3}$,

$$E_n = \frac{L}{2}[\varepsilon_{m1}(2\varepsilon_{m2} - \varepsilon_{m1}) - \frac{E_1}{E}(\varepsilon_{m2} - \varepsilon_{m1})^2] + \frac{1}{2}[\frac{E_3}{E_1}\sqrt{\frac{E_3}{E_1}}(\tan L_1 - L_1) + L_3 - \tanh L_3]\frac{T^2 - \tau_{m2}^2}{E_3\sqrt{E_3}\sqrt{E}}$$

To determine the solution for given E, E_1 and wire length l_0, we first solve for L_1, L_3 or l_1, l_3 from (B.14) and (B.19). Note that Δl is linearly proportional to T. The ranges of Δl or T are limited by the condition that $\varepsilon_{m1} \le \varepsilon \le \varepsilon_{m2}$ in zone 1 and $\varepsilon \le \varepsilon_{m1}$ in zone 2.

Equation (B.19) is independent of Δl or T, which implies the sizes of zones 1 and 3 are constant. There is no non-zero solution for l_3 from (B.14) and (B.19) if $l_0\sqrt{E/\gamma} < \pi\sqrt{E/E_1}$. Evaluating Eq. (B.15) at z = l_1, we find

$$T = \tau_{m2} + E_1(\varepsilon_{m2} - \varepsilon_{l_1})\cos L_1 / (\cos L_1 - 1),$$

where ε_{l_1} is the value of ε at $z = l_1$. Since $\varepsilon_{m1} \le \varepsilon_{l_1} \le \varepsilon_{m2}$, and $\cos L_1 / (\cos L_1 - 1)$ is non-negative for $\pi - \tan^{-1}(\sqrt{E_1 / E}) \le L_1 (= l_1\sqrt{E_1/\gamma}) \le \pi$, we have

(B.22) $\tau_{m2} \le T \le \tau_{m2} + E_1(\varepsilon_{m2} - \varepsilon_{m1})\cos L_1 / \cos L_1 - 1 = T_{b2}$.

The bounds for T_{b2} are given in Eq. (18). One finds the range of Δl from (B.20).

For long wires, we can approximate the strain and size of zone 3 [(B.17), (B.19)] as

(B.23) $\varepsilon = \varepsilon_{m2} + [1 - \exp(z\sqrt{E_3/\gamma})](T - \tau_{m2}) / E_3$

(B.24) $\tan L_1 = -\sqrt{E_1 / E_3}$ or $L_1 = \pi - \tan^{-1}\sqrt{E_1 / E_3}$.

Solution with zones 1, 2 and 3. In this case

(B.25) $l_1 + l_2 + l_3 = l_0$.

The strain and elongation ratio in zone 3 are the same as those given by (B.17) and (B.18) with $d\varepsilon / dz = 0$ at $z = -l_3$ and $\varepsilon = \varepsilon_{m2}$ at z = 0. The solution in zone 2 is the same as that given by (B.7) and (B.8) with $d\varepsilon / dz = 0$ at $z = l_1 + l_2$ and equals to $\varepsilon = \varepsilon_{m1}$ at $z = l_1$. The solution in zone 1 can be written in the form as

(B.26) $\varepsilon = \varepsilon_{m1} + \dfrac{\tau_{m1} - T}{E_1} + \dfrac{T - \tau_{m2}}{E_1}[\cos(\sqrt{\dfrac{E_1}{\gamma}}z) - \sqrt{\dfrac{E_1}{E_3}}\tan L_3 \sin(\sqrt{\dfrac{E_1}{\gamma}}z)]$,

(B.27) $\Delta l_1 / l_1 = \varepsilon_{m2} + [\sqrt{E_1 / E_3}\tanh L_3(\cos L_1 - 1)/L_1 + (\sin L_1)/L_1 - 1](T - \tau_{m2}) / E_1$,

$$E_{n1} = \frac{L_1\sqrt{E}\varepsilon_{m1}^2}{2\sqrt{E_1}} + (\varepsilon_{m2} - \varepsilon_{m1})\frac{2\tau_{m2} - (\tau_{m2} + \tau_{m1})\cos L_1}{2\sqrt{E_1 E}\sin L_1} + \frac{2\cos L_1 + L_1 \sin L_1 - 2}{2E_1\sqrt{E_1 E}\sin L_1}(\tau_{m1}^2 - T^2)$$

in which the strain equals to ε_{m2} and the strain gradient equals to that of zone 1 at z = 0. The total elongation and strain energy of the three zones are

(B.28) $\Delta l / l_0 = (\Delta l_1 + \Delta l_2 + \Delta l_3) / l_0$,

(B.29) $E_n = E_{n1} + E_{n2} + E_{n3}$,

where $\Delta l_2, \Delta l_3$ and E_{n2}, E_{n3} are the same as given before. Making the strain and the strain gradient match those of zone 2 at $z = l_1$, we obtain

(B.30) $(\tau_{m1} - T)\sqrt{E_3} = (T - \tau_{m2})(\sqrt{E_1}\tanh L_3 \sin L_1 - \sqrt{E_3}\cos L_1)$

(B.31) $(E_1 \tanh L_2 \tanh L_3 - \sqrt{E}\sqrt{E_3})\tan L_1 = \sqrt{E_1}(\sqrt{E_3}\tanh L_2 + \sqrt{E}\tanh L_3)$

To determine the solution for given E, E_1, E_3 and wire length l_0, we can determine $\Delta l, l_1, l_2, l_3$ in terms of T or T, l_1, l_2, l_3 in terms Δl from (B.25) and (B.28)-(B.31). Unlike the two-zone solutions, the sizes of different zones depend on Δl or T and Δl is no longer linearly proportional to T. If $L(= l_0\sqrt{E/\gamma}) < \pi\sqrt{E/E_1}$, no non-zero l_2 and l_3 exist. The solution is subjected to the constraints that $\varepsilon \le \varepsilon_{m1}$ in zone 2 and that $\varepsilon_{m2} \le \varepsilon$ in zone 3. From (B.11) and (B.22), the constraints require that

(B.32) $$\tau_{m2} + \frac{E_1 \cos L_{12}}{\cos L_{12} - 1}(\varepsilon_{m2} - \varepsilon_{m1}) \le T \le \tau_{m1} - \frac{E_1 \cos L_{11}}{\cos L_{11} - 1}(\varepsilon_{m2} - \varepsilon_{m1}),$$

in which L_{11}, L_{12} are the values of L_1 defined in Eqs. (B.9) and (B.19), respectively. The upper and lower limits are close to $(\tau_{m1} + \tau_{m2})/2$ if E_1 is much less than E and E_3.

The solution can be simplified for long wires that (B.30) and (B.31) reduce to

(B.33) $$T = \frac{\tau_{m1}\sqrt{E_3}\sqrt{E + E_1} + \tau_{m2}\sqrt{E}\sqrt{E_3 + E_1}}{\sqrt{E_3}\sqrt{E + E_1} + \sqrt{E}\sqrt{E_3 + E_1}},$$

(B.34) $\tan L_1 = -\sqrt{E_1}(\sqrt{E_3} + \sqrt{E})/(\sqrt{E}\sqrt{E_3} - E_1)$,

that T equals to the Maxwell stress and the size of the transition zone (zone 1) is constant. Similar to the case of the two-zone solutions, zones of appropriate constant strain or repeated patterns can be added to form solutions for long wires.

The solution corresponds to the minimum potential energy and is stable for fixed elongation. It can be shown that the load-controlled case is not stable for $\varepsilon_{m1} \le \varepsilon \le \varepsilon_{m2}$.

References

1. Shaw, J.A. and Kyriakides, S., Thermomechanical Aspects of NiTi, *J. Mech. Phys. Solids*, **43** (1995), pp. 1243-1281.
2. Sun, Q. P., and Hwang, K. C., Micromechanics modeling for the constitutive behaviors of polycrystalline shape memory alloys, *J. Mech. Phys. Solids*, **41** (1993) 1-33.
3. Coleman, B. D., Necking and drawing in polymeric fibers under tensions, *Arch. Ration. Mech. Analysis,* **83** (1983), pp. 115-137.
4. Sun, Q. P., Li, Z. Q. and Tse, K. K. (2000), On superelastic deformation of NiTi shape memory alloy micro-tubes and wires, in *Proceedings of IUTAM Symposium on Smart Structures and Structronic Systems (Gabbert and Tzou Eds.)*, Kluwer Academic Publisher, pp. 113-120.
5. Zhong, Z., Sun, Q. P. and Tong, P.: On the elastic axisymmetric deformation of a rod containing a single cylindrical inclusion, *Int. J. of Solids and Struct*., **37** (2000), pp. 5943-5955.

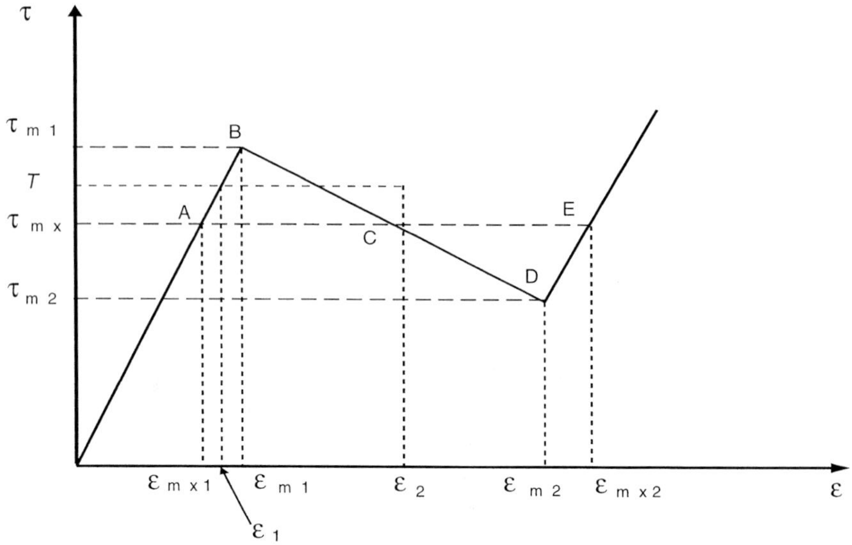

Fig. 1. Trilinear phase transformation stress strain relations

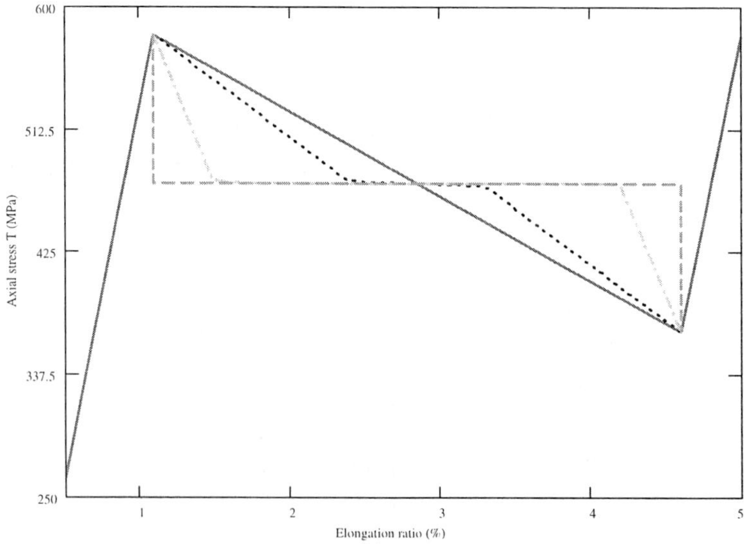

Fig. 2. Axial stress vs. elongation ratio for $l_0\sqrt{E/\gamma} = 9$ (solid), 12 (dot), 30 (dash-dot), $\propto$ (dash).

UNCONVENTIONAL DOMAIN BANDS NEAR A CRACK TIP IN FERROELECTRICS

W. Yang, F. Fang, H.T. Wang and Y.Q. Cui
Department of Engineering Mechanics, Tsinghua University, Beijing 100084, P.R. China

Abstract

Domain polarization switch near the tip of a crack plays a critical role in the fracture or toughening of ferroelectric ceramics. The intensive electric field near a crack tip stimulates local domain switching. Experiment indicates that the domain band structure in front of an indentation crack is unconventional, attributed to the highly localized crack tip electric field. The partially switched ferroelectric grain resembles a banded Eshelby inclusion embedded in a polycrystalline ferroelectric matrix. The domain wall energy for unconventional domain structures is estimated via arrays of misfit dislocations. The analysis prescribed in this paper predicts important aspects including the volume fraction, the thickness, and the orientation of switched domain bands ahead of the crack.

1. Introduction

Ferroelectric ceramics exhibit peculiar behaviors such as fracture near a defect or an electrode under electric load [1]. The intensive electric field near a crack tip stimulates local domain polarization switching that plays a critical role in electric fracture. The switched domains generate incompatible strain under the constraint of un-switched material and consequently alter the stress distribution near the crack. Experiment [2] indicates that unconventional domain band structure appears ahead of an indentation crack tip when subjected to lateral electric field. The band structure is caused by highly localized crack tip electric field. The elastic mismatch energy for a partially switched ferroelectric grain embedded in a polycrystalline ferroelectric matrix is evaluated as a banded Eshelby inclusion. The domain wall energy is derived for unconventional domain structures via arrays of misfit dislocations. The analysis quantifies various parameters of unconventional domain bands near a crack tip, including the volume

Q.P. Sun (ed.), IUTAM Symposium on Mechanics of Martensitic Phase Transformation in Solids, 233–240.

fraction, the thickness, and the orientation of switched domain bands.

2. Microscopy of Crack Tip Unconventional Domain Bands

Experimental observations for domain bands were mostly reported under uniform fields. It was the work in [2] that provided direct evidence for the domain switching near a crack tip subjected to an electric field concentration. The PLZT ceramic samples were mechanically cut and grounded into specimens with the size of 3.5mm×5mm×25mm. XRD revealed that they adopt a tetragonal perovskite structure with $a = b = 0.4055$ nm, and $c = 0.4109$nm. Consequently the aspect ratio c/a is 1.013, and the spontaneous domain switching strain is 1.3%. The samples were poled under an electric field of 2300V/mm at 120°C from gold electrodes sprayed on two opposing 5mm×25mm surfaces. During poling, ferroelectric domains switch and align with a polarization normal to the plane. After poling, we removed the Au electrodes, and indented the polished surface of 5mm×25mm under 5kg load. The two opposing surfaces of 3.5mm× 25mm were then sprayed with Au electrodes and lateral electric field of $0.6E_c$ (E_c= 1100V/mm) was applied.

Indentation followed by lateral field application altered the domain structures near the indenting cracks. Etching process was carried out for the polished top surface of $5\text{mm}\times 25\text{mm}$ with Vickers indents to reveal the domain bands. Fig. 1 provided a SEM image near an indentation crack tip [2]. The actual crack originates far away from the viewing frame, enters the view from the upper-left, and runs down at about 45° degrees. Lamellar 90° domain structure appeared under $0.6E_c$ lateral electric field, as evidenced by the black and write strips shown in Fig. 1. The observed 90° domain switching is attributed to the intense electric field near the crack tip.

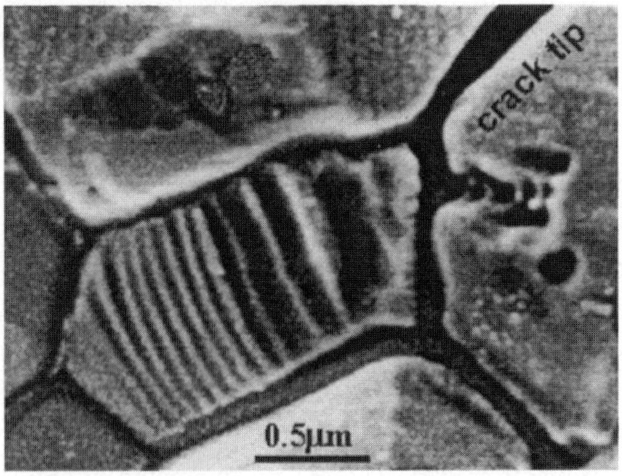

Fig. 1 SEM shows 90° domain switching bands near a crack tip.

Directly ahead of the crack tip, the electric field concentration is intense and aligned normal to the crack extension line. The intensified local electric field, regardless its direction, forms an angle almost normal to the poling direction. Accordingly, 90° domain switching takes place in several strips in the grain ahead of the crack tip, leading to marks of straight 90° domain walls.

3. Energetics of Domain Bands

Early studies of the assembly patterns can be traced back to the pioneer works of Kittel [3] for ferromagnetics. Based on the electrostatics equation and depolarization energy of domain formation, ferroelectric structures of plate-like and needle-like geometries were analyzed [4]. Studied in [5] is the relationship between the domain assemblies and the grain size by taking into account the release of elastic energy during a switch. The equilibrium domain assemblies will be quantified here by free energy minimization.

Consider a ferroelectric grain of volume D^3 embedded within a homogenized ferroelectric matrix. Remote stress $\overline{\sigma}_{ij}$ and electric field $\overline{E}_i$ are applied. The formation of domain band structure in the ferroelectric grain causes the following change in system free energy:

$$\Delta U = \Delta U_{\text{Elastic}} + \Delta U_{\text{Electric}} + \Delta U_{\text{Wall}} - \Delta U_{\text{M}}^{\infty} - \Delta U_{\text{E}}^{\infty}, \tag{1}$$

where $\Delta U_{\text{Elastic}}$ denotes the elastic mismatch energy, $\Delta U_{\text{Electric}}$ the electric depolarization energy, ΔU_{Wall} the domain wall energy, and $\Delta U_{\text{M}}^{\infty}$ and $\Delta U_{\text{E}}^{\infty}$ correspond to the works done by the remote stress and electric fields. All terms bear the meaning of energy densities per unit volume. Denote the spacing of two adjacent domain plates as $2t$. For the case of an instantaneous 90° domain switch, the depolarization energy can be computed under the principle outlined in [4-5] as

$$\Delta U_{\text{Electric}} = \frac{2P_s^2 t}{\pi^3 \varepsilon D} f(V_{90}); \quad f(V_{90}) = \sum_{k=1}^{\infty} \frac{\sin^2(V_{90} k\pi)}{k^3}, \tag{2}$$

where ε denotes the mean square root average of the dielectric constants, P_s the spontaneous polarization, and V_{90} the volume fraction of the switched domain. The domain wall energy has a form of $\Delta U_{\text{Wall}} = \Gamma_{90}/t$, where the 90° domain wall energy Γ_{90} is composed of the coherent domain wall energy Γ_{coh} and the energy Γ_{misfit} of misfit dislocations along the domain wall. The quantity Γ_{coh} assumes a value of $\Gamma_{\text{coh}} = 0.002\text{J/m}^2$ for PZT. The mechanical and electrical works done by remote fields due to switching are given by $\Delta U_{\text{M}}^{\infty} = V_{90}\overline{\sigma}_{ij}\Delta\gamma_{ij}$ and $\Delta U_{\text{E}}^{\infty} = V_{90}\overline{E}_i\Delta P_i$.

4. Banded Eshelby Inclusion

Consider an infinite region Ω_∞ with boundary free of remote loading. Embedded within is a transformable ferroelectric grain Ω. Remove the ferroelectric region Ω from Ω_∞, and let it undergo an unconstrained transformation $\hat{\gamma}_{ij}^*$. Due to the domain band structure, $\hat{\gamma}_{ij}^*$ is not uniform within the region Ω but has a band-like feature. When the ferroelectric region Ω is inserted to the infinite region Ω_∞ as a banded Eshelby inclusion, stress and the strain fields are induced and they are denoted by $\hat{\sigma}_{ij}$ and $\hat{\gamma}_{ij}$. The elastic mismatch energy due to the insertion process is $\Delta U_{\text{Elastic}} = -\dfrac{1}{2D^3}\int_{\partial\Omega}\hat{t}_i\hat{u}_i^*\text{d}S$ [6]. Let us decompose the unconstrained transform $\hat{\gamma}_{ij}^*$ to its domain average γ_{ij}^* and fluctuation $\tilde{\gamma}_{ij}^*$. The self-equilibrium feature of the residual mismatch stress renders the decomposition of the elastic mismatch energy as $\Delta U_{\text{Elastic}} = \Delta U_{\text{LR}} + \Delta U_{\text{SR}}$. The first term signifies the long-range elastic mismatch energy for a uniform transform of γ_{ij}^* in region Ω. Dimensional consideration dictates that:

$$\Delta U_{\text{LR}} = \alpha(\nu)Y\left(V_{90}\frac{c-a}{a}\right)^2. \tag{3}$$

The transformation strain solution of Eshelby [7] leads to the dimensionless quantity $\alpha(\nu)$ as $(7-5\nu)/[15(1-\nu^2)]$ for a spherical shape, and the value of $3/[16(1-\nu^2)]$

for a cylindrical grain unconstrained in axial direction.

The short-range mismatch energy due to the fluctuating mismatch $\tilde{u}_i^*$ along the grain boundary can be computed as

$$\Delta U_{\mathrm{SR}} = -\frac{1}{2D^3}\int_{\partial\Omega} \tilde{t}_i \tilde{u}_i^* \, \mathrm{d}S \, . \tag{4}$$

5. Domain Wall Energy for Unconventional Bands

The band geometry shown in Fig. 2 is used to reveal the misfit dislocations caused by misfit strains. In Fig. 2(a), the shaded bands are the domains which have switched, and the blank ones the un-switched domains. If the domain bands are sliced along the domain walls, then misfit strain of $\Delta\gamma$ is observed for every shaded bands. Apply compressive stress $-[Y/2(1-\nu^2)]\Delta\gamma$ to the switched domains and tensile stress $[Y/2(1-\nu^2)]\Delta\gamma$ to the unswitched ones. Under plane strain, the mechanical strains produced by two sets of stresses exactly cancel the misfit strain $\Delta\gamma$. Thus, two domain bands can be coherently adhered with coherent domain wall energy Γ_{coh}, Fig. 2(b). After long time of relaxation, the long-range stress is released by forming arrays of misfit dislocations along the domain walls, Fig. 2(c). The misfit dislocations have a Burgers vector of b and a spacing of $l = b/\Delta\gamma$. For a particular array of dislocations in Fig. 2(c), their stresses can be computed by summing the infinite series for the stress field of a single edge dislocation. The final stress field is obtained by the superposition of that solution and the band residual stresses $\mp[Y/2(1-\nu^2)]\Delta\gamma$.

Let the lattice mismatches of different orientations along the domain wall be represented by the respective misfit dislocations along the domain boundary. Their interaction is weak since the associated stress fields decay exponentially from the misfit dislocation arrays. The domain misfit energy can be obtained by integrating the strain energy density in a domain layer then dividing it to the domain area as

$$\Gamma_{\mathrm{misfit}} = \frac{Yl}{8\pi(1-\nu^2)}\left(\frac{c-a}{a}\right)^2\left[\ln\frac{l}{4\pi r_0} - \frac{1}{4(1-\nu)}\right]\frac{\sin^2\phi(1+3\cos^2\phi)}{(1+\cos^2\phi)^2} \, . \tag{5}$$

where ϕ denotes the top view angle of domain wall. That angle is measured by microscopy of etched specimen as $70°$.

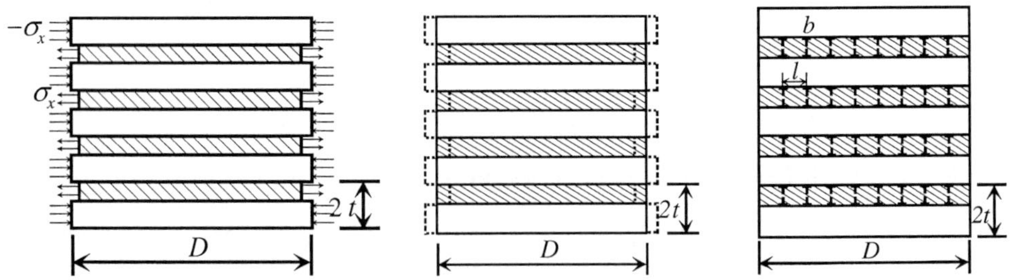

Fig. 2 Misfit dislocations caused by inter-band misfit strain $\Delta\gamma$, (a) two sets of domain bands with misfit strain $\Delta\gamma$; (b) adherence after applying respective tensile and compression stresses; (c) misfit dislocations are formed after relaxation

6. Band Parameters near a Crack Tip

The mismatch along the grain boundary can be written as

$$\begin{Bmatrix} \tilde{u}_1^*(\xi_1) \\ \tilde{u}_2^*(\xi_1) \end{Bmatrix} = \frac{c-a}{a} \begin{Bmatrix} -\dfrac{\sin^2\phi}{1+\cos^2\phi} \\ 2\sqrt{2}\dfrac{\cos\phi}{\sqrt{1+\cos^2\phi}} \end{Bmatrix} \times \begin{Bmatrix} V_{90}(-t-\xi_1) & -t \le \xi_1 < -V_{90}t \\ (1-V_{90})\xi_1 & -V_{90}t \le \xi_1 < V_{90}t \\ V_{90}(t-\xi_1) & V_{90}t \le \xi_1 < t \end{Bmatrix}. \tag{6}$$

Expanding $\tilde{u}_i^*(\xi_1)$ in the interval $[-t,t]$ by sinusoidal Fourier series, then substituting into (4), one has

$$\Delta U_{\mathrm{SR}} = \frac{4(1-\nu)Yf(V_{90})}{(1+\nu)(3-4\nu)\pi^3} \frac{(c-a)^2}{a^2} \frac{t}{D} \frac{1+2\cos^2\phi+5\cos^4\phi}{(1+\cos^2\phi)^2}. \tag{7}$$

That leads to a band thickness of

$$t = \sqrt{\frac{\pi^3\Gamma_{90}D}{2f(V_{90})\left[\dfrac{P_s^2}{\varepsilon} + \dfrac{2(1-\nu)Y}{(1+\nu)(3-4\nu)} \dfrac{(c-a)^2}{a^2} \dfrac{1+2\cos^2\phi+5\cos^4\phi}{(1+\cos^2\phi)^2}\right]}}. \tag{8}$$

Substituting (2), (3), (7) and (8) into (1), one finds the change in system energy as

$$\Delta U = \frac{Y(7-5\nu)}{15(1-\nu^2)}\left(V_{90}\frac{c-a}{a}\right)^2 - V_{90}\left[\frac{c-a}{a}(\bar{\sigma}_{22}-\bar{\sigma}_{33})+P_s(\bar{E}_2-\bar{E}_3)\right]$$
$$+\frac{2}{\pi}\sqrt{\frac{2f(V_{90})\Gamma_{90}}{\pi D}\left[\frac{P_s^2}{\varepsilon}+\frac{2(1-\nu)Y}{(1+\nu)(3-4\nu)}\frac{(c-a)^2}{a^2}\frac{1+2\cos^2\phi+5\cos^4\phi}{(1+\cos^2\phi)^2}\right]}\ . \quad (9)$$

Numerical calculations indicate that the last term only offers an effect of about a few percent in determining the switched volume fraction V_{90}. If that term is neglected, the following explicit expression is derived for V_{90}:

$$V_{90} = \frac{1}{2Y\alpha(\nu)}\frac{a}{c-a}\left[\bar{\sigma}_{22}-\bar{\sigma}_{33}+\frac{a}{c-a}(\bar{E}_2-\bar{E}_3)P_s\right]\ . \quad (10)$$

Fig. 1 indicates that the switched ferroelectric grain ahead of the crack has a shape of $1\times 1.5\mu\text{m}$ rectangle, and the length D along the crack extension line is approximately $1.5\mu\text{m}$. Integrate the electric field along the crack extension line from 0 to D, then divide it by D for a grain average measure of the applied field, this gives $\bar{E}_{\text{app}} = E_{\text{app}}\cos\omega\sqrt{2a/D}$. The applied field E_{app} in the experiment is $0.6E_c = 660\text{V/mm}$, with an angle ω of $45°$ to the crack. The non-contacted portion of the crack length is measured approximately as $2a = 60\mu\text{m}$. These data give the average field $\bar{E}_{\text{app}}$ applied on the grain ahead of the crack tip of $2\sqrt{5}E_{\text{app}}$. Neither remote stress nor remote strain exists in the test. Substituting only $\bar{E}_2 \approx 2\sqrt{5}E_{\text{app}}$ and $\bar{E}_3 = 0$ into (10), one finds that the volume fraction for switched bands is $\frac{\sqrt{5}}{Y\alpha(\nu)}\left(\frac{a}{c-a}\right)^2 P_s E_{\text{app}}$. For the "soft" ferroelectric ceramics used in this test, the relevant physical constants are $Y = 33\text{GPa}$, $\nu = 1/3$, $c/a = 1.013$, $E_c = 1.1\times 10^6\text{V/m}$

and $P_s = 0.3787\text{C/m}^2$. The calculation gives $V_{90} = 0.249$ for fully constrained spherical grain, and $V_{90} = 0.474$ for cylindrical grain unconstrained in the x_3-direction. They provide bounds for actual value of V_{90}. This prediction agrees with the measured volume fraction of domain switch. The latter can be assessed from Fig. 1 as $V_{90} \approx 0.35$.

Substituting $Y = 33\text{GPa}$, $\nu = 1/3$, $c/a = 1.013$, $r_0 = b = 0.4\text{nm}$ and $\phi = 70°$ into (5), the misfit energy is obtained for the particular domain wall in Fig. 1 as $\Gamma_{\text{misfit}} \approx 0.021\text{J/m}^2$. Consequently, Γ_{90} is about 0.023J/m^2. For at a volume fraction of $V_{90} \approx 0.35$, it is found that $f(0.35) \approx 0.902$ from (2). The substitution of the above data, plus $P_s = 0.3787\text{C/m}^2$ and $\varepsilon = 1800\varepsilon_0 = 1.593 \times 10^{-8}\text{F/m}$, into (8) gives a predicted domain wall spacing of $t = 0.2178\mu\text{m}$. That value is of the same order, but somewhat larger, than the experiment measurement from Fig. 1.

Acknowledgement: The present work is sponsored by the National Natural Science Foundation of China.

References

1. Winzer, S.R., Shankar, N. and Ritter, A.P., Designing cofired multilayer electrostrictive actuators for reliability, *J. Am. Ceramic Soc.* **72**(1989), 2246-2257.
2. Fang, F., Yang, W. and Zhu, T., Crack tip 90^0 switching in tetragonal lanthanum-modified lead zirconia titanate under an electric field. *Journal of Materials Research*, **14** (1999), 2940-2944.
3. Kittel, C., Theory of the structure of Ferromagnetic domains in films and small particles, *Physical. Review*, **70** (1946), 965-971.
4. Mitsui, T. and Furuichi, J., Domain structure of Rochelle salt and KH_2PO_4. *Physical Review*, **90** (1953), 193-202.
5. Arlt, G., Twinning in ferroelectric and ferroelastic ceramics. *J. Mater. Sci.*, **25**(1990), 2655-2666.
6. Mura, T., *Micromechanics of Defects in Solids*, Amsterdam: Nijhoff, 1987.
7. Eshelby, J.D., Determination of the elastic field of an ellipsoidal inclusion and related problems. Proc. Roy. Society London Ser. **A241**(1957), 376-396.

A MACROSCOPIC MODEL FOR MICROSCOPICALLY HETEROGENEOUS SHAPE MEMORY MATERIALS

DAVIDE BERNARDINI

Dipartimento di Ingegneria Strutturale e Geotecnica
Università di Roma "La Sapienza"
Via Eudossiana 18, 00184 - Roma Italy

1. Introduction

The macroscopic scale behavior of Shape Memory Materials (SMM) is the result of microstructural rearrangements induced by solid phase transformations among Austenite (**A**) and Martensite variants (**M**) [1]. While the phases exhibit different material properties such as elastic moduli, thermal expansion and specific heat, most of the available models neglect this issue [2]. In [3], some expressions of the macroscopic free energy functions for SMM have been derived via the study of a representative volume composed of heterogeneous thermoelastic phases subjected to a prescribed eigenstrain field. In this paper, such free energies are complemented by suitable phase transformation criteria to build a model for pseudoelasticity. A double-inclusion embedding scheme [4, 5] is implemented as underlying micromechanical model. By endowing the comparison material with a suitable physical interpretation, an approximate model is given for the strain energy relaxation effects that take place around the inclusions during phase accommodation. In this way a correct prediction of the order of magnitude of the transformation stresses is obtained.

2. Basic equations

The model is composed by three groups of equations: (i) balance equations and entropy inequality, (ii) constitutive equations for strain and entropy, (iii) phase transformation criteria that enable to determine the microstructure evolution.

In the case of small displacement gradients and neglecting heat conduction, *balance equations* for mass, momenta, internal energy and entropy, together with

Q.P. Sun (ed.), IUTAM Symposium on Mechanics of Martensitic Phase Transformation in Solids, 241–248.

Clausius-Duhem inequality are assumed to hold in the following local form

$$\begin{array}{lll} \dot{\rho} = 0, & \rho \dot{\mathbf{v}} = \operatorname{div} \boldsymbol{\Sigma} + \mathbf{b}, & \boldsymbol{\Sigma} = \boldsymbol{\Sigma}^T, \\ \dot{\eta} = \Gamma / \vartheta + r / \vartheta, & \dot{e} = \boldsymbol{\Sigma} \cdot \dot{\mathbf{E}} + r, & \Gamma \geq 0, \end{array} \tag{1}$$

where: $\rho, \mathbf{v}, \vartheta$ denote mass density, velocity and temperature; $\mathbf{E}, \boldsymbol{\Sigma}$ infinitesimal strain and stress; η entropy; $\mathbf{b}, r$ body force and rate of heat supply; e, Γ internal energy and rate of energy dissipation (extensive quantities per unit volume).

Constitutive equations are given for $\mathbf{E}$ and η in terms of $\mathcal{S} := \{\boldsymbol{\Sigma}, \vartheta, \boldsymbol{\alpha}\}$ where $\boldsymbol{\alpha} \in \mathbb{R}^n$ is a suitable set of microstructural descriptors (internal variables) representing a class of microstructures. Standard thermodynamic arguments can then be used to show, or to postulate, the following relations

$$\mathbf{E} = \frac{\partial \widehat{\Psi}}{\partial \boldsymbol{\Sigma}}, \quad \eta = -\frac{\partial \widehat{\Psi}}{\partial \vartheta} \tag{2}$$

where $\widehat{\Psi}(\mathcal{S})$ is a constitutive function expressing the value assumed by the free energy $\Psi := e - \vartheta\eta - \boldsymbol{\Sigma} \cdot \mathbf{E}$. The substitution of (2) as well as of the energy and the entropy balances into Clausius-Duhem inequality yields

$$\boldsymbol{\Pi} \cdot \dot{\boldsymbol{\alpha}} \geq 0 \quad \text{with} \quad \boldsymbol{\Pi} := \partial \widehat{\Psi} / \partial \boldsymbol{\alpha}. \tag{3}$$

The Forward $\mathbf{A} \rightarrow \mathbf{M}$ (FwT) and the Reverse $\mathbf{M} \rightarrow \mathbf{A}$ (RvT) phase transformations are driven respectively by Π and $-\Pi$ and are assumed to occur, in a time interval $[t, t_1]$, if and only if the corresponding *transformation criterion* holds, i.e.

$$\Pi(\tau) = \boldsymbol{\Lambda}_F(\tau) \quad \text{or} \quad \Pi(\tau) = \boldsymbol{\Lambda}_R(\tau) \qquad \text{for any } \tau \in [t, t_1] \tag{4}$$

where $\boldsymbol{\Lambda}_F$ and $\boldsymbol{\Lambda}_R$ are constitutive functions expressing the thresholds attained by the driving forces during the transformations. As shown below, (4) yield kinetic equations for $\boldsymbol{\alpha}$ describing the microstructure evolution. By virtue of (3) and of $\Gamma \geq 0$ the functions $\boldsymbol{\Lambda}_F, \boldsymbol{\Lambda}_R$ must obey, for any choice of their arguments,

$$\Lambda_F \geq 0, \quad \Lambda_R \leq 0. \tag{5}$$

In this framework the development of a model for SMM thus requires three steps: the choice of microstructural descriptors $\boldsymbol{\alpha}$, the specification of the free energy function $\widehat{\Psi}(\mathcal{S})$ and of the driving force threshold functions $\boldsymbol{\Lambda}_F(\mathcal{S})$, $\boldsymbol{\Lambda}_R(\mathcal{S})$.

3. Underlying model for the microstructure

Each point of the material, seen at the macroscopic scale, is associated with a representative volume composed by two isotropic thermoelastic phases with elastic modulus (compliance) $\mathbf{L}_i$ ($\mathbf{M}_i$), thermal expansion $\mathbf{l}_i$ and specific heat c_i where $i = 1, 2$ denote respectively $\mathbf{A}$ and $\mathbf{M}$. At the microscopic scale, the effect of the phase transformations is modeled by the occurrence of a piecewise-constant *local* *transformation eigenstrain* field assuming the values γ_i^*. This provides a close model for the microstructure whenever a single family of martensitic inclusions is expected to form, as for single crystals under uniaxial stress. In this case the γ_i^*'s may be obtained either from experiments or computed from lattice parameters. In case of more complex microstructures, it becomes a coarser model and γ_i^*'s have to be understood as effective values prescribed by specific constitutive equations [3, 6]. Here, as explained later, γ_i^*'s are regarded as known and only the *phase fractions* ξ_1, ξ_2 are retained as internal variables. Hence, denoting $\xi := \xi_2$ and $\xi_1 = 1 - \xi$, the vector $\boldsymbol{\alpha}$ becomes $\boldsymbol{\alpha} = \{\xi\}$. During FwT starting from pure **A**, the phase 1 is the parent ($\gamma_1^* = \mathbf{0}$) and phase 2 forms as product with $\gamma_2^* = \gamma^*$. Conversely, during RvT starting from pure **M**, the previously formed phase 2 is the parent and converts into phase 1 with $\gamma_1^* = -\gamma^*$. The other details on the microstructure, are reflected by the expression of the free energy.

4. Free energy function

A free energy function with the following structure is considered [3]

$$\widehat{\Psi} := \frac{1}{2}\Sigma \cdot \overline{\mathbf{M}}\Sigma + \Sigma \cdot \left(\bar{\mathbf{l}}\Theta + \overline{\gamma}^*\right) - W^* - \overline{E}^{ch} \tag{6}$$

where Θ is the temperature difference with respect to a reference state, $\overline{\mathbf{M}}$ and $\bar{\mathbf{l}}$ are the *effective elastic compliance* and *effective thermal expansion* tensors, $\overline{\gamma}^*$ the *macroscopic transformation strain*, W^* the *interaction energy* and $\overline{E}^{ch}$ the *overall chemical free energy*. The reference state is: $\boldsymbol{\Sigma} = \mathbf{0}$, $\vartheta = \vartheta_0 \geq A_f$, $\xi = 0$, that is a, pure-**A**, macroscopic and microscopic stress-free state. While the complete model with heterogeneous thermal expansion and specific heat can be obtained from (6) by straightforward substitution of the expressions reported in [3], here, for the sake of simplicity, only the heterogeneity of the elastic modulus is retained. Accordingly, the quantities characterizing $\widehat{\Psi}$ are assumed to be given by [3]

$$\overline{\gamma}^* := \xi\gamma^* + (\delta\mathbf{M})\,\mathbf{R}\gamma^*, \quad W^* := -\frac{1}{2}\gamma^* \cdot \mathbf{R}\gamma^*$$

$$\overline{E}^{ch} := c\left(\vartheta - \vartheta_0 - \vartheta \ln\frac{\vartheta}{\vartheta_0}\right) - \eta_{1,0}\vartheta - \delta\eta_0\,(\vartheta - \vartheta_e)\,\xi + e_{1,0} \tag{7}$$

where $\eta_{i,0}$ and $e_{i,0}$ are the entropy and the internal energy of the pure phases evaluated at ϑ_0, $\delta\mathbf{M} := \mathbf{M}_2 - \mathbf{M}_1$ and $\delta\eta_0 := \eta_{2,0} - \eta_{1,0}$. The parameter ϑ_e is the temperature at which the pure phases have the same chemical free energy and

$$\mathbf{R} := (\delta\mathbf{M})^{-1}\left[\overline{\mathbf{M}} - \langle\mathbf{M}\rangle\right](\delta\mathbf{M})^{-1} \tag{8}$$

where $\langle\mathbf{M}\rangle$ is the average elastic compliance. Taking into account (7) and (8), the free energy is completely determined whenever an estimate of $\overline{\mathbf{M}}$ is known. Consequently, the constitutive equations that relate the strain and driving force to the state are also determined since they follow from (2)

$$\begin{aligned} \mathbf{E} &= \overline{\mathbf{M}}\boldsymbol{\Sigma} + \xi\gamma^* + (\delta\mathbf{M})\,\mathbf{R}\gamma^* + 1\Theta \\ \Pi &= \Sigma \cdot \left(\gamma^* + \delta\mathbf{M}\frac{\partial\mathbf{R}}{\partial\xi}\gamma^*\right) + \frac{1}{2}\gamma^* \cdot \frac{\partial\mathbf{R}}{\partial\xi}\gamma^* + \delta\eta_0\,(\vartheta - \vartheta_e) + \frac{1}{2}\Sigma \cdot \frac{\partial\overline{\mathbf{M}}}{\partial\xi}\Sigma. \quad (9) \end{aligned}$$

Any specific choice of $\overline{\mathbf{M}}$ reflects the underlying micromechanical model for the microstructure. As shown by Dvorak and Srinivas [4], several existing models for the estimate of the effective elastic modulus of composite materials, as well as new ones, can be obtained from a common *double-inclusion embedding scheme* [5], in which, the phases are regarded as inclusions in a volume of a comparison material $\mathbf{L}_0$ which, in turn, is an inclusion in the effective medium. Each model is then characterized by a choice of $\mathbf{L}_0$. For example, if the comparison medium coincides with the matrix then, the Mori-Tanaka estimate is obtained, whereas Hashin-Shtrikman bounds follow when the comparison material is identified, respectively, with the phases with upper and lower moduli, in an appropriate sense. While the double-inclusion scheme is usually employed to approximate the solution of the heterogeneous elasticity problem arising from composite mechanics, here, it will be used to model a different situation relevant for SMM modeling.

It is known that the computation of the phase interaction energy by means of elastic micromechanical models can lead to large overestimates with respect to the values obtained from experiments [7, 8, 9]. This has been explained by the fact that purely elastic models neglect the relaxation mechanisms that take place, within the martensite regions, by twinning and, in the region around the inclusions, by dislocation motions and other inelastic accommodation phenomena. While the effect of twinning can be taken into account in the computation of the local transformation strain [8, 10, 11], the modeling of the other mechanisms seems to be an open problem. In Ref. [9] a plastic strain acting within the martensite inclusion is introduced to reduce the local transformation strain. Here, a different possibility to model such effects is proposed. In the double-inclusion embedding, $\mathbf{L}_0$ represents the elastic modulus of the region that surrounds the inclusions. Since this region is interested by the relaxation mechanisms, it seems reasonable to think that its equivalent stiffness will be lower than the virgin matrix, as much as the relaxation is relevant. Hence we propose to use to double-inclusion method with a suitable value of $\mathbf{L}_0$ reduced to take into account the dislocation motions around inclusions. Under this new interpretation of $\mathbf{L}_0$ the name "comparison material", suitable for composites where it represents a fictitious medium used for the approximation of the elasticity problem, will be replaced henceforth by "surrounding material".

By calculations analogous to those in [4], the stress concentration tensor $\mathbf{B}_2$ of the phase 2 consistent with the double-inclusion scheme turns out to be

$$\mathbf{B}_2 = \mathbf{B}_2^0\left[\mathbf{B}_1^0 + \xi\left(\mathbf{B}_2^0 - \mathbf{B}_1^0\right)\right]^{-1} \quad (10)$$

where $\mathbf{B}_i^0$ denotes the stress concentration tensor of the phase i in dilute conditions evaluated in $\mathbf{L}_0$, i.e.

$$\mathbf{B}_1^0 := \left[\mathbf{I} + \mathbf{Q}_0 \left(\mathbf{M}_1 - \mathbf{M}_0\right)\right]^{-1}, \quad \mathbf{B}_2^0 := \left[\mathbf{I} + \mathbf{Q}_0 \left(\mathbf{M}_2 - \mathbf{M}_0\right)\right]^{-1} \tag{11}$$

with $\mathbf{Q}_0 := \mathbf{L}_0 \left(\mathbf{I} - \mathbf{S}_0\right)$ and $\mathbf{S}_0$ being the Eshelby tensor in $\mathbf{L}_0$. If $\mathbf{B}_2$ is known then so are the effective compliance

$$\overline{\mathbf{M}} = \mathbf{M}_1 + \xi \left(\delta \mathbf{M}\right) \mathbf{B}_2, \tag{12}$$

the $\mathbf{R}$ tensor

$$\mathbf{R} = \xi \left(\mathbf{B}_2 - \mathbf{I}\right) \left(\delta \mathbf{M}\right)^{-1} \tag{13}$$

and, finally, the free energy.

5. Driving force thresholds and transformation kinetic

According to (4), FwT occurs in a time interval if and only if Π equals Λ_F at any time of the interval. Consequently, FwT will occur at the time t if and only if $\Pi(t) = \Lambda_F(t)$ and

$$\lim_{\tau \downarrow t} \Pi(\tau) - \Lambda_F(\tau) = 0 \tag{14}$$

Summarizing, whenever Π and Λ possess such a right limit,

$$\text{FwT} \quad \Longleftrightarrow \quad \Pi(t) = \Lambda_F(t) \quad \text{and} \quad \dot{\Pi}(t) = \dot{\Lambda}_F(t), \tag{15}$$

where superposed dot here denotes right-derivative. Similarly

$$\text{RvT} \quad \Longleftrightarrow \quad \Pi(t) = \Lambda_R(t) \quad \text{and} \quad \dot{\Pi}(t) = \dot{\Lambda}_R(t). \tag{16}$$

Taking into account the state dependence of both Π and Λ's, the chain rule, (15) and (16), the following transformation kinetic is obtained

$$\dot{\xi} = \begin{cases} \max\left\{ \left(\frac{\partial(\Lambda_F - \Pi)}{\partial \xi}\right)^{-1} \left[\frac{\partial(\Pi - \Lambda_F)}{\partial \boldsymbol{\Sigma}} \cdot \dot{\boldsymbol{\Sigma}} + \frac{\partial(\Pi - \Lambda_F)}{\partial \vartheta} \dot{\vartheta}\right], 0\right\} & \text{if } \Pi = \Lambda_F \\ \min\left\{ \left(\frac{\partial(\Lambda_R - \Pi)}{\partial \xi}\right)^{-1} \left[\frac{\partial(\Pi - \Lambda_R)}{\partial \boldsymbol{\Sigma}} \cdot \dot{\boldsymbol{\Sigma}} + \frac{\partial(\Pi - \Lambda_R)}{\partial \boldsymbol{\Sigma}} \dot{\vartheta}\right], 0\right\} & \text{if } \Pi = \Lambda_R \\ 0 & \text{otherwise} \end{cases} \tag{17}$$

Having already introduced $\widehat{\Psi}$, whence Π, the transformation kinetic is determined as soon as constitutive expressions are given for Λ_F and Λ_R that represent the resistance encountered by the progress of the phase transformations. In the ideal case of a single inclusion growing in absence of defects, such resistance could be thought, roughly, proportional to the surface of the phase interface so that a linear dependence on the product phase fraction could be appropriate. However, in the multiple inclusions case and in presence of defects, additional resistance contributions should arise from the coalescence and the interaction with defects. The detailed modeling of such contributions would require the knowledge of microstructural information that is beyond the possibilities of macroscopic models based on the phase fractions. However, in the present framework, it is possible to implement the intuitive fact that coalescence and interactions with defects should reasonably increase with the product phase fraction. This is done by assuming the following simple expressions for the driving force thresholds:

$$\begin{aligned}\Lambda_F &= \Lambda_{M_s} + D_{F,1}\xi + \frac{D_{F,2}\left(e^{F\xi}-1\right)}{e^F-1} \\ \Lambda_R &= \Lambda_{A_s} - D_{R,1}\left(1-\xi\right) - \frac{D_{R,2}\left(e^{R(1-\xi)}-1\right)}{e^R-1}.\end{aligned} \tag{18}$$

All the constitutive parameters in (18) have to be identified from experimental or theoretical data but *must* agree with the restrictions induced by the Clausius-Duhem inequality (5). The importance of such restrictions cannot be underestimated during the parameters identification since, otherwise, violations of the second law of the thermodynamics can arise as in the Ref. [12].

More specifically, the parameters $D_{F,1}, D_{F,2}$ and their RvT counterparts, are identified by prescribing the differences among the transformation temperatures $h_M := M_f - M_s$, $h_A := A_f - A_s$ and the equilibrium temperature ϑ_e. Thus, the transformation temperatures and stresses are predicted as consequences of the modeling assumptions. The parameters F, R regulate the steepness of the resistance increase with the product phase fraction and are set heuristically (they influence the shape of the stress-strain curves near the end of transformation). The consistency with (5) is then assured by

$$\Lambda_{M_s}, D_{F,1}, D_{F,2}, D_{R,1}, D_{R,2} \geq 0 \quad \text{and} \quad \Lambda_{A_s} \leq 0. \tag{19}$$

6. Local transformation strain

The attention is focused on cubic-to-monoclinic stress-induced martensitic transformations in NiTi single crystals under uniaxial stress. According to [11, 10], the Martensite forms as *Correspondence Variant Pairs* (CVP) composed of twinned Martensite crystallographic variants. Under the hypothesis of invariant plane deformation, each CVP is identified, in the parent phase coordinate system, by the habit plane normal $\mathbf{n}$, the transformation direction $\mathbf{m}$ and the transformation displacement modulus g. According to the phenomenological theory, in such conditions, 24 CVP's can arise and the local transformation strain of the I-th CVP is $\boldsymbol{\gamma}^*_{(I)} = g\left[\mathbf{m}^{(I)} \otimes \mathbf{n}^{(I)} + \mathbf{n}^{(I)} \otimes \mathbf{m}^{(I)}\right]/2$.

Once all the local transformation strains $\boldsymbol{\gamma}^*_{(I)}$ have been computed from $\mathbf{n}, \mathbf{m}, g$, then a basic issue is the determination of the active CVP and of the aspect ratio of the inclusions. Under uniaxial stress $\boldsymbol{\Sigma} = \sigma \mathbf{e}_i \otimes \mathbf{e}_j$ this is done by computing the FwT start stress, σ^{M_s}, for all variants and for a set of aspect ratios, spanning the spheroidal geometries from the oblate to the prolate case, and selecting the CVP-aspect ratio pair that gives rise to the lower value of σ^{M_s}. The transformation stress stems from the uniaxial version of $\Pi = \Lambda_F$ that depends on the aspect ratio via the Eshelby tensor $\mathbf{S}_0$ and yields a quadratic equation in σ with one positive and one negative solution. The quadratic term is due to the heterogeneity of the elastic moduli of the phases and then is absent in the models that neglect this aspect. When dealing with tensile stress, this enables to take into account the tension/compression asymmetry since different CVP's can be activated for the positive and negative values of σ^{M_s}.

7. An example

As an illustrative example, the isothermal stress-strain behavior under uniaxial tension and compression is considered. The material parameters are partly taken from literature data pertinent to NiTi single crystals and partly identified as indicated above. The crystallographic parameters $\mathbf{n}, \mathbf{m}, g$ are taken equal to the following values, computed in [11]:

$$\mathbf{n} = \{-0.88888, 0.40442, 0.21523\}\,, \ \mathbf{m} = \{0.43449, 0.48739, 0.75741\}\,, \ g = 0.13079$$

while the other parameters are taken as follows

$$\begin{array}{llll} \delta\eta_0 = 0.2 \text{ MPa/K} & \vartheta_e = 297 \text{ K} & h_M = -13 \text{ K} & h_A = 13 \text{ K} \\ \Lambda_{M_s} = 0.5 \text{ MPa} & \Lambda_{A_s} = -0.5 \text{ MPa} & D_{F,1} = 2.6 \text{ MPa} & D_{F,2} = 1.3 \text{ MPa} \\ D_{R,1} = 2.6 \text{ MPa} & D_{R,2} = 1.3 \text{ MPa} & F = 30.0 & R = 30.0 \end{array}$$

Figure 1 shows uniaxial tension and compression isothermal behavior for two different values of the Martensite Young modulus E_2 ($\nu_1 = \nu_2 = 0.3$).

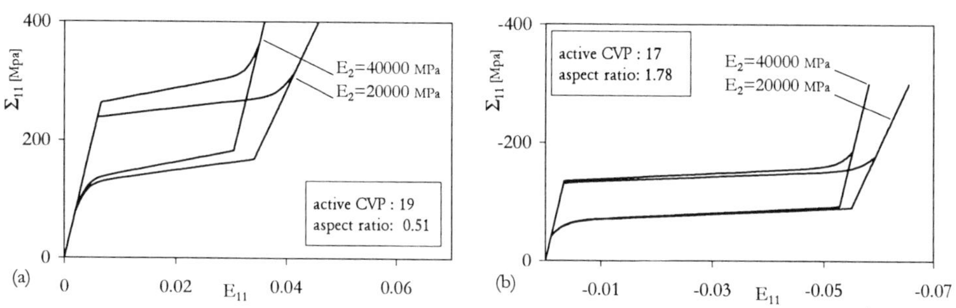

Fig. 1. Isothermal stress-strain curves (10 K above A_f) for uniaxial tension (a) and compression (b) and different Martensite Young modulus (E_1=40000 MPa)

In these simulations $E_0 = 200\,\text{MPa}$ and, remarkably, the order of magnitude of the transformation stresses is the same of the one observed experimentally. For the sake of comparison, let us mention that if $E_0 = 20000\,\text{MPa}$, so as to obtain the purely elastic Hashin-Shtrikman lower bound, the corresponding FwT start stress is about 3 times greater. It is to note that the consideration of the heterogeneity of the elastic modulus leads to a reduction of the transformation stresses. Relevant tension/compression asymmetry is obtained due to the different active CVP and moreover, different aspect ratios are found in tension and compression.

8. Conclusions

A double-inclusion embedding scheme for the Martensite inclusions has been implemented as micromechanical model underlying a free energy function consistent with [3]. By introducing expressions for the driving force thresholds that comply with Clausius-Duhem inequality, a model for pseudoelasticity allowing for the heterogeneity of the material properties of the phases is obtained. The double-inclusion scheme provides a way to model, in an approximate way, the relaxation mechanisms occurring during the phase accommodation process and enables to predict values for the transformation stresses of the same order of those observed experimentally. The proposed example under uniaxial tension and compression shows that not only different CVP's but also different aspect ratios can be activated.

9. References

[1] Fischer, F.D., Berveiller, M., Tanaka, K., Oberaigner, E.R. (1994) Continuum mechanical aspects of phase transformations in solids, *Arch. Appl. Mech.*, **64**, 54-85

[2] Bernardini D., Pence T. J. (2001) Shape memory materials: modeling, in *Encyclopedia of Smart Materials*, John Wiley and Sons, New York, in press

[3] Bernardini D. (2001) On the macroscopic free energy functions for shape memory alloys, *J. Mech. Phys. Solids*, **49**, 813-837

[4] Dvorak, G.J., Srinivas, M.V. (1999) New estimates of overall properties of heterogeneous solids, *J. Mech. Phys. Solids*, **47**, 899-920 and *J. Mech. Phys. Solids* **47**, 2207-2208

[5] Nemat-Nasser S., Hori M. (1993) *Micromechanics: overall properties of heterogeneous materials*, North-Holland, Amsterdam

[6] Sun, Q.P., Hwang, K.C. (1994) Micromechanics constitutive description of thermoelastic martensitic transformation, *Adv. Appl. Mech.*, **31**, 249-298

[7] Mura T., Mori T., Kato M. (1976) The elastic field caused by a general ellipsoidal inclusion and the application to martensite formation, *J. Mech. Phys. Solids*, **24**, 305-318

[8] Huang, M., Brinson, L.C. (1998) A multivariant model for single crystal shape memory alloy behavior, *J. Mech. Phys. Solids*, **46**, 1379-1409

[9] Zhang S., McCormick P. G. (2000) Thermodynamic analysis of shape memory phenomena - I. Effect of transformation plasticity on elastic strain energy", *Acta Mater.*, **48**, 3081-3089

[10] Gall K., Sehitoglu H., Chumlyakov Y. I., Kireeva I. V. (1999) Tension-compression asymmetry of the stress-strain response in aged single crystal and polycrystalline NiTi, *Acta Mater.*, **47**, 1203-1217

[11] Matsumoto O., Miyazaki S., Otsuka K., Tamura H. (1987) Crystallography of martensitic transformation in Ti-Ni single crystals, *Acta Metall.*, **35**, 2137-2144

[12] Lu, Z.K., Weng, G.J. (1997) Martensitic transformation and stress-strain relations of shape-memory alloys, *J. Mech. Phys. Solids*, **45**, 1905-1928

Author Index

Mechanics

SOLID MECHANICS AND ITS APPLICATIONS

Series Editor: G.M.L. Gladwell

Aims and Scope of the Series

The fundamental questions arising in mechanics are: *Why?*, *How?*, and *How much?* The aim of this series is to provide lucid accounts written by authoritative researchers giving vision and insight in answering these questions on the subject of mechanics as it relates to solids. The scope of the series covers the entire spectrum of solid mechanics. Thus it includes the foundation of mechanics; variational formulations; computational mechanics; statics, kinematics and dynamics of rigid and elastic bodies; vibrations of solids and structures; dynamical systems and chaos; the theories of elasticity, plasticity and viscoelasticity; composite materials; rods, beams, shells and membranes; structural control and stability; soils, rocks and geomechanics; fracture; tribology; experimental mechanics; biomechanics and machine design.

1. R.T. Haftka, Z. Gürdal and M.P. Kamat: *Elements of Structural Optimization*. 2nd rev.ed., 1990 ISBN 0-7923-0608-2
2. J.J. Kalker: *Three-Dimensional Elastic Bodies in Rolling Contact*. 1990 ISBN 0-7923-0712-7
3. P. Karasudhi: *Foundations of Solid Mechanics*. 1991 ISBN 0-7923-0772-0
4. *Not published*
5. *Not published.*
6. J.F. Doyle: *Static and Dynamic Analysis of Structures*. With an Emphasis on Mechanics and Computer Matrix Methods. 1991 ISBN 0-7923-1124-8; Pb 0-7923-1208-2
7. O.O. Ochoa and J.N. Reddy: *Finite Element Analysis of Composite Laminates*. ISBN 0-7923-1125-6
8. M.H. Aliabadi and D.P. Rooke: *Numerical Fracture Mechanics*. ISBN 0-7923-1175-2
9. J. Angeles and C.S. López-Cajún: *Optimization of Cam Mechanisms*. 1991 ISBN 0-7923-1355-0
10. D.E. Grierson, A. Franchi and P. Riva (eds.): *Progress in Structural Engineering*. 1991 ISBN 0-7923-1396-8
11. R.T. Haftka and Z. Gürdal: *Elements of Structural Optimization*. 3rd rev. and exp. ed. 1992 ISBN 0-7923-1504-9; Pb 0-7923-1505-7
12. J.R. Barber: *Elasticity*. 1992 ISBN 0-7923-1609-6; Pb 0-7923-1610-X
13. H.S. Tzou and G.L. Anderson (eds.): *Intelligent Structural Systems*. 1992 ISBN 0-7923-1920-6
14. E.E. Gdoutos: *Fracture Mechanics*. An Introduction. 1993 ISBN 0-7923-1932-X
15. J.P. Ward: *Solid Mechanics*. An Introduction. 1992 ISBN 0-7923-1949-4
16. M. Farshad: *Design and Analysis of Shell Structures*. 1992 ISBN 0-7923-1950-8
17. H.S. Tzou and T. Fukuda (eds.): *Precision Sensors, Actuators and Systems*. 1992 ISBN 0-7923-2015-8
18. J.R. Vinson: *The Behavior of Shells Composed of Isotropic and Composite Materials*. 1993 ISBN 0-7923-2113-8
19. H.S. Tzou: *Piezoelectric Shells*. Distributed Sensing and Control of Continua. 1993 ISBN 0-7923-2186-3
20. W. Schiehlen (ed.): *Advanced Multibody System Dynamics*. Simulation and Software Tools. 1993 ISBN 0-7923-2192-8
21. C.-W. Lee: *Vibration Analysis of Rotors*. 1993 ISBN 0-7923-2300-9
22. D.R. Smith: *An Introduction to Continuum Mechanics*. 1993 ISBN 0-7923-2454-4
23. G.M.L. Gladwell: *Inverse Problems in Scattering*. An Introduction. 1993 ISBN 0-7923-2478-1

Mechanics

SOLID MECHANICS AND ITS APPLICATIONS

Series Editor: G.M.L. Gladwell

24. G. Prathap: *The Finite Element Method in Structural Mechanics*. 1993 ISBN 0-7923-2492-7
25. J. Herskovits (ed.): *Advances in Structural Optimization*. 1995 ISBN 0-7923-2510-9
26. M.A. González-Palacios and J. Angeles: *Cam Synthesis*. 1993 ISBN 0-7923-2536-2
27. W.S. Hall: *The Boundary Element Method*. 1993 ISBN 0-7923-2580-X
28. J. Angeles, G. Hommel and P. Kovács (eds.): *Computational Kinematics*. 1993 ISBN 0-7923-2585-0
29. A. Curnier: *Computational Methods in Solid Mechanics*. 1994 ISBN 0-7923-2761-6
30. D.A. Hills and D. Nowell: *Mechanics of Fretting Fatigue*. 1994 ISBN 0-7923-2866-3
31. B. Tabarrok and F.P.J. Rimrott: *Variational Methods and Complementary Formulations in Dynamics*. 1994 ISBN 0-7923-2923-6
32. E.H. Dowell (ed.), E.F. Crawley, H.C. Curtiss Jr., D.A. Peters, R. H. Scanlan and F. Sisto: *A Modern Course in Aeroelasticity*. Third Revised and Enlarged Edition. 1995 ISBN 0-7923-2788-8; Pb: 0-7923-2789-6
33. A. Preumont: *Random Vibration and Spectral Analysis*. 1994 ISBN 0-7923-3036-6
34. J.N. Reddy (ed.): *Mechanics of Composite Materials*. Selected works of Nicholas J. Pagano. 1994 ISBN 0-7923-3041-2
35. A.P.S. Selvadurai (ed.): *Mechanics of Poroelastic Media*. 1996 ISBN 0-7923-3329-2
36. Z. Mróz, D. Weichert, S. Dorosz (eds.): *Inelastic Behaviour of Structures under Variable Loads*. 1995 ISBN 0-7923-3397-7
37. R. Pyrz (ed.): *IUTAM Symposium on Microstructure-Property Interactions in Composite Materials*. Proceedings of the IUTAM Symposium held in Aalborg, Denmark. 1995 ISBN 0-7923-3427-2
38. M.I. Friswell and J.E. Mottershead: *Finite Element Model Updating in Structural Dynamics*. 1995 ISBN 0-7923-3431-0
39. D.F. Parker and A.H. England (eds.): *IUTAM Symposium on Anisotropy, Inhomogeneity and Nonlinearity in Solid Mechanics*. Proceedings of the IUTAM Symposium held in Nottingham, U.K. 1995 ISBN 0-7923-3594-5
40. J.-P. Merlet and B. Ravani (eds.): *Computational Kinematics '95*. 1995 ISBN 0-7923-3673-9
41. L.P. Lebedev, I.I. Vorovich and G.M.L. Gladwell: *Functional Analysis*. Applications in Mechanics and Inverse Problems. 1996 ISBN 0-7923-3849-9
42. J. Menčik: *Mechanics of Components with Treated or Coated Surfaces*. 1996 ISBN 0-7923-3700-X
43. D. Bestle and W. Schiehlen (eds.): *IUTAM Symposium on Optimization of Mechanical Systems*. Proceedings of the IUTAM Symposium held in Stuttgart, Germany. 1996 ISBN 0-7923-3830-8
44. D.A. Hills, P.A. Kelly, D.N. Dai and A.M. Korsunsky: *Solution of Crack Problems*. The Distributed Dislocation Technique. 1996 ISBN 0-7923-3848-0
45. V.A. Squire, R.J. Hosking, A.D. Kerr and P.J. Langhorne: *Moving Loads on Ice Plates*. 1996 ISBN 0-7923-3953-3
46. A. Pineau and A. Zaoui (eds.): *IUTAM Symposium on Micromechanics of Plasticity and Damage of Multiphase Materials*. Proceedings of the IUTAM Symposium held in Sèvres, Paris, France. 1996 ISBN 0-7923-4188-0
47. A. Naess and S. Krenk (eds.): *IUTAM Symposium on Advances in Nonlinear Stochastic Mechanics*. Proceedings of the IUTAM Symposium held in Trondheim, Norway. 1996 ISBN 0-7923-4193-7
48. D. Ieşan and A. Scalia: *Thermoelastic Deformations*. 1996 ISBN 0-7923-4230-5

Mechanics

SOLID MECHANICS AND ITS APPLICATIONS

Series Editor: G.M.L. Gladwell

49. J.R. Willis (ed.): *IUTAM Symposium on Nonlinear Analysis of Fracture*. Proceedings of the IUTAM Symposium held in Cambridge, U.K. 1997 ISBN 0-7923-4378-6
50. A. Preumont: *Vibration Control of Active Structures*. An Introduction. 1997 ISBN 0-7923-4392-1
51. G.P. Cherepanov: *Methods of Fracture Mechanics: Solid Matter Physics*. 1997 ISBN 0-7923-4408-1
52. D.H. van Campen (ed.): *IUTAM Symposium on Interaction between Dynamics and Control in Advanced Mechanical Systems*. Proceedings of the IUTAM Symposium held in Eindhoven, The Netherlands. 1997 ISBN 0-7923-4429-4
53. N.A. Fleck and A.C.F. Cocks (eds.): *IUTAM Symposium on Mechanics of Granular and Porous Materials*. Proceedings of the IUTAM Symposium held in Cambridge, U.K. 1997 ISBN 0-7923-4553-3
54. J. Roorda and N.K. Srivastava (eds.): *Trends in Structural Mechanics*. Theory, Practice, Education. 1997 ISBN 0-7923-4603-3
55. Yu.A. Mitropolskii and N. Van Dao: *Applied Asymptotic Methods in Nonlinear Oscillations*. 1997 ISBN 0-7923-4605-X
56. C. Guedes Soares (ed.): *Probabilistic Methods for Structural Design*. 1997 ISBN 0-7923-4670-X
57. D. François, A. Pineau and A. Zaoui: *Mechanical Behaviour of Materials*. Volume I: Elasticity and Plasticity. 1998 ISBN 0-7923-4894-X
58. D. François, A. Pineau and A. Zaoui: *Mechanical Behaviour of Materials*. Volume II: Viscoplasticity, Damage, Fracture and Contact Mechanics. 1998 ISBN 0-7923-4895-8
59. L.T. Tenek and J. Argyris: *Finite Element Analysis for Composite Structures*. 1998 ISBN 0-7923-4899-0
60. Y.A. Bahei-El-Din and G.J. Dvorak (eds.): *IUTAM Symposium on Transformation Problems in Composite and Active Materials*. Proceedings of the IUTAM Symposium held in Cairo, Egypt. 1998 ISBN 0-7923-5122-3
61. I.G. Goryacheva: *Contact Mechanics in Tribology*. 1998 ISBN 0-7923-5257-2
62. O.T. Bruhns and E. Stein (eds.): *IUTAM Symposium on Micro- and Macrostructural Aspects of Thermoplasticity*. Proceedings of the IUTAM Symposium held in Bochum, Germany. 1999 ISBN 0-7923-5265-3
63. F.C. Moon: *IUTAM Symposium on New Applications of Nonlinear and Chaotic Dynamics in Mechanics*. Proceedings of the IUTAM Symposium held in Ithaca, NY, USA. 1998 ISBN 0-7923-5276-9
64. R. Wang: *IUTAM Symposium on Rheology of Bodies with Defects*. Proceedings of the IUTAM Symposium held in Beijing, China. 1999 ISBN 0-7923-5297-1
65. Yu.I. Dimitrienko: *Thermomechanics of Composites under High Temperatures*. 1999 ISBN 0-7923-4899-0
66. P. Argoul, M. Frémond and Q.S. Nguyen (eds.): *IUTAM Symposium on Variations of Domains and Free-Boundary Problems in Solid Mechanics*. Proceedings of the IUTAM Symposium held in Paris, France. 1999 ISBN 0-7923-5450-8
67. F.J. Fahy and W.G. Price (eds.): *IUTAM Symposium on Statistical Energy Analysis*. Proceedings of the IUTAM Symposium held in Southampton, U.K. 1999 ISBN 0-7923-5457-5
68. H.A. Mang and F.G. Rammerstorfer (eds.): *IUTAM Symposium on Discretization Methods in Structural Mechanics*. Proceedings of the IUTAM Symposium held in Vienna, Austria. 1999 ISBN 0-7923-5591-1

Mechanics

SOLID MECHANICS AND ITS APPLICATIONS

Series Editor: G.M.L. Gladwell

69. P. Pedersen and M.P. Bendsøe (eds.): *IUTAM Symposium on Synthesis in Bio Solid Mechanics.* Proceedings of the IUTAM Symposium held in Copenhagen, Denmark. 1999 ISBN 0-7923-5615-2
70. S.K. Agrawal and B.C. Fabien: *Optimization of Dynamic Systems.* 1999 ISBN 0-7923-5681-0
71. A. Carpinteri: *Nonlinear Crack Models for Nonmetallic Materials.* 1999 ISBN 0-7923-5750-7
72. F. Pfeifer (ed.): *IUTAM Symposium on Unilateral Multibody Contacts.* Proceedings of the IUTAM Symposium held in Munich, Germany. 1999 ISBN 0-7923-6030-3
73. E. Lavendelis and M. Zakrzhevsky (eds.): *IUTAM/IFToMM Symposium on Synthesis of Nonlinear Dynamical Systems.* Proceedings of the IUTAM/IFToMM Symposium held in Riga, Latvia. 2000 ISBN 0-7923-6106-7
74. J.-P. Merlet: *Parallel Robots.* 2000 ISBN 0-7923-6308-6
75. J.T. Pindera: *Techniques of Tomographic Isodyne Stress Analysis.* 2000 ISBN 0-7923-6388-4
76. G.A. Maugin, R. Drouot and F. Sidoroff (eds.): *Continuum Thermomechanics.* The Art and Science of Modelling Material Behaviour. 2000 ISBN 0-7923-6407-4
77. N. Van Dao and E.J. Kreuzer (eds.): *IUTAM Symposium on Recent Developments in Non-linear Oscillations of Mechanical Systems.* 2000 ISBN 0-7923-6470-8
78. S.D. Akbarov and A.N. Guz: *Mechanics of Curved Composites.* 2000 ISBN 0-7923-6477-5
79. M.B. Rubin: *Cosserat Theories: Shells, Rods and Points.* 2000 ISBN 0-7923-6489-9
80. S. Pellegrino and S.D. Guest (eds.): *IUTAM-IASS Symposium on Deployable Structures: Theory and Applications.* Proceedings of the IUTAM-IASS Symposium held in Cambridge, U.K., 6–9 September 1998. 2000 ISBN 0-7923-6516-X
81. A.D. Rosato and D.L. Blackmore (eds.): *IUTAM Symposium on Segregation in Granular Flows.* Proceedings of the IUTAM Symposium held in Cape May, NJ, U.S.A., June 5–10, 1999. 2000 ISBN 0-7923-6547-X
82. A. Lagarde (ed.): *IUTAM Symposium on Advanced Optical Methods and Applications in Solid Mechanics.* Proceedings of the IUTAM Symposium held in Futuroscope, Poitiers, France, August 31–September 4, 1998. 2000 ISBN 0-7923-6604-2
83. D. Weichert and G. Maier (eds.): *Inelastic Analysis of Structures under Variable Loads.* Theory and Engineering Applications. 2000 ISBN 0-7923-6645-X
84. T.-J. Chuang and J.W. Rudnicki (eds.): *Multiscale Deformation and Fracture in Materials and Structures.* The James R. Rice 60th Anniversary Volume. 2001 ISBN 0-7923-6718-9
85. S. Narayanan and R.N. Iyengar (eds.): *IUTAM Symposium on Nonlinearity and Stochastic Structural Dynamics.* Proceedings of the IUTAM Symposium held in Madras, Chennai, India, 4–8 January 1999 ISBN 0-7923-6733-2
86. S. Murakami and N. Ohno (eds.): *IUTAM Symposium on Creep in Structures.* Proceedings of the IUTAM Symposium held in Nagoya, Japan, 3-7 April 2000. 2001 ISBN 0-7923-6737-5
87. W. Ehlers (ed.): *IUTAM Symposium on Theoretical and Numerical Methods in Continuum Mechanics of Porous Materials.* Proceedings of the IUTAM Symposium held at the University of Stuttgart, Germany, September 5-10, 1999. 2001 ISBN 0-7923-6766-9
88. D. Durban, D. Givoli and J.G. Simmonds (eds.): *Advances in the Mechanis of Plates and Shells The Avinoam Libai Anniversary Volume.* 2001 ISBN 0-7923-6785-5
89. U. Gabbert and H.-S. Tzou (eds.): *IUTAM Symposium on Smart Structures and Structonic Systems.* Proceedings of the IUTAM Symposium held in Magdeburg, Germany, 26–29 September 2000. 2001 ISBN 0-7923-6968-8

Mechanics

SOLID MECHANICS AND ITS APPLICATIONS

Series Editor: G.M.L. Gladwell

90. Y. Ivanov, V. Cheshkov and M. Natova: *Polymer Composite Materials – Interface Phenomena & Processes*. 2001 ISBN 0-7923-7008-2
91. R.C. McPhedran, L.C. Botten and N.A. Nicorovici (eds.): *IUTAM Symposium on Mechanical and Electromagnetic Waves in Structured Media*. Proceedings of the IUTAM Symposium held in Sydney, NSW, Australia, 18-22 Januari 1999. 2001 ISBN 0-7923-7038-4
92. D.A. Sotiropoulos (ed.): *IUTAM Symposium on Mechanical Waves for Composite Structures Characterization*. Proceedings of the IUTAM Symposium held in Chania, Crete, Greece, June 14-17, 2000. 2001 ISBN 0-7923-7164-X
93. V.M. Alexandrov and D.A. Pozharskii: *Three-Dimensional Contact Problems*. 2001 ISBN 0-7923-7165-8
94. J.P. Dempsey and H.H. Shen (eds.): *IUTAM Symposium on Scaling Laws in Ice Mechanics and Ice Dynamics*. Proceedings of the IUTAM Symposium held in Fairbanks, Alaska, U.S.A., 13-16 June 2000. 2001 ISBN 1-4020-0171-1
95. U. Kirsch: *Design-Oriented Analysis of Structures*. A Unified Approach. 2002 ISBN 1-4020-0443-5
96. A. Preumont: *Vibration Control of Active Structures*. An Introduction (2nd Edition). 2002 ISBN 1-4020-0496-6
97. B.L. Karihaloo (ed.): *IUTAM Symposium on Analytical and Computational Fracture Mechanics of Non-Homogeneous Materials*. Proceedings of the IUTAM Symposium held in Cardiff, U.K., 18-22 June 2001. 2002 ISBN 1-4020-0510-5
98. S.M. Han and H. Benaroya: *Nonlinear and Stochastic Dynamics of Compliant Offshore Structures*. 2002 ISBN 1-4020-0573-3
99. A.M. Linkov: *Boundary Integral Equations in Elasticity Theory*. 2002 ISBN 1-4020-0574-1
100. L.P. Lebedev, I.I. Vorovich and G.M.L. Gladwell: *Functional Analysis*. Applications in Mechanics and Inverse Problems (2nd Edition). 2002 ISBN 1-4020-0667-5; Pb: 1-4020-0756-6
101. Q.P. Sun (ed.): *IUTAM Symposium on Mechanics of Martensitic Phase Transformation in Solids*. Proceedings of the IUTAM Symposium held in Hong Kong, China, 11-15 June 2001. 2002 ISBN 1-4020-0741-8
102. M.L. Munjal (ed.): *IUTAM Symposium on Designing for Quietness*. Proceedings of the IUTAM Symposium held in Bangkok, India, 12-14 December 2000. 2002 ISBN 1-4020-0765-5
103. J.A.C. Martins and M.D.P. Monteiro Marques (eds.): *Contact Mechanics*. Proceedings of the 3rd Contact Mechanics International Symposium, Praia da Consolação, Peniche, Portugal, 17-21 June 2001. 2002 ISBN 1-4020-0811-2
104. H.R. Drew and S. Pellegrino (eds.): *New Approaches to Structural Mechanics, Shells and Biological Structures*. 2002 ISBN 1-4020-0862-7

Kluwer Academic Publishers – Dordrecht / Boston / London

ICASE/LaRC Interdisciplinary Series in Science and Engineering

1. J. Buckmaster, T.L. Jackson and A. Kumar (eds.): *Combustion in High-Speed Flows.* 1994 ISBN 0-7923-2086-X
2. M.Y. Hussaini, T.B. Gatski and T.L. Jackson (eds.): *Transition, Turbulence and Combustion.* Volume I: Transition. 1994 ISBN 0-7923-3084-6; set 0-7923-3086-2
3. M.Y. Hussaini, T.B. Gatski and T.L. Jackson (eds.): *Transition, Turbulence and Combustion.* Volume II: Turbulence and Combustion. 1994 ISBN 0-7923-3085-4; set 0-7923-3086-2
4. D.E. Keyes, A. Sameh and V. Venkatakrishnan (eds): *Parallel Numerical Algorithms.* 1997 ISBN 0-7923-4282-8
5. T.G. Campbell, R.A. Nicolaides and M.D. Salas (eds.): *Computational Electromagnetics and Its Applications.* 1997 ISBN 0-7923-4733-1
6. V. Venkatakrishnan, M.D. Salas and S.R. Chakravarthy (eds.): *Barriers and Challenges in Computational Fluid Dynamics.* 1998 ISBN 0-7923-4855-9
7. M.D. Salas, J.N. Hefner and L. Sakell (eds.): *Modeling Complex Turbulent Flows.* 1999 ISBN 0-7923-5590-3

KLUWER ACADEMIC PUBLISHERS – DORDRECHT / BOSTON / LONDON